Archives of Virology

Supplementum 8

W. H. Gerlich (ed.)

**Research in Chronic
Viral Hepatitis**

Springer-Verlag Wien New York

Prof. Dr. W.H. Gerlich
Institut für Medizinische Virologie, Justus-Liebig-Universität Giessen, Giessen,
Federal Republic of Germany

© 1993 Springer-Verlag/Wien

Typesetting: Best-set Typesetter Ltd., Hong Kong

Printed on acid-free and chlorine-free bleached paper

With 46 partly coloured Figures

ISSN 0939-1983
ISBN-13:978-3-211-82497-9   e-ISBN-13:978-3-7091-9312-9
DOI: 10.1007/978-3-7091-9312-9

Preface

Chronic viral hepatitis remains one of the major medical problems worldwide. Neither a cure nor eradication of this disease is in sight. The chronic disease caused by hepatitis viruses type B, C and D is a much greater problem than the acute disease caused by the same viruses or by hepatitis viruses type A and E. Chronic viral hepatitis often remains unrecognized until the patient develops decompensated liver cirrhosis or hepatocellular carcinoma. Furthermore, unrecognized chronic virus carriers are a persistent source of infection by sexual and other close contacts as well as during many medical procedures. The viruses of chronic hepatitis are very different from each other from a taxonomical point of view, but they share many common pathogenic properties and they often coinfect individuals.

Six years ago Carlo De Bac, Gloria Taliani (Rome) and I undertook an effort to bring together, under the auspices of the European Society against Virus Diseases, clinicians, laboratory physicians, epidemiologists, pathologists and molecular biologists whose primary research interest is chronic viral hepatitis. The contributions from these quite divergent participants to a meeting devoted solely to chronic viral hepatitis were most stimulating and valuable. As a result of the success of the first meeting in Fiuggi (Italy), a second followed in Siena (Italy) 1990 and the recent third meeting was held in Pisa (Italy). Most of the speakers expressed interest in publishing their contributions in the form of a proceedings volume, as was done in the case of the Siena meeting.

Prof. Klenk and Springer-Verlag kindly agreed again to publish the proceedings as a Supplement Volume of Archives of Virology. Some of the authors preferred to give a brief review on their field, others submitted original articles. My personal feeling is that reading this collection of articles is a must for all those who have a specific interest in viral hepatitis, may it be from a practical or a theoretical point of view. Most of the information published in this volume cannot be found anywhere else or is widely spread throughout the literature. The contents of this proceedings volume is organized in seven parts.

Therapy. There is still no reliable cure for chronic hepatitis C. The articles on interferon therapy from Budillon et al. and Piccinino et al. make this very clear. They do, however, find that younger patients without cirrhosis respond more often to interferon. Sometimes a change from recombinant E.coli-derived interferon to more natural lymphoblast-derived interferon brings an improvement. The latter interferon is less immunogenic than the recombinant interferon which often induces neutralizing interferon antibodies. The two available recombinant interferons also have different immunogenicity (Antonelli and Dianzani). But as reported by Seta et al. there is some hope that those who have a potential to respond to interferon will be recognized more and more reliably. Viruses other than HBV, HCV and HDV may also cause severe chronic hepatitis, most notably cytomegalovirus, but this is only true for immunocompromised patients (Nigro et al.).

Transplantation. A last resort to "cure" chronic hepatitis is liver transplantation. Zignego et al. and Caccamo et al. report some important findings on the rate and clinical consequences of reinfection of the graft. While reinfection by HBV can be prevented by passive immunization with hepatitis B hyperimmune globulin, reinfection by HDV and HCV cannot be prevented, but fortunately this is not as detrimental as one would have anticipated.

Diagnosis. More encouraging is the progress in the diagnosis of chronic viral hepatitis. It appears that antiviral IgM antibodies may turn out to be the disease marker for which virologists have so long searched. Both in HBV (Colloredo et al.) and HCV infection (Brillanti et al.), presence of antiviral IgM is a marker of inflammation. Moreover, there is an interesting time shift between reactivation of HBV, appearence of serum transaminase and IgM antiHBc (Colloredo et al.). Isolated antibodies against the core protein of HCV may be more significant than was previously believed. Taliani et al. found that about one half of the patients with such an "indeterminate result" have in fact liver disease, HCV RNA in the serum and antibodies against the NS5 protein of HCV. One last enigma in the diagnosis of HCV remains: the question of whether HCV is transmitted from mother to child pre- or perinatally (Tanzi et al.). Co-infection with HIV seems to greatly enhance the infectivity of HCV for the child, but transmission of HCV in the absence of HIV is probably rare.

Impressive, as usual, was the progress in *molecular biology* of HBV. Carman and Thomas, as well as Miska and Will, reported on a large number of *HBV variants* which may play a major role in the chronification of hepatitis B. The significance of such variants for the response against interferon is possibly not as high as previously believed (Santantonio et al.). Furthermore, some elements of HBV cannot be deleted as the preS1 attachment site (Petit et al.). Besides the clinical importance of genetic variation, genotype analysis allows for fascinating studies on evolution of HBV and its geographical distribution throughout the human population (Norder et al.).

An important selective force for variants and genotypes of HBV is the host's *immune response*. Ferrari et al. summarize their land-mark experiments on cytotoxic T-cells against the core protein of HBV, and show clearly that this type of immune reaction is both pathogenic and curative. Repp et al. demonstrated in their group of children with HBV-infection during drug-induced immune paralysis leads to persistent infection without any disease. Even hepatitis B vaccine may induce specific-immune tolerance if it is given during multidrug cancer therapy. As studied over 7 years in Senegal, vaccination is successful in children. It prevents HBV infection if antiHBs is present (Chabaud et al.). It is, however, a pity that current HBV vaccines do not contain preS antigen. Experiments in ducks show that the preS antigen of the duck virus DHBV is an optimal target for immune protection (Chassot et al.), and there is much reason to believe that this would also be the case for human HBV.

Oncogenesis by HBV is still not well understood. DHBV is apparently not oncogenic. Those cases of liver cancer which have previously been found in DHBV-infected Chinese ducks are obviously caused by aflatoxin (Cova et al.). Human HBV is certainly oncogenic but the mechanism is still enigmatic. The X protein of HBV is an activator of gene expression even in transgenic mice, but such mice obviously do not always develop liver cancer (Balsano et al.). HDV has been recently incriminated to be oncogenic as well. Tappero et al. have found a surprising interaction between HDAg and c-myc, but it is not clear what the function of this interaction is.

The role of *HBeAg* also remains unsolved. Schlicht et al. have unraveled the mechanism of how the pre-core sequence alters the conformation of the core protein so that it exposes HBeAg, but unfortunately in this case molecular biology does not give a clue to possible function. Also engimatic is the *protein kinase* of HBV, studied by Kann et al. The authors suggest that this kinase is most likely an activated derivative of the cellular protein kinase C.

Replication of hepatitis viruses in cell culture is still a problem but significant progress has been made. One of the pioneers in this field, C. Sureau, presented his latest cell culture system which can be efficiently infected by HDV in a preS1 and S dependent way, but surprisingly HBV (which has the same envelope proteins) cannot infect these cells. Ciccaglioni et al. have induced persistent HDV infection in WHV carrier woodchucks by transfection with cloned cDNA of the HDV genome. HCV can now be grown in primary fetal hepatocytes (Carloni et al.) and it also infects lymphocytes in vitro (Artini et al.).

In summary, chronic viral hepatitis is a rapidly moving field with many, not yet reached, goals. I thank all the authors for their great efforts to provide the manuscripts, Dr. Bruce Boschek (Giessen) and Dr. Michael Kann for critical reading of the manuscripts, and Springer-Verlag for publishing this book.

W.H. Gerlich

Contents

I. Replication of hepatitis viruses

II. Molecular biology of hepatitis B virus

III. Pathogenic and protective immune responses against hepatitis viruses

IV. Variability of hepatitis B virus

V. Diagnosis of chronic viral hepatitis

VI. Clinical course and therapy of chronic viral hepatitis

VII. Liver transplantation and chronic viral hepatitis

Listed in Current Contents

I. Replication of hepatitis viruses

Arch Virol (1993) [Suppl] 8: 3–14

Archives of Virology
© Springer-Verlag 1993

In vitro culture systems for hepatitis B and delta viruses

Camille Sureau

Department of Virology and Immunology, Southwest Foundation for Biomedical Research, San Antonio, Texas, U.S.A.

Summary. The development of tissue culture technology has led to invaluable information in many fields of modern virology. Until recently, the lack of an in vitro culture system for the hepatitis B virus (HBV) was a considerable impediment to the study of its life cycle at the cellular and molecular levels. However, it did not prevent its isolation and molecular cloning. Such has been the case also for the hepatitis delta virus (HDV), the genome of which was cloned and sequenced before its replication could be observed in cultured cells. In recent years, tissue culture systems for HBV and HDV have been developed progressively by the identification of permissive, established cell lines for production of virions and susceptible primary hepatocyte cultures for infection assays. I will briefly review here the recent experiments that have contributed to replicate HBV and HDV in cell culture systems.

Hepatitis B viruses

The lack of an in vitro culture system did not hamper the initial investigations of HBV. Its identification, isolation and characterization were possible in part because of the peculiar properties of this virus such as the abundance of viral particles in an infectious serum, the persistence of viremia in chronically infected individuals, and the stability of the viral particle [17, 51]. The rapid progress made in molecular biology allowed the cloning and sequencing of the HBV genome soon after the identification of the viral particle [51]. In addition, the discovery of HBV related viruses, namely the duck hepatitis B virus (DHBV), the woodchuck hepatitis virus (WHV) and the ground squirrel hepatitis virus (GSHV), has contributed to the understanding of the hepadnavirus family [17]. Each of these animal species represents therefore an important experimental system. The human hepadnavirus HBV can be studied in the chimpanzee, the only animal that can be experimentally infected [50]. The chimpanzee model has been used in a limited number of experiments aimed mainly at the evaluation of vaccines and antiviral drugs.

The establishment of an in vitro system for HBV has been hampered primarily by its very narrow host range. HBV infects humans and chimpanzees exclusively, and the liver is the only tissue in which productive infection can be observed. Unfortunately, primary human or chimpanzee hepatocytes are difficult to obtain and to grow in vitro, and the isolation of a liver-derived cell line susceptible to infection with HBV has so far been unsuccessful.

A tissue culture system for human hepatitis viruses should have the following characteristics: i) the cells should be of human liver origin, ii) they should be immortalized to ensure an unlimited supply, iii) the cells should retain the characteristics and functions of differentiated hepatocytes, iv) they should be both susceptible to and permissive for HBV, and v) they should sustain replication in quantities sufficient for most biochemical and molecular analysis procedures.

Most of the early experiments to develop a tissue culture system for HBV involved either inoculation of animal or human cells, such as adult and fetal hepatocytes, with infectious sera or short-term cultures of cells obtained by liver biopsies from infected patients (for a review see [20]). These initial attempts resulted in somewhat inconclusive observations. Replication of HBV DNA could not be demonstrated and production of viral particles was not thoroughly evaluated. Therefore these systems were not practical or reliable.

Soon after the cloning of HBV DNA, envelope and capsid proteins were expressed by transfection of mammalian cells from different species and phenotypes, but none of these experiments resulted in the synthesis and release of complete HBV virions [20, 21]. Recently, several in vitro culture systems which fulfill only some of the requirements listed above, have been established using the following criteria: i) permissivity was demonstrated by transfection of transformed cells where the inoculum was a cloned viral DNA and the readout product was an infectious viral particle released into the culture medium of the transfected cell; ii) susceptibility was demonstrated by inoculation of primary hepatocytes in culture where the inoculum was a viral particle and the readout product was replicative intermediates of the viral genome in the infected cell.

Susceptibility

In a landmark experiment Tuttleman et al. [53] established a system susceptible to DHBV infection in vitro using primary cultures of duck hepatocytes. Susceptibility of these cells to DHBV was shown to last for few days after isolation. Infection was demonstrated by the appearance of intracellular replicative forms of DHBV DNA which were not present in the inoculum. De novo production of infectious virions was also documented which clearly indicated that this system was also permissive. This represented the first practical system in which all stages of hepadnavirus replication could be studied in vitro. However, there were some limitations to this model: i) primary hepatocytes were by definition in limited supply, ii) among the hepadnavirus members, DHBV is the most distantly related to HBV, and iii) after isolation, the cells remain susceptible for a very limited period of time, probably due to the loss of cell differentiation after a few days in culture. However the period of susceptibility was later extended by complementing the culture medium with dimethyl sulfoxide, a solvent known to help maintain cells at a differentiated stage in culture [15, 33].

Following the description of the DHBV system, primary woodchuck hepatocytes were shown to be susceptible to in vitro infection with WHV and GSHV for a limited period of time after isolation [2]. Interestingly, replicative intermediates of the viral genome were not detected in substantial amounts until 7–10 days post-infection as opposed to 3–4 days in the DHBV system.

Gripon et al. [18] and Ochiya et al. [29] described a system of primary cultures of human adult or fetal hepatocytes for in vitro infection. Both systems demonstrated susceptibility and permissivity over a few days after isolation. The appearance of intracellular replicative intermediates of HBV DNA and the detection of Dane particles in the culture medium were documented. Both systems should be valuable, if not very practical, for the propagation of HBV in vitro. Their limitations reside mainly in the short supply of cells and in the reproducibility of the infection assays. Recently, Jacob et al. [23] and Sureau et al. [44] have shown that primary cultures of chimpanzee hepatocytes maintained in a serum-free medium are permissive for production of HBV, but, in spite of numerous attempts, susceptibility was not demonstrated.

Permissivity

An important contribution to the search for a permissive tissue culture system for HBV was the demonstration by Will et al. [58] that cloned viral DNA could induce an infection in the chimpanzee when introduced directly into the liver. These authors showed that a full-length circular monomer or a head to tail multimer of cloned DNA was capable of inducing the replication of HBV DNA and the production of virus particles when forced into a hepatocyte. It was then theoretically possible to bypass the infection event of the HBV replication cycle to produce viral particles in vitro using transfection techniques.

Prior to this study, transfection of cultured cells with cloned HBV DNA had been employed to express HBV proteins in a variety of cell lines but had not resulted in substantial production of Dane particles or even measurable viral DNA replication. By transfection with cloned circular HBV DNA, Sureau et al. [46] demonstrated that a differentiated human hepatoma cell line (HepG2) was permissive for the production of human HBV. HepG2 cells were cotransfected with cloned HBV DNA and a plasmid coding for neomycin resistance. Following transfection and selection with neomycin, a clone called HepG2T14 was isolated. HepG2T14 cells sustained the replication of HBV DNA, the production of HBV proteins, and the production of Dane particles. Those particles were later proven to be infectious in the chimpanzee. The infected animal underwent a typical course of type B hepatitis infection and molecularly cloned virus was recovered from the serum [42]. This experiment established a system that could be amenable to a variety of applications including genetic analysis, biochemical analysis, and screening for antiviral drugs. Similar results have been obtained by other groups using different human hepatoma cell lines including HepG2, Huh6 and Huh7, and using either stable or transient transfection procedures [1, 7, 39, 52, 55]. Although permissivity was proven, susceptibility of the cells to direct infection was not demonstrated with the exception of one report where weak viral DNA replication was detected following in vitro exposure of HepG2 cells to HBV particles [3].

Following these initial experiments, cell lines of different origins were shown to be permissive for hepadnaviruses by transfection with cloned viral DNA. Infectious DHBV was produced in Huh7 cells and in a chicken hepatoma cell line [12, 14, 19, 34]. Infectious HBV was produced in a rat hepatoma cell line [40, 41], and HBV-like particles were produced in murine fibroblasts, in transformed mouse hepatocytes [8], in a human monocytic cell line [27], in differentiated adult rat hepatocytes [13], and other non-hepatocyte cells [16, 38]. However, the system of choice remains the human hepatoma cell lines such as HepG2 and Huh7 which are easily transfectable and which release large amounts of infectious Dane particles in the culture medium after transfection. After the demonstration of their permissivity, these two cell lines were used extensively to analyze the replication mechanism of HBV. Production of infectious WHV was also obtained in HepG2 and murine or avian fibroblasts [37]. In this latter experiment, replication of WHV DNA was induced by transcription of pregenomic RNA, a replication intermediate of the viral DNA, from the cytomegalovirus immediate-early promoter. Surprisingly, when the WHV promoter was used to direct the transcription of the pregenome, replication of WHV DNA was observed at a much lower level.

Therefore, the entire life cycle of HBV can be examined using two sources of cultured cells: i) permissive cell lines, preferably derived from human hepatoma, in which the events of transcription, reverse transcription, translation, and morphogenesis and release of viral particles can be studied, and ii) susceptible primary cultures of human hepatocytes for in vitro infection, where attachment, penetration, and uncoating of the particles can be examined.

Hepatitis delta virus

HDV and HBV particles have in common the protein composition of their envelope. Both viruses use the large (L), middle (M), and small (S) HBV envelope proteins to package their respective genomes in a spherical particle of 42 and 36 nm in diameter, respectively [17, 35, 48]. In the case of HBV, the viral DNA and DNA polymerase are contained in a capsid made of HBV core proteins. This capsid interacts with the envelope proteins S, M, and L to produce mature enveloped virions also referred to as Dane particles. In the case of HDV, the viral RNA interacts with HDV-encoded proteins which bear the delta antigen (HDAg) to form a ribonucleoprotein complex which is in turn packaged in an envelope consisting of the HBV-encoded envelope proteins S, M, and L [4, 5]. The implications of this latter observation are that HDV needs the presence of HBV to acquire its envelope to ensure its propagation, that both viruses have the same host range, and that productive HDV infection will take place only in tissues susceptible to HBV infection.

Natural infection with HDV has been observed only in the human and experimental infection can been achieved only in the chimpanzee. However a woodchuck HDV pseudotype was produced experimentally in woodchucks by exposing a chronically WHV-infected animal to HDV virions derived from a chimpanzee [32]. A productive infection occurred where the progeny pseudotype virions were coated with WHV instead of the HBV envelope proteins. As a result, the woodchuck system has become a practical model where the full cycle of HDV can be examined as well as its interaction with helper WHV. In view of these results, one would assume that other hepadnaviruses would also support the replication of HDV, but there has been no indication that ducks infected with DHBV, another member of the hepadnavirus family, can be infected with HDV to produce a duck HDV pseudotype coated with the DHBV envelope proteins.

Susceptibility

The initial description of an in vitro culture system was made by J. Taylor et al. [49] using the woodchuck pseudotype virus. In this study, primary cultures of woodchuck hepatocytes were shown to support the early events of HDV infection, from adsorption and penetration of the viral particle to replication of the genome. Genome replication following infection was demonstrated by detection of antigenomic HDV RNA by RNA blot analysis and in situ hybridization. This system was later used to analyze the effect of drugs such as suramin, ribavirin, alpha amanitin, and acyclovir on HDV [10, 11, 30].

Using primary cultures of chimpanzee or human hepatocytes, Sureau et al. [44] developed a system for infection with human HDV in vitro. Primate hepatocytes isolated from uninfected livers were grown in a serum-free medium. They were shown to conserve a differentiated status for several weeks in culture, including the production of several liver-specific proteins [23, 26] and the susceptibility to HDV. Cells remained susceptible to infection for at least 3 weeks in culture as demonstrated by the detection of increasing amounts of intracellular genomic and antigenomic HDV RNA following infection. However, susceptibility to infection with HBV under the same culture conditions was not demonstrated. As a result, HDV infections of non HBV-infected hepatocytes in vitro were abortive, since the envelope proteins necessary to assemble HDV virions were not supplied. When hepatocytes were derived from an HBV-infected animal, replication of HBV in vitro was maintained, and when the cells were exposed to HDV, superinfection occurred as shown by the appearance of replicative forms of the HDV genome and the synthesis and release in the culture medium of progeny HDV. Primary cultures of HBV-infected hepatocytes were therefore both susceptible to and permissive for HDV. The amounts of HDV RNA replication following in vitro infection were similar to those observed in vivo. This can be as high as 300 000 molecules of HDV RNA per infected hepatocyte, including 10–50% of antigenomic molecules which are absent in the inoculum [9]. Typically 5–10% of the cells exposed to HDV in vitro became infected [44]. Therefore, the measurement of intracellular antigenomic HDV RNA in hepatocytes exposed to HDV in vitro represents a very sensitive and reliable assay. This system has enabled us to conduct several experiments such as infectivity assays of recombinant HDV particles produced in Huh7 cells by transfection [43], and in vitro neutralization assays with antibodies to HBV envelope proteins. Infection was prevented when particles were preincubated with antibodies directed against the preS1 and preS2 domains of the HBV envelope proteins. These results indicated that the infectious agent was a particle coated with envelope proteins containing the preS1 and preS2 antigens [45].

Permissivity

Permissivity of hepatoma cell lines for HDV was also demonstrated by transfection following important preliminary experiments. They include: i) the cloning of full-length HDV cDNA and assembly of a head-to-tail multimer construct [25, 57], ii) the demonstration that transfection of cultured cells with a cloned head-to-tail multimer HDV cDNA could lead to replication of HDV RNA [24], iii) the demonstration that the same HDV cDNA clone could lead to a characteristic HDV infection when injected directly into the liver of an HBV-infected chimpanzee [47], and iv) the prior demonstration of permissivity for HBV of several human hepatoma cell lines which thus became candidates for production of HDV as well.

It was thus proven that the HDV cDNA clone was biologically active and that the replication mechanism of HDV RNA was totally independent of HBV. As a consequence, adding a means for HBV production to the transfection of human hepatoma cell lines with HDV cDNA would most certainly lead to production of HDV particles. This was achieved in Huh7 cells by cotransfection with cloned HDV cDNA and HBV DNA [45, 59]. Both HBV and HDV replication could be demonstrated by the detection of replicative intermediates of the viral genome in quantities sufficient for most experimental procedures, and both HBV and HDV particles were released in the culture medium. In addition, it was shown that HDV replication could suppress the expression of HBV [59]. Such a phenomenon was previously described in experimentally infected animals or in patients superinfected with HDV [22, 47]. Furthermore, the transfection experiments indicated that suppression of HBV expression could result from the action of the delta antigen on HBV transcription [59]. HDV particles produced in transfected Huh7 cells were proven to be infectious by inoculation of primary cultures of chimpanzee hepatocytes [45].

The Huh7 system was also useful to demonstrate that the full cycle of HBV replication was not required for HDV production. The provision of the envelope proteins S, M and L only, or more interestingly the S protein alone was sufficient to package HDV RNA and HDAg proteins in an HDV-like particle [36, 43, 56]. This was in contrast to the mechanism of HBV particle assembly which requires the presence of at least the L and perhaps the M protein in addition to S to assemble a Dane particle [6, 54]. The fact that HDV particles produced in Huh7 cells in the presence of HBV envelope proteins but in the absence of HBV replication were also infectious in the primate hepatocyte cultures [43] demonstrated that HBV had no helper function in the entry events of HDV and therefore need not be present in an infectious HDV inoculum. Recently, it was also demonstrated that the envelope of HDV must include the HBV L envelope protein for infectivity [43]. Recombinant HDV particles with an envelope devoid of the L protein were not infectious in the chimpanzee hepatocyte cultures. This result was in agreement with the prior description that the preS1 domain of the L protein contains a site for binding to the hepatocyte as a means of entry for the viral particle [28, 31]. However, other mechanisms than direct binding through the preS1 domain can be proposed which involve the L protein for infection.

Conclusion

Permissive cell lines such as Huh7 cells have proven to be very practical and reliable for the production of HBV or HDV particles by transfection. This system has already been very useful to study many aspects of HBV replication and to examine the assembly mechanism and infectivity of HDV particles. Infection assays in primate hepatocyte cultures were proven to be reliable and very sensitive although entirely dependent on the availability of human or chimpanzee liver tissue. This problem may be resolved in the future by the increasing number of human liver surgeries including transplants where sections of liver otherwise discarded are made available to research. The progress made in live tissue preservation and in surgical procedures may also contribute to increase the supply of liver tissue. Nonetheless, primate hepatocyte isolation will remain a laborious procedure which requires special equipment and training. Liver wedge biopsies must be processed rapidly after explant by extensive washings, followed by perfusion with collagenase, seeding of the hepatocytes on collagen treated plates, and maintenance in a hormone supplemented medium.

The need for a permanent primate hepatocyte cell line susceptible to productive infection with HBV and HDV remains. The generation of cell lines derived from experimentally immortalized chimpanzee or human hepatocytes is a promising prospect that hopefully will resolve the quest for a simple tissue culture model for HBV and HDV. Alternatively, cloning of the HBV and HDV receptors could allow for their expression in a permissive cell line which would thereby become susceptible as well.

Acknowledgements

I thank R.E. Lanford for helpful comments on the text and J. Fletcher for preparation of the manuscript.

References

1. Acs G, Sells MA, Purcell RH, Price P, Engle R, Shapiro M, Poper H (1987) Hepatitis B virus produced by transfected Hep G2 cells causes hepatitis in chimpanzees. Proc Natl Acad Sci USA 84: 4641–4644
2. Aldrich CE, Coates L, Wu T-T, Newbold J, Tennant BC, Summers J, Seeger C, Mason WS (1989) In vitro infection of woodchuck hepatocytes with woodchuck hepatitis virus and ground squirrel hepatitis virus. Virology 172: 247–252
3. Bchini R, Capel F, Dauguet C, Dubanchet S, Petit M-A (1990) In vitro infection of human hepatoma (HepG2) cells with hepatitis B virus. J Virol 64: 3025–3032
4. Bonino, F, Heermann, KH, Rizzetto M, Gerlich WH (1986) Hepatitis delta virus: protein composition of delta antigen and its hepatitis B virus-derived envelope. J Virol 58: 945–950
5. Bonino F, Hoyer B, Shih JW-K, Rizzetto M, Purcell RH, Gerin JL (1984) Delta hepatitis agent: structural and antigenic properties of the delta-associated particle. Infect Immun 43: 1000–1005
6. Bruss V, Ganem D (1991) The role of envelope proteins in hepatitis B virus assembly. Proc Natl Acad Sci USA 88: 1059–1063
7. Chang C, Jeng K-S, Hu C-P, Lo SJ, Su T-S, Ting L-P, Chou C-K, Han S-H, Pfaff E, Salfeld J, Schaller H (1987) Production of hepatitis B virus in vitro by transient expression of cloned HBV DNA in a hepatoma cell line. EMBO J 6: 675–680
8. Chen S-H, Hu C-P, Chang C (1992) Hepatitis B virus replication in well differentiated mouse hepatocyte cell lines immortalized by plasmid DNA. Cancer Res 52: 1329–1335
9. Chen P-J, Kalpana G, Goldberg J, Mason W, Werner B, Gerin J, Taylor J (1986) Structure and replication of the genome of the hepatitis ∂ virus. Proc Natl Acad Sci USA 83: 8774–8778
10. Choi S-S, Rasshofer R, Roggendorf M (1988) Propagation of woodchuck hepatitis delta virus in primary woodchuck hepatocytes. Virology 167: 451–45
11. Choi S-S, Rasshofer R, Roggendorf M (1989) Inhibition of hepatitis delta virus RNA replication in primary woodchuck hepatocytes. Antiviral Res 12: 213–222
12. Condreay LD, Aldrich CE, Coates L, Mason WS, Wu T-T (1990) Efficient duck hepatitis B virus production by an avian liver tumor cell line. J Virol 64: 3249–3258
13. Diot C, Gripon P, Rissel M, Guguen-Guillouzo C (1992) Replication of hepatitis B virus in differentiated adult rat hepatocytes transfected with cloned viral DNA. J Med Virol 36: 93–100

14. Galle PR, Schlicht HJ, Fischer M, Schaller H (1988) Production of infectious duck hepatitis B virus in a human hepatoma cell line. J Virol 62: 1736–1740
15. Galle PR, Schlicht H-J, Kuhn C, Schaller H (1989) Replication of duck hepatitis B virus in primary duck hepatocytes and its dependence on the state of differentiation of the host cell. Hepatology 10: 459–465
16. Galun E, Offensperger W-B, von Weizsäcker F, Offensperger S, Wands JR, Blum HE (1992) Human non-hepatocytes support hepadnaviral replication and virion production. J Gen Virol 73: 173–178
17. Ganem D, Varmus HE (1987) The molecular biology of the hepatitis B viruses. Annu Rev Biochem 56: 651–693
18. Gripon P, Diot C, Thézé N, Fourel I, Loreal O, Bréchot C, Guguen-Guillouzo C (1988) Hepatitis B virus infection of adult human hepatocytes cultured in the presence of dimethyl sulfoxide. J Virol 62: 4136–4143
19. Hirsh R, Colgrove R, Ganem D (1988) Replication of duck hepatitis B virus in two differentiated human hepatoma cell lines after transfection with cloned viral DNA. Virology 167: 136–142
20. Hirschman SZ (1984) Replication of hepatitis B virus in cell culture systems. In: Chisari FV (ed) Advances in hepatitis research. Masson Publ., New York, pp 54–61
21. Hirschman SZ, Price P, Garfinkel E, Christman J, Acs G (1980) Expression of cloned hepatitis B virus DNA in human cell cultures. Proc Natl Acad Sci USA 77: 5507–5511
22. Ichimura H, Tamura I, Tsubakio T, Kurimura O, Kurimura T (1988) Influence of hepatitis delta virus superinfection on the clearance of hepatitis B virus (HBV) markers in HBV carriers in Japan. J Med Virol 26: 49–55
23. Jacob JR, Eichberg JW, Lanford RE (1989) In vitro replication and expression of hepatitis B virus from chronically infected primary chimpanzee hepatocytes. Hepatology 10: 921–927
24. Kuo MY-P, Chao M, Taylor J (1989) Initiation of replication of the human hepatitis delta virus genome from cloned DNA: role of delta antigen. J Virol 63: 1945–1950
25. Kuo MYP, Goldberg J, Coates L, Mason W, Gerin J, Taylor J (1988) Molecular cloning of hepatitis delta virus RNA from an infected woodchuck liver: sequence, structure, and applications. J Virol 62: 1855–1861
26. Lanford RE, Carey KD, Estlack LE, Smith GC, Hay RV (1989) Analysis of plasma protein and lipoprotein synthesis in long-term primary cultures of baboon hepatocytes maintained in serum-free medium. In Vitro Cell Dev Biol 25: 174–182
27. Müller C, Bergmann KF, Gerin JL, Korba BE (1992) Production of hepatitis B virus by stably transfected monocytic cell line U-937: a model for extrahepatic hepatitis B virus replication. J Infect Dis 165: 929–933
28. Neurath AR, Kent SBH, Strick N, Parker K (1986) Identification and chemical synthesis of a host cell receptor binding site on hepatitis B virus. Cell 46: 429–436
29. Ochiya T, Tsurimoto T, Ueda K, Okubo K, Shiozawa M, Matsubara K (1989) An in vitro system for infection with hepatitis B virus that uses primary human fetal hepatocytes. Proc Natl Acad Sci USA 86: 1875–1879
30. Petcu DJ, Aldrich CE, Coates L, Taylor JM, Mason WS (1988) Suramin inhibits in vitro infection by duck hepatitis B virus, Rous sarcoma virus, and hepatitis delta virus. Virology 167: 385–392
31. Petit M-A, Dubanchet S, Capel F, Voet P, Dauguet C, Hauser P (1991) HepG2 cell binding activities of different hepatitis B virus isolates: inhibitory effect of anti-HBs and anti-preS1 (21–47). Virology 180: 483–491

32. Ponzetto A, Cote PJ, Poper H, Hoyer BH, London WT, Ford EC, Bonino F, Purcell RH, Gerin JL (1984) Transmission of the hepatitis B virus-associated ∂ agent to the eastern woodchuck. Proc Natl Acad Sci USA 81: 2208–2212

33. Pugh JC, Summers JW (1989) Infection and uptake of duck hepatitis B virus by duck hepatocytes maintained in the presence of dimethyl sulfoxide. Virology 172: 564–572

34. Pugh JC, Yaginuma K, Koike K, Summers J (1988) Duck hepatitis B virus (DHBV) particles produced by transient expression of DHBV DNA in a human hepatoma cell line are infectious in vitro. J Virol 62: 3513–3516

35. Rizzetto M (1983) The delta agent. Hepatology 3: 729–737

36. Ryu W-S, Bayer M, Taylor J (1992) Assembly of hepatitis delta virus particles. J Virol 66: 2310–2315

37. Seeger C, Baldwin B, Tennant BC (1989) Expression of infectious woodchuck hepatitis virus in murine and avian fibroblasts. J Virol 63: 4665–4669

38. Seifer M, Heermann KH, Gerlich WH (1990) Replication of hepatitis B virus in transfected nonhepatic cells. Virology 179: 300–311

39. Sells MA, Chen M-L, Acs G (1987) Production of hepatitis B virus particles in Hep G2 cells transfected with cloned hepatitis B virus DNA. Proc Natl Acad Sci USA 84: 1005–1009

40. Shih C, Li L-S, Roychoudhury S, Ho M-H (1989) In vitro propagation of human hepatitis B virus in a rat hepatoma cell line. Proc Natl Acad Sci USA 86: 6323–6327

41. Shih C, Yu M-YW, Li LS, Shih W-K (1990) Hepatitis B virus propagated in a rat hepatoma cell line is infectious in a primate model. Virology 179: 871–873

42. Sureau C, Eichberg JW, Hubbard GB, Romet-Lemonne JL, Essex M (1988) A molecularly cloned hepatitis B virus produced in vitro is infectious in a chimpanzee. J Virol 62: 3064–3067

43. Sureau C, Guerra B, Lanford RE (1993) Role of the large hepatitis B virus envelope protein in infectivity of the hepatitis delta virion. J Virol 67: 366–372

44. Sureau C, Jacob JR, Eichberg JW, Lanford RE (1991) Tissue culture system for infection with human hepatitis delta virus. J Virol 65: 3443–3450

45. Sureau C, Moriarty AM, Thornton GB, Lanford RE (1992) Production of infectious hepatitis delta virus in vitro and neutralization with antibodies directed against hepatitis B virus pre-S antigens. J Virol 66: 1241–1245

46. Sureau C, Romet-Lemonne J-L, Mullins JI, Essex M (1986) Production of hepatitis B virus by a differentiated human hepatoma cell line after transfection with cloned circular HBV DNA. Cell 47: 37–47

47. Sureau C, Taylor J, Chao M, Eichberg JW, Lanford RE (1989) Cloned hepatitis delta virus cDNA is infectious in the chimpanzee. J Virol 63: 4292–4297

48. Taylor J (1991) Human hepatitis delta virus. Curr Top Microbiol Immunol 168: 141–166

49. Taylor J, Mason W, Summers J, Goldberg J, Aldrich C, Coates L, Gerin J, Gowans E (1987) Replication of human hepatitis delta virus in primary cultures of woodchuck hepatocytes. J Virol 61: 2891–2895

50. Thung, SN, Gerber, MA, Purcell RH, London WT, Mihalik KB, Popper H (1981) Animal model of human disease, chimpanzee carriers of hepatitis B virus. Am J Pathol 105: 328–332

51. Tiollais P, Pourcel C, Dejean A (1985) The hepatitis B virus. Nature 317: 489–495

52. Tsurimoto T, Fujiyama A, Matsubara K (1987) Stable expression and replication of hepatitis B virus genome in an integrated state in a human hepatoma cell line transfected with the cloned viral DNA. Proc Natl Acad Sci USA 84: 444–448

53. Tuttleman JS, Pugh JC, Summers JW (1986) In vitro experimental infection of primary duck hepatocyte cultures with duck hepatitis B virus. J Virol 58: 17–25
54. Ueda K, Tsurimoto T, Matsubara K (1991) Three envelope proteins of hepatitis B virus: large S, middle S and major S proteins needed for the formation of Dane particles. J Virol 65: 3521–3529
55. Yaginuma K, Shirakata Y, Kobayashi M, Koike K (1987) Hepatitis B virus (HBV) particles are produced in a cell culture system by transient expression of transfected HBV DNA. Proc Natl Acad Sci USA 84: 2678–2682
56. Wang C-J, Chen P-J, Wu J-C, Patel D, Chen D-S (1991) Small-form hepatitis B surface antigen is sufficient to help in the assembly of hepatitis delta virus-like particles. J Virol 65: 6630–6636
57. Wang K-S, Choo Q-L, Weiner AJ, Ou J-H, Najarian RC, Thayer RM, Mullenbach GT, Denniston KJ, Gerin JL, Houghton M (1986) Structure, sequence and expression of the hepatitis delta (∂) viral genome. Nature 323: 508–514
58. Will H, Cattaneo R, Koch H-G, Darai G, Schaller H, Schellekens H, van Eerd PMCA, Deinhardt F (1982) Cloned HBV DNA causes hepatitis in chimpanzees. Nature 299: 740–742
59. Wu J-C, Chen P-J, Kuo MYP, Lee S-D, Chen D-S, Ting L-P (1991) Production of hepatitis delta virus and suppression of helper hepatitis B virus in a human hepatoma cell line. J Virol 65: 1099–1104

Author's address: Dr. Camille Sureau, Department of Virology and Immunology, Southwest Foundation for Biomedical Research, P.O. Box 28147, San Antonio, TX 78228-0147, U.S.A.

Arch Virol (1993) [Suppl] 8: 15–21

Chronic infection in woodchucks infected by a cloned hepatitis delta virus

Anna R. Ciccaglione[1], **Maria Rapicetta**[1], **Antonia Fabiano**[2], **C. Argentini**[1],
Maria Silvestro[2], **R. Giuseppetti**[1], **F. Varano**[1], **Nicoletta D'Urso**[2], **L. Dinolfo**[2],
Anna Morgando[2], **R. Bruni**[1], and **A. Ponzetto**[2]

[1] Department of Virology, Istituto Superiore di Sanità, Rome
[2] Department of Gastroenterology, Ospedale Le Molinette, Torino, Italy

Summary. Two woodchucks (*Marmota monax*) intrahepatically inoculated with hepatitis delta virus (HDV) complementary DNA clones pSVL-D3 and pSVL-Ag showed virological and pathological signs of acute and chronic HDV infection. HDV-RNA and hepatitis delta antigen (HDAg) were detected in serum by slot-blot hybridization and by western blot five weeks after inoculation. Liver biopsy specimens collected at 8th week post inoculum were positive for HDV-RNA. Anti-HDV antibodies were detected at the 11th and 9th weeks, respectively. Histological finding of hepatocarcinoma and persistence of circulating HDV-RNA and anti-HDV were observed up to the 10th month. Both woodchucks produced "small" and "large" HDAg antigen, although the inoculated cloned DNA bears the coding capability solely for the small antigen. A transient decrease of woodchuck hepatitis virus DNA (WHV-DNA) level was observed during the peak of HDV infection. Successive inoculation of acute-phase serum in three woodchucks resulted in a successful infection in one of the animals.

Introduction

HDV is a defective human pathogen that causes severe liver disease [3]. The human hepatitis B virus and the related woodchuck hepatitis virus (WHV) have been shown to provide the necessary helper function(s) for infection and replication of HDV in man and animals [3].

HDV replication can also occur in the absence of the helper hepadnavirus, after introduction of HDV encoding recombinant cDNA into cells. In vitro transcribed HDV-RNA can also initiate HDV replication within transfected cells [1], even if these are not hepatocytes.

However, HDV replication is strongly activated by expression of the sole protein encoded by HDV, HDAg [2]. Introduction of a well-defined full-length cDNA copy of the HDV genome into woodchucks might help in the understanding of HDV replication in vivo, pathogenicity and interaction with the helper hepadnavirus.

In this study two chronic WHV carrier woodchucks were inoculated by direct injection into the liver with two recombinant plasmids, one containing a trimeric form of a full-length cDNA copy of the HDV genome, and the other containing the HDAg open reading frame only.

Materials and methods

Experimental animals

Five chronic WHV carrier woodchucks (*Marmota monax*) were used: all were wild-caught from the mid-Atlantic United States. Two animals (W 597, W 598) were experimentally infected by intrahepatic route with HDV-DNA molecular clone pSVLD3 (kindly provided by Dr. J. Taylor) that contains three head-to-tail HDV genomes [2], and another derivative of pSVL (pSVL-Ag) that contains the open reading frame for the small form of HDAg inserted so as to be capable of directing the synthesis of HDAg-specific messanger RNA (mRNA). A mixture of 25 µl of both the trimer and the antigen clones was dissolved in 5 ml of Williams medium E containing 250 µg of DEAE dextran and was injected at five different locations into the liver of W 597 and W 598. Serum samples were collected weekly, prior to and during the study. Liver autopsy and biopsy tissues (week 4 and 8) were obtained.

Three chronic WHV carrier woodchucks (W 164, W 231, W 2040) were inoculated intravenously with 0.2 ml of the serum collected during the peak of HDV viremia from W 598.

Serological and virological studies

Anti-HDV was determined in sera by a commercial immunoenzymatic assay (Anti delta Eia-Abbott Il.).

Detection of WHV-DNA in serum and liver

DNA extraction was performed following standard procedures [4] from 20 µl of serum or 10–20 mg of liver tissue.

DNA pellets were resuspended in 20 µl of distilled water, and were analyzed by slot and Southern Blotting standard procedures [4]. Hybridization was performed with 10×10^6 cpm of nick-translated ^{32}P-WHV-DNA (specific activity 2×10^8 cpm/µg of DNA; gift of Dr. J. Summers, Philadelphia). WHV-DNA was quantitated by using 0.4 ng/µl cloned WHV-DNA blotted in parallel to the test sample. Autoradiographic reading was performed by Scanning Densitometer (Hoefer).

Detection of HDV-RNA in serum and liver

RNA extraction was performed from 100 µl of serum or 10–20 mg of liver tissue by standard procedures [4].

The RNA pellets were resuspended in 20 µl of distilled water and analyzed by slot and northern Blotting standard procedures [4]. Hybridization was performed using as a probe ^{32}P-cDNA purified from pGBZ-HDV plasmid (gift of Dr. J. Taylor, Philadelphia).

HDV-RNA was quantified by densitometric analysis using cloned HDV-DNA (12 pg/µl) as a reference sample (Scanning Densitometer Hoefer).

Immunoblot assay

150 µl of serum was sedimented in an Airfuge (Beckmann). Pellets were suspended in a buffer containing 2% sodium dodecyl sulfate and 2% β-mercaptoethanol heated to 100°C for 5 min prior to analysis by the immunoblot procedure as described [7].

PCR assay

RT-PCR assay was performed as previously described [6].

Results

A cloned cDNA copy of HDV (pSVL-D3) was infectious when injected intrahepatically into two chronic WHV-carrier *Marmota monax* (W 597, W 598). Follow-up sera at weekly intervals and liver samples from autopsies performed 17 (W 598) and 38 (W 597) weeks after inoculation, were tested for the presence of WHV and HDV serological markers.

Figure 1 presents the quantity of WHV-DNA and HDV-RNA detected during the entire observation period. HDV-RNA was first detected in serum by slot-blot hybridization after 4 (W 597) and 7 (W 598) weeks after inoculation. Persistent positivity was observed up to the 17th and 38th week, respectively, with a wide fluctuation in the levels of RNA. Anti-HDV antibodies appeared in circulation at 9th (W 598) and 10th (W 597) week post-inoculation. In all animals, a fluctuating decrease of WHV-DNA signals was demonstrated in coincidence with HDV-RNA positivity.

The results of immunoblot analysis performed on W 597 and W 598 sera, collected at 5 weeks post-inoculation, are shown in Fig. 2.

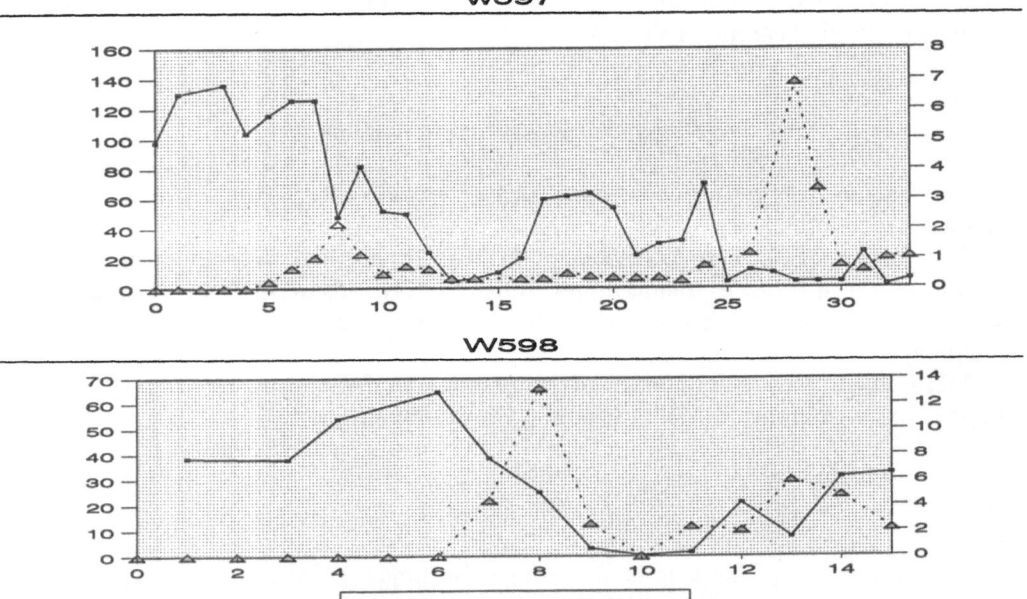

HDV-RNA pg/0.2ml ; WHV-DNA ng/0,02ml ;

Fig. 1. Follow-up sera of woodchuck 597 (W 597) and woodchuck 598 (W 598) were hybridized with WHV and HDV cDNA probes. The detection of WHV–DNA (■–■) in ng/μl and HDV-RNA (△–△) in pg/200 μl is shown

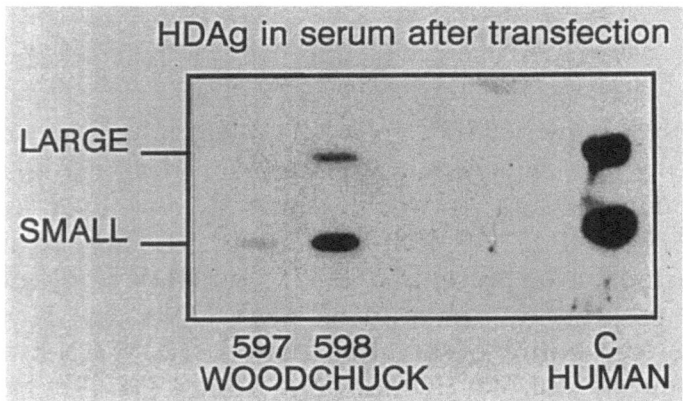

Fig. 2. 150 μl of W 597 and W 598 sera were analyzed by immunoblot procedures, showing the large and small forms of HDV antigen

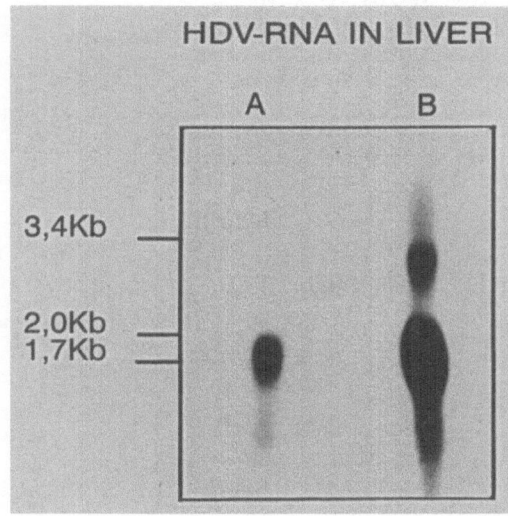

Fig. 3. HDV-RNA extracted from livers of W 597 (*A*) and W 598 (*B*) was analyzed by electrophoresis and blotted. The hybridization with the appropriate probe detects monomeric and dimeric forms of HDV genome

Two HDAg-reactive polypeptides with molecular weights of 24 and 27 kd, as is usually seen in serum of patients or after experimental infection in woodchucks or chimpanzees, [7] are present.

Figure 3 shows the results of northern blots performed with HDV-RNA liver extracts. High levels of genomic HDV-RNA, corresponding to 1.7 kb, as well as low levels of 3.6–3.8 and 5.4 kb dimeric and trimeric forms were observed in the livers of all animals.

Furthermore, 0.8 kb forms which may include candidates for the HDAg mRNA were present. The various species of HDV-RNA observed in livers are consistent with a complex rolling circle replication model.

W 598 survived 4 months after inoculation and showed histological findings of hepatocellular carcinoma at autopsy. Markers of HDV infection were still present. W 597 survived until the 8th month and was persistently positive for HDV-RNA in serum and liver.

Table 1 shows the results relative to three other chronic WHV carrier woodchucks (W 164, W 231, W 2040), who were intravenously inoculated with serum from the acute phase of HDV infection of W 598, to assess the capability of transfected cDNA to generate a true HDV infection. Only one (W 164) of these three woodchucks (W 164, W 231, W 2040) was infected and showed a typical HDV infection, with presence of HDV-RNA in serum and liver, that persisted throughout a follow up of 17 weeks.

W 231 and W 2040 were negative for serum and liver HDV-RNA. However, PCR analysis showed a low level viremia 13 weeks after inoculation in W 231. Anti-HDV appeared early, at week 9, only in W 164.

Table 1. HDV-RNA in W 164, W 231, W 2040
intravenously inoculated with acute-phase W 598 serum

Animal No.	Weeks 2	4	6	8	13	17	
W 164	+	+	+	+	+	+	slot hybrid.
	ND	ND	ND	ND	ND	ND	PCR
W 231	−	−	−	−	−	−	slot hybrid.
	−	−	−	−	+	ND	PCR
W 2040	−	−	−	−	−	−	slot hybrid.
	−	−	−	−	−	−	PCR

ND Not determined; − negative, + positive

Discussion

Intrahepatic injection of a cloned full-length HDV cDNA trimer caused a typical acute HDV hepatitis in both the chronic WHV carrier woodchucks inoculated.

The resulting infection was identical to an experimental infection transmitted by parenteral inoculation of viral particles: incubation time was identical in the two animals (five weeks post-inoculation), and it was similar to that of a virus-transmitted infection. In addition, the appearance of high levels of HDV-RNA in the liver and serum followed kinetics typical of an acute infection.

The cDNA clone [2] transfected has a nucleotide sequence predicting the presence of the sole small HDAg. However, the two 24 and 27 kd HDAg forms were present in sera of the inoculated animals. This finding confirms a similar observation made in the chimpanzee [5]. We also observed a marked transient suppression of the synthesis of WHV-DNA during the active phase of HDV replication. This phenomenon has been described previously as a characteristic feature of HDV superinfection in both human and chimpanzee. This observation seems to indicate a perfect analogy of the primate and the woodchuck model.

One main new point established by the present study is that the cloned HDV cDNA is able to induce chronic HDV infection and disease. The outcome of infection was less efficient in animals infected with serum from a transfected animal than in woodchucks inoculated with a standard HDV infectious pool. The reasons for such an incomplete outcome are under investigation.

In conclusion, this study demonstrates the relevance of using a cloned cDNA from HDV for functional analysis of the HDV genome. Recombinant plasmid pSVL-D3 contains a biologically competent clone that should be useful in mutagenesis studies aimed at determining the role of specific regions of the HDV genome in the molecular biology and pathogenicity of this peculiar viral agent.

References

1. Glenn JS, Taylor J, White JM (1990) In vitro-synthesis of hepatitis delta virus RNA initiates genome replication in cultured cells. J Virol 64: 3104–3107
2. Kuo MY, Chao M, Taylor J (1989) Initiation of replication of the human hepatitis delta virus genome cloned DNA: role of delta antigen. J Virol 63: 1945–1950
3. Ponzetto A (1991) Hepatitis delta virus: a clinical outlook. In: Hollinger FB, Lemon SM, Margolis H (eds) Viral hepatitis and liver disease. Wilkins, Baltimore, pp 882–884
4. Sambrook J, Fritsch EF, Maniatis T (1989) Molecular cloning: a laboratory manual, 2nd edn. Cold Spring Harbor Laboratory, Cold Spring Harbor
5. Sureau C, Taylor J, Chao M, Eichberg JW, Lanford RE (1989) Cloned hepatitis delta virus cDNA is infectious in the chimpanzee. J Virol 63: 4292–4297
6. Zignego AL, Deny P, Feray C, Ponzetto A, Gentilini P, Tiollais P, Brechot C (1990) Amplification of hepatitis delta virus RNA sequences by polymerase chain reaction: a tool for viral detection and cloning. Mol Cell Probes 4: 43–51
7. Zyzik E, Ponzetto A, Forzani B, Hele C, Heermann KH, Gerlich WH (1987) Proteins of Hepatitis delta virus in serum and liver. In: Robinson W, Koike K, Will H (eds) Hepadnaviruses. Alan R. Liss, New York, pp 565–577

Authors' address: Dr. M. Rapicetta, Department of Virology, Istituto Superiore di Sanità, Viale Regina Elena, 299, I-00161 Rome, Italy.

Arch Virol (1993) [Suppl] 8: 23–29

Detection of replicative intermediates of viral RNA in peripheral blood mononuclear cells from chronic hepatitis C virus carriers

M. Artini, G. Natoli, Maria Laura Avantaggiati, Clara Balsano, P. Chirillo, A. Costanzo, Maria Simona Bonavita, and **M. Levrero**

I Clinica Medica and Fondazione Andrea Cesalpino, Universita' La Sapienza, Rome, Italy

Summary. Clinical and experimental evidence suggests the possible existence of one or more extrahepatic sites of HCV infection. In order to demonstrate the "in vivo" infection of lymphoid cells by HCV, we applied a nested PCR to total cytoplasmic RNA extracted from fresh or cultured peripheral blood mononuclear cells (PBMCs) of HCV chronically infected patients, using primers derived from the highly conserved 5' untranslated region of the HCV genome. The presence of virions in PBMCs occurs frequently, if not always, and is often accompanied by active viral replication. Moreover, the appearance of replicative intermediates after stimulation of cellular growth with mitogens suggests that latent genomes could undergo replication upon cellular activation and/or proliferation.

Introduction

Hepatitis C virus (HCV) is considered a major aetiologic agent of NANB hepatitis, chronic liver disease and hepatocellular carcinoma (HCC) around the world. It is a positive-stranded RNA virus, the genome of which was recently cloned and entirely sequenced. Comparative sequence analysis indicates that the virus is distantly related to the flaviviruses and pestiviruses [2]. As a result of this similarity it has been assumed that HCV might replicate through the synthesis of an antigenomic RNA strand complementary to the genomic RNA, used as a template for the synthesis of new viral genomes. Due to the relatively small quantity of HCV-RNA found in serum and tissues of infected individuals, the only presently available method of detecting both genomic and antigenomic viral RNA is the polymerase chain reaction (PCR) method performed after reverse transcription of the viral genome.

The frequent recurrence of hepatic infection after liver transplant-ation, even in patients that are not viremic after grafting [3, 6, 10], has suggested the possible existence of one or more extrahepatic sites of HCV infection that might act as viral reservoirs. In particular, the possibility of lymphoid-cell infection by HCV is suggested by the ability of flaviviruses and pestiviruses to infect cells belonging to the mononuclear-phagocyte system [7]. This possibility is further suggested by the re-ported experimental transmission of NANB virus (possibly HCV) after injection of leucocyte preparations from patients with NANB hepatitis into healthy chimpanzees [4].

The aim of the present study was to search for in vivo evidence of infection and active replication of HCV by cDNA-PCR in fresh PBMCs as well as in short- and medium-term lymphocyte cultures obtained from chronically HCV infected patients.

Materials and methods

Patients

Peripheral blood mononuclear cells and sera were collected from 16 chronically HCV-infected patients (22 to 61 years old, 8 males and 8 females), 8 chronic HBV carriers and 10 healthy subjects. Percutaneous liver biopsies were obtained from all the patients affected by chronic HCV and HBV infection. Liver histology showed the features of chronic persistent hepatitis (CPH) in two cases, chronic active hepatitis (CAH) in 10 cases and CAH with liver cirrhosis (CAH/K) in the remaining 4 cases of the HCV patients (Table 1).

Serological tests

Serum samples were collected from total blood samples obtained on the same day as PBMCs extraction and were quickly stored at −70°C. The presence of anti-HCV antibodies was assayed for by a commercially available ELISA test (ORTHO diag-nostic Systems Inc., Westwood, MA, U.S.A.), and confirmed using a recombinant immunoblot assay RIBA-2 (Ortho diagnostic Systems Inc., Westwood, MA, U.S.A.). Serum HBsAg, HBeAg, antiHBs, antiHBc and antiHBe were measured by means of radioimmunoassay using standard kits (Abbott Laboratories, Chicago, IL, U.S.A.).

Preparation of fresh PBMC and short-term cultures

Heparinized blood (40 ml) from patients was diluted with Hanks' balanced salt solution (HBSS vol/vol), separated on a density gradient (Lymphoprep, Nycomed Pharma AS, Oslo, Norway) by centrifugation at 350 × g for 30 min. PBMC were collected and washed four times with about 300 ml of HBSS; a sample of the last washing medium was collected for subsequent testing by nested PCR. The final cell pellet was resuspended in 1 ml of HBSS, prior to RNA extraction.

Ten PBMC samples obtained from selected patients were cultured in RPMI supplemented with 10% FCS (GIBCO Lab., Grand Island, NY, U.S.A.) in the presence or absence of phorbol myristate acetate (PMA) and phytohemagglutinin (PHA) for 24 to 72 h (patients 3 and 5), or for 15 days in presence of PHA (patient 14).

RNA extraction from sera and cells

Sera (120 μl) were incubated in a lysis buffer consisting of 100 mM TRIS-HCl (pH 8), 1% sodium dodecyl sulfate and 50 μg/ml proteinase K, for 1 h at 37°C. After phenol/chloroform extraction, RNA was precipitated in ethanol, redissolved in 20 μl of sterile distilled water, and stored at $-70°C$. RNA was prepared from fresh or cultured PBMC using standard procedures [8]. Briefly, cells were lysed in cold isotonic buffer (30 mM TRIS-HCl pH 8.3, 100 mM NaCl, 5 mM $MgCl_2$, 5 mM $CaCl_2$, containing 0.5% Triton X and 0.1% diethylpyrocarbonate); after pelleting nuclei by centrifugation, the supernatant was extracted twice with phenol/chloroform and RNA was recovered by ethanol precipitation, resuspended in sterile water in aliquots of 1 μg and stored at $-70°C$.

Reverse transcription for HCV-RNA cDNA synthesis

Ten microliters of RNA solution and 10 pmol of downstream primer were incubated for 5 min at 65°C and rapidly cooled on ice. cDNA synthesis was carried out at 37°C for 45 min after adjustment of the mixture to contain 50 mM TRIS-HCl (pH 8.3), 75 mM potassum chloride, 3 mM magnesium chloride, 10 mM DTT, 2 mM dNTP and 200 IU of MMLV reverse transcriptase (Gibco, BRL), in a final volume of 25 μl. To detect either the positive or the negative strand of viral RNA, reverse transcription of the RNA samples was performed in the presence of the antisense primer SR1 (5'GG TGC ACG GTC TAC GAG ACC 3' spanning from position −21 to −1) for the genomic HCV-RNA, and of the sense primer SF1 (5'GCC ATG GGC GTT AGT ATG AG 3' nucleotides −259 to −239) for the antigenomic RNA. After RNase digestion the reverse transcriptase was inactivated by heating at 95°C for at least 20 min and then quickly chilled on ice; to rule out residual RT activity after heat inactivation of RT, a reaction was done using all components except the primer during the RT step. The cDNAs were stored at $-20°C$.

Nested PCR

For the detection of both genomic and antigenomic HCV-RNA a nested PCR was performed: 10 μl of cDNA were added to 40 μl of first PCR mix (dNTP: 250 μM each, KCl 50 mM, $MgCl_2$ 1.5 mM, 0.01% gelatine, 8 pmol of each outer primer and 2.5 U of cloned Taq polymerase (Cetus, Emeryville, CA, U.S.A.) were added; after carrying out a first amplification (25 cycles: 94°C for 45 sec, 45°C for 45 sec and 72°C for 55 sec) 150 μl of the second PCR mix were added containing 100 picomoles of each inner primer but no Taq polymerase: the second cycle of amplification (30 rounds of 94°C for 45 sec, 45°C for 45 sec and 72°C for 30 sec) was carried out in a final volume of 200 μl with an inner/outer ratio of 12/1. The oligonucleotide primers used were from the highly conserved 5' untranslated region of the prototype HCV cDNA nucleotide sequence of HCV(1): for the first cycle of amplification primers SR1 and SF1 were used, while primer antisense SR2 (5'ACG GGT GAG GTA GTA GAC CC 3' spanning from −24 to −44) and primer sense SF2 (5'GTG CAG CCT CCA GGA CCC CCGA 3' from −236 to °216) were used for the second step of the nested PCR. The expected size of the final product from the second amplification was 212 bp long.

Analysis of PCR products

Twenty microliters of each amplified product were analyzed by electrophoresis on 2% agarose gels. Bands were visualized by ethidium bromide staining, transferred to nylon membranes (Hybond-N+, Amersham, U.K.), and hybridized with a 32P-labeled probe (5'CCG GAA TTG CCA GGA CCG GGT CCT TTC TTG 3' spanning from −76 to −46).

Specificity and sensitivity controls

Due to the risk of contamination in the PCR procedure we used the following general precautions: the different steps in the PCR procedure were performed in distinct rooms; only disposable reagents were used and each RNA was processed with and without reverse transcriptase. Moreover, to detect carry-over, HCV-PCR was performed on serum-free lysis buffer for the detection of contamination at any step of procedure, on sera and PBMCs from blood donors who were HBV and HCV negative, and on the PCR mix alone. Finally, to detect contamination of the cell preparations with circulating virions, the last washing medium of fresh and cultured PBMCs was subjected to cDNA PCR.

Results

HCV-RNA was detected in PBMCs from 12 out of the 16 patients; 8 out of the 12 patients HCV-RNA positive in PBMCs were viremic, while 4 had no detectable serum HCV-RNA at the moment of the analysis (Table 1). HBV patients and healthy blood donors were all HCV-RNA negative. It is noteworthy that, since we excluded from the study about 30 more HCV patients on the basis of positivity for HCV-RNA in the PBMCs last washing medium, our results suggest that HCV-RNA can be frequently found in lymphocytes, but do not allow estimation of the actual prevalence of lymphocytic HCV-RNA in vivo.

Since four out of the eight patients harboring virions in their PBMCs were not viremic at the time of the study, the possibility of a false positive result due to the contamination of the cellular samples by the viremic sera can be ruled out in these cases.

Table 1. Detection of genomic and anti-genomic
HCV-RNA in PBMCs from chronic HCV carriers

No.	Sex	Hist.	Serum	PBMCs	
				genom.	anti-genomic
1.	F	CAH	+	+	−
2.	M	CAH/K	−	+	−
3.	M	CAH	−	+	−
4.	M	CPH	+	+	−
5.	F	CAH	+	+	+
6.	M	CAH	+	+	+
7.	F	CAH	+	+	+
8.	F	CAH	+	+	+
9.	F	CAH	+	−	NT.
10.	F	CAH/K	+	−	NT.
11.	M	CAH/K	+	+	NT.
12.	M	CPH	+	−	NT.
13.	M	CAH/K	−	+	NT.
14.	F	CAH	−	−	−
15.	F	CAH	−	+	NT.
16.	M	CAH	+	+	NT.

In 8 out of the 12 patients that were positive for genomic HCV-RNA in PBMCs we also searched for the presence of minus-strand HCV-RNA, as evidence for active viral replication. Antigenomic HCV-RNA was detected in 4 out of the 8 tested patients.

PBMCs from patients 3 and 5 showed the presence of genomic HCV-RNA before and after 24–72 h in culture in the presence of mitogens. Antigenomic HCV-RNA was also detected in the PBMCs from both patients, but, interestingly, in patient 3 antigenomic RNA was positive only after lymphocyte stimulation. In patient 14, both genomic and antigenomic HCV-RNA were detectable only after stimulation with PHA for 15 days (Table 2).

Discussion

In this study we confirm the possibility of the infection of cells other than hepatocytes (in particular lymphocytes) by HCV in vivo. Clinical observations in liver-transplant patients [3] and experiments on chimpanzees [4] had already suggested this possibility, which was recently confirmed in vitro by demonstration of the infection and active replica-

Table 2. Detection of HCV-RNA in short- and medium-term PBMCs cultures from chronic HCV carriers

	Fresh	24 h	72 h
Pz. 3			
genomic	+	+	+
antigenomic	−	+	+
Pz. 5			
genomic	+	+	−
antigenomic	+	+	−
	fresh	unstim.[a]	stim.[a]
Pz. 14			
genomic	−	−	+
antigenomic	−	−	+

[a] PBMCs were cultured in presence or absence of PHA for 15 days

tion of HCV in the human T cell line [9], MOLT 4. Our observations demonstrate that in patients with chronic HCV infection, the presence of virions in PBMCs is a frequent, but not constant event, often accompanied by active viral replication. Moreover, the appearance of replicative intermediates after stimulation of the cellular growth with mitogens, suggests that latent genomes could undergo replication upon cellular activation and/or proliferation. As already described for other viral infections of lymphocytes [5] the status of cellular growth seems to supply a more suitable environment for viral replication.

These observations indicate that the lymphoid compartment of chronic carriers can be a reservoir for latent HCV that in some instances can be induced to replicate by cellular activation. Whether the presence and/or replication of HCV in lymphocytes might be responsable for altered lymphocyte trafficking or for the immunologic deletion of virus-infected clones of lymphocytes still remains to be determined.

References

1. Choo QL, Richman KH, Han JH, Berger K, Lee C, Dong C, Gallegos C, Coit D, Medina-Selby A, Barr PJ, Weiner AJ, Bradley DW, Kuo G, Houghton M (1991) Genetic organization and diversity of the hepatitis C virus. Proc Natl Acad Sci USA 88: 2451–2455

2. Choo QL, Kuo G, Weiner AJ, Overby LR, Bradley DW, Houghton M (1989) Isolation of a cDNA clone derived from a blood-borne nonA-nonB viral hepatitis genome. Science 244: 359–362

3. Feray C, Samuel D, Thiers V, Gigou M, Pichon F, Bismuth A, Reynes M, Maisonneuve P, Bismuth H, Brechot C (1992) Reinfection of liver graft by hepatitis C virus after liver transplantation. J Clin Invest 89: 1361–1365

4. Hellings JA, van der Veen-du Prie J, Boender P (1988) Transmission of non-A non-B hepatitis by leukocyte preparations. In: Zuckerman AJ (ed) Viral hepatitis and liver disease. Alan R. Liss, New York, pp 543–549

5. Korba BE, Cote PJ, Gerin JL (1988) Mitogen-induced replication of woodchuck hepatitis virus in cultured peripheral blood lymphocytes. Science 241: 1213–1216

6. Martin P, Munoz SJ, Di Bisceglie AM, Rubin R, Waggoner JG, Armenti VT, Moritz MJ, Jarell BE, Maddrey WC (1991) Recurrence of hepatitis C virus infection after orthotopic liver transplantation. Hepatology 13: 719–721

7. Moennig V, Plagemann P (1992) The pestiviruses. Adv Virus Res 41: 53–98

8. Sambrook J, Fritsch EF, Maniatis T (1989) Molecular cloning, a laboratory manual, 2nd edn. Cold Spring Harbor Laboratory, Cold Spring Harbor

9. Shimizu YK, Iwamoto A, Hijikata M, Purcell RH, Yoshikura H (1992) Evidence for in vitro replication of hepatitis C virus genome in a human T-cell line. Proc Natl Acad Sci USA 89: 5477–5481

10. Wright TL, Donegan E, Hsu HH, Ferrel L, Lake JR, Kim M, Combs C, Fennessy S, Roberts JP, Ascher NL, Greenberg HB (1992) Recurrent and acquired hepatitis C viral infection in liver transplant recipients. Gastroenterology 103: 317–322

Authors' address: Dr. M. Artini, Istituto di I Clinica Medica, Policlinico Umberto I, Viale del Policlinico 155, I-00161 Rome, Italy.

Arch Virol (1993) [Suppl] 8: 31–39

Susceptibility of human liver cell cultures to hepatitis C virus infection

G. Carloni[1], S. Iacovacci[1], M. Sargiacomo[2], G. Ravagnan[1], A. Ponzetto[3], C. Peschle[2], and M. Battaglia[1]

[1] Institute of Experimental Medicine, C.N.R., Rome
[2] Laboratory of Hematology-Oncology, Instituto Superiore di Sanità, Rome
[3] Division of Gastroenterology, Le Molinette Hospital, Turin, Italy

Summary. To develop a cell culture system susceptible to infection by hepatitis C virus (HCV), human fetal hepatocytes, grown in serum-free medium, were inoculated with serum samples from two HCV-infected patients. Viral RNA sequences were detected by polymerase chain reaction, using primers specific for the 5′ noncoding region of HCV, in extracts prepared from the hepatocyte cultures as early as 5 days after inoculation. Virus was also released from the infected cells into the medium. The HCV strains could be serially passaged three times into fresh liver cell cultures using intracellular virus as inoculum. Evidence that HCV replication really took place in primary human fetal hepatocytes was also obtained by detection of minus-strand viral RNA (replication intermediate) in cell extracts and of viral antigens in the infected cells.

Introduction

Until recently, progress in research on the hepatitis viruses was hampered by the lack of suitable in vitro models [15]. Nonetheless, by recombinant DNA technology important knowledge has been acquired on the biology of these viruses [4]. In the case of hepatitis C virus (HCV), it was possible to identify by molecular cloning its genome from plasma of an experimentally infected chimpanzee [1]. This also enabled the development of serological assays for detection of HCV-specific antibody, which greatly expanded our knowledge of the epidemiology and on the natural history of non-A, non-B hepatitis [2, 16].

It has been recently demonstrated that hepatocytes from liver biopsies obtained from experimentally infected chimpanzees, could be maintained in vitro in serum-free medium [7]. HCV-specific antigens were demonstrated by immunocytochemical methods in the cell cultures derived from the biopsies. The detection of virus-like particles with a diameter of 40–60 nm, by electron microscopy, and of HCV-specific RNA sequences, by polymerase chain reaction (PCR), in cells and culture media, revealed that long-term liver cell cultures were able to support HCV replication.

This prompted us to prepare in vitro cultures of primary human fetal hepatocytes, and to test them for their ability to support HCV replication. Evidence for HCV replication was obtained, based on the detection of structural and non-structural HCV antigens, on the presence of genomic and antigenomic viral RNA strands in the infected cells, and on serial propagation of virus.

Materials and methods

Fetal hepatocyte cultures

Primary human fetal hepatocytes were isolated from livers of 9 week-gestation fetuses, obtained from therapeutic abortions, by the method of Salas-Prato et al. [14], with the following modifications: embryos were soaked at 4°C in extracellular matrix and cells were dissociated by repeated flushing through a 22 gauge needle. One ml of cell suspension in Hanks balanced salt solution was centrifuged for 20 min at 200 g at 4°C. The liver cell pellets were planted onto plastic substrates coated with basement membrane matrix (Matrigel, ICN Biochemicals Inc., Costa Mesa, CA, U.S.A.) in Dulbecco's modified Eagle medium supplemented with epidermal growth factor, insulin and other metabolites, as previously described for baboon's hepatocytes [9]. The cultures were incubated at 37°C in a humidified atmosphere of 95% air and 5% CO_2. Monolayers of viable human fetal hepatocytes were maintained in their differentiated state for over three months, based on their ability to synthesize liver-specific proteins [9] and on their typical morphology by phase contrast microscopy (Fig. 1). Secretion of albumin and alpha-fetoprotein into culture medium, as determined by commercially available nephelometric and immunoradiometric assays, respectively, remained constant throughout long-term cultivation of liver cells.

Virus inoculation

Cultures of human fetal hepatocytes, maintained as described above in 25 cm^2 polystyrene flasks, were inoculated with sera from two patients, GF and MM, who were chronically infected with HCV. The specimens contained approximately 1×10^5 HCV genomes per ml, as determined by polymerase chain reaction (PCR) in end-point dilution assays (data not shown). Each flask received 0.5 ml of inoculum diluted in 5 ml of culture medium. The cell monolayers were washed six times with fresh medium 18 h

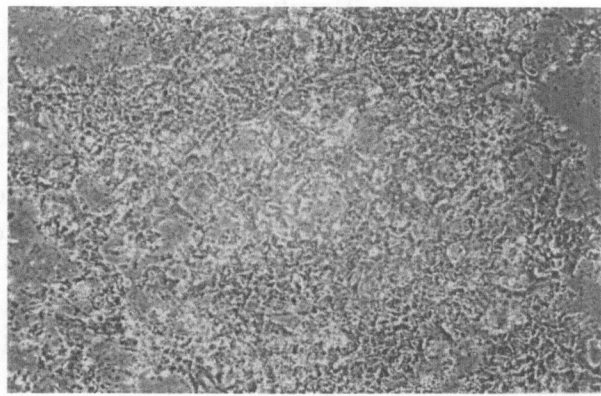

Fig. 1. Phase contrast microscopy of primary human fetal hepatocytes cultured for 21 days in serum-free medium. Magnification ×25

after inoculation and were maintained for over 4 weeks at 37°C. Serum specimens from the infected patients were also inoculated into the following continuous human liver cell lines: Li7A, HepG2, and HepG2.2.15.

RNA extraction and PCR analysis

The RNAs were extracted from cytoplasmic fractions [8] and from ultracentrifuged medium pellets of inoculated and uninoculated cultures as described in [11]. The presence of HCV genome was determined by nested PCR following a cDNA reverse transcription step [12]. Primers, product sizes, and the PCR reaction conditions for the 5' noncoding region and the NS4 nonstructural C-100 region have been extensively described [11, 15]. The minus strand of viral RNA was detected after reverse transcription in the presence of the outer sense primer for the 5' noncoding region [11]. The reaction was carried out as described by Fong et al. [3], except that reverse transcriptase was inactivated by heating at 95°C for 60 min. PCR products were analyzed by electrophoresis in 2% agarose gels and ethidium bromide staining followed by Southern blotting on nylon membranes, hybridization with a ^{32}P-probe, and autoradiography. The probes were: 5 'CCATAGTGGTCTGCGGAACCGTGAGTACACCGGAAT 3' at positions 122–158 of 5' noncoding region and 5'CCTACGGATTCCAATACTCACC AGG 3' at positions 6481-6505 of the C-100 nonstructural region.

Immunofluorescence for antigens in infected cultures

Infected and uninfected cells were fixed for 15 min in absolute ethanol at −20°C, and then examined for the presence of viral antigens by indirect immunofluorescence. The following mouse monoclonal antibodies, all of IgG1 subclass, were employed: i) the TORDJI-22 clone (Biosoft, Paris, France), directed to the C-100-3 nonstructural HCV protein, and, ii) clone No. 1851 (Virostat, Portland, MN, U.S.A.) directed to viral capsid protein. The binding of antibody to infected cells was revealed by using fluorescein-conjugated affinity-purified goat antibody specific for mouse IgG (Sigma, St. Louis, MO, U.S.A.).

Results

The infected culture fluids were harvested at different times after inoculation and examined for the presence of genomic HCV RNA by PCR using primers specific for the 5′ noncoding region of the HCV genome. As shown in Table 1, small amounts of virus were released into the media of fetal hepatocyte cultures as early as 8 days after inoculation, as demonstrated by their PCR positivity only when the samples were ultracentrifuged. The experiments with HCV strains GF and MM gave comparable results. Viral sequences were not detected in the media from uninoculated hepatocyte cultures or from inoculated cultures of the continuous human liver cell lines Li7A, HepG2, and 2.2.15. Fetal liver cells were harvested at 5, 10, 12 and 24 days after inoculation, and were examined for the presence of genomic HCV RNA by nested PCR (Table 2). HCV RNA was detected in cell extracts for as long as 4 weeks after inoculation. Infection of hepatocytes at different times after establishment of the cultures indicated that they remained susceptible to infection for at least 3 weeks after preparation. Reduction of the adsorption time

Table 1. Time course of HCV infection in human liver cells

Cell culture	PCR on the RNAs extracted from pellets of ultracentrifuged medium at the indicated days after inoculation[a]					
	0[b]	1	5	8	12	24
Fetal hepatocytes[c]	−	−	−	−	−	−
Fetal hepatocytes + HCV	+	+/−[d]	−	+	+	+
Li7A hepatoma cells + HCV	−	−	−	−	−	−
HepG2 hepatoblastoma cells + HCV	−	−	−	−	−	−
HepG 2.2.15 cell clone + HCV	−	−	−	−	−	−

[a] HCV RNA was assayed by PCR and primer pairs for the 5′ noncoding region. The results are the summation of several duplicate experiments using the MM virus strain

[b] The day 0 colum shows results of the ultracentrifuged pellets from the last wash after removal of unadsorbed inoculum

[c] Media harvested from uninoculated control cultures

[d] Discordant results

Table 2. Detection of HCV RNA in human fetal
hepatocyte cell cultures

Cell culture	PCR on the RNAs extracted from cell pellets at the indicated days after inoculation[a]			
	5	10	12	24
Fetal hepatocytes[b]	–	–	–	–
Fetal hepatocytes + HCV	+	+	+	+

[a] PCR was performed as in Table 1
[b] Cell pellets from uninoculated control cultures

from 18 h, used in all the experiments described above, to 2 h did not seem to appreciably reduce the efficiency of viral multiplication. Viral sequences were not detectable in uninoculated control cultures, or in the cultures of the continuous human liver cell lines inoculated with serum samples from the HCV-infected patients.

The two HCV strains could be serially passed three times into fresh fetal liver cell cultures using as inoculum the cells of the isolation cultures frozen and thawed once. Approximately 3×10^6 cells were inoculated with the extracts of infected cultures containing about 1×10^6 cells. Viral sequences were not detected in Li7A, HepG2 and 2.2.15 cultures which, in control experiments, were inoculated with the same materials (data not shown). In addition, intracellular minus-strand viral RNA, putative replication intermediate, undetectable in the virions pelleted from the serum inocula, was detected in the hepatocyte cultures from 2 to 4 weeks after inoculation. The cells were examined for HCV core and C-100-3 antigen by indirect immunofluorescence with mouse monoclonal antibodies. At 12 days after inoculation, a small proportion of cells (1%) were positive for both antigens (Fig. 2).

Discussion

A system for the in vitro propagation of HCV was developed, using monolayers of primary human fetal hepatocytes that retained their liver-specific differentiated function for more than three months. The cells

G. Carloni et al.

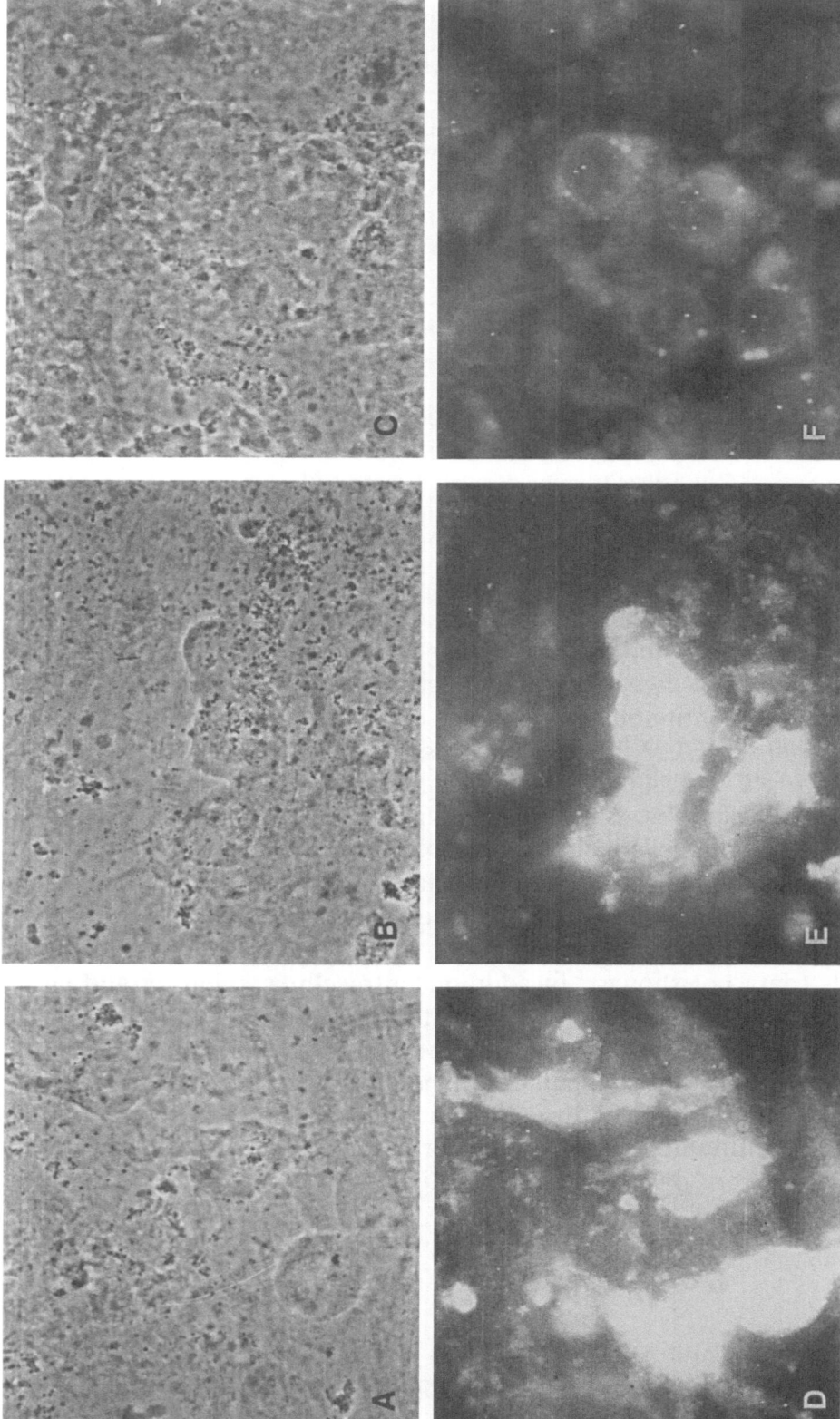

Fig. 2. Indirect immunofluorescent staining of HCV-infected human fetal liver cells 12 days after inoculation using mouse monoclonal antibodies directed to HCV capsid protein (**D**), and to the nonstructural protein C-100-3 (**E**). Lack of staining with a mixture of the same antibodies in uninoculated cells (**F**). **A–C** Phase contrast micrographs of the same fields shown in **D–F**, respectively. Magnification ×500

were maintained in serum-free medium supplemented with epidermal growth factor and insulin, employing an extracellular matrix to improve the persistence of the differentiated state of cultured hepatocytes [10]. Primary human fetal [13] and adult [6] hepatocytes were previously demonstrated to support in vitro replication of hepatitis B virus. This suggests that human liver cell cultures might be a valuable model for the study of several hepatotropic viruses. However, persistence of the differentiated state of these cell systems was limited to a few weeks, probably because the method of cell culture preparation (i.e., the use of collagenase for dissociation of tissue, and of bovine serum for growing the cells).

Active replication of HCV was suggested by the release of detectable virus from the inoculated cultures into the medium, and by the detection of positive and negative strands of HCV RNA in the extracts of infected hepatocyte cultures. In this study, we could not perform a quantitative PCR, however, semiquantitative end-point dilution assays (i.e., PCR with serial dilutions of replicate samples) revealed that effective viral replication takes place in fetal hepatocytes (unpubl. data).

Although the fluids of the hepatocyte cultures harvested 1 day after inoculation were positive for detection of HCV genome by PCR, probably because of the persistence of small amounts of the input virus, we believe for several reasons that our findings are not the consequence of a carry over effect. First, at virus isolation and at each passage, the inocula were removed and the cell monolayers were extensively washed after virus adsorption; passages were done, using frozen and thawed cells, 2–4 weeks after inoculation. It seems unlikely that HCV would be so stable, after adsorption to cells, to be detected by PCR after as long as 4 weeks at 37°C. A passive carry-over effect seems unlikely also because viral sequences were not detectable in non-susceptible cells inoculated, in control experiments, with HCV-infected hepatocytes frozen and thawed once. Second, negative strand HCV RNA (replicative intermediate) was detected in the hepatocyte cultures 12 days after inoculation, but not in the serum specimens used as inocula. Thus, we believe that our cultures are positive for HCV detection by PCR due to viral infection and replication. Finally, using monoclonal antibodies from two different sources, we showed by immunofluorescence the presence of antigens related to the viral capsid and to the nonstructural protein C-100-3 in the inoculated cultures. This further supports that HCV genome replication and expression really takes place in fetal liver cells. It would be of interest to investigate whether the viral progeny produced by infected human fetal liver cultures show an altered cell or organ tropism in vivo, and to determine whether mutations in the HCV genome might be responsible for such changes.

In conclusion, primary human fetal hepatocyte cultures might prove useful for the study of HCV replication as well as for the production and assay of infectious virus, with important implications for diagnosis and prevention of HCV infection.

Acknowledgements

During this study S.I. was recipient of fellowships from Clonit and from Consiglio Nazionale delle Ricerche, BTBS target project. This work was partially supported by grants from Consiglio Nazionale delle Ricerche (CNR) "Target projects FATMA and BTBS".

References

1. Choo QL, Kou G, Weiner AJ, Overby LR, Bradley DW, Houghthon M (1969) Isolation of cDNA clone from a blood-borne non-A, non-B viral hepatitis genome. Science 244: 359–361
2. Esteban JI, Gonzalez A, Hernandez JM, Viladomiu L, Sanchez C, Lopez-Tavalera JC, Lucea D, Martin-Vega C, Vidal X, Esteban R, Guardia J (1990) Evaluation of antibodies to hepatitis C virus in a study of transfusion-associated hepatitis. N Engl J Med 323: 1107–1112
3. Fong TL, Shindo M, Feinstone SM, Hoofnagle JH, Di Bisceglie AM (1991) Detection of replicative intermediate of hepatitis C viral RNA in liver and serum of patients with chronic hepatitis C. J Clin Invest 88: 1058–1060
4. Ganem D, Varmus HE (1987) Molecular biology of the hepatitis viruses. Annu Rev Biochem 56: 651–693
5. Garson JA, Tedder RS, Briggs M, Tuke A, Parker D, Barbara JAJ, Contreras M, Aloysius A (1990) Detection of hepatitis C viral sequences in blood donations by "nested" polymerase chain reaction and prediction of infectivity. Lancet 325: 1419–1422
6. Gripon P, Diot C, Thézé N, Fourel I, Loreal O, Brechot C, Guguen-Guillouzo C (1988) Hepatitis B virus infection of adult human hepatocytes cultured in the presence of dimethyl sulfoxide. J Virol 62: 4136–4143
7. Jacob JR, Burk KH, Eichberg JW, Dreesman GR, Lanford RE (1990) Expression of infectious viral particles by primary chimpanzee hepatocytes isolated during the acute phase of non-A, non-B hepatitis. J Infect Dis 161: 1121–1127
8. Kawasaki ES, Clark SS, Coyne MY, Smith SD, Champlin R, Witte ON, McCormick FP (1988) Diagnosis of chronic myeloid and acute lymphocytic leukemias by detection of leukemia-specific mRNA sequences amplified in vitro. Proc Natl Acad Sci USA 85: 5698–5702
9. Lanford RE, Carey KD, Estlack LE, Smith GC, Hay B (1989) Analysis of plasma protein and lipoprotein synthesis in long-term primary cultures of baboon hepatocytes maintained in serum-free medium. In Vitro Cell Dev Biol 25: 174–182
10. Moshage H, Yap SH (1992) Primary cultures of human hepatocytes: a unique system for studies in toxicology, virology parasitology and liver pathophysiology in man. J Hepatol 15: 404–411

11. Nalpas B, Thiers V, Pol S, Driss F, Thepot V, Berthelot P, Brechot C (1992) Hepatitis C viremia and anti-HCV antibodies in alcoholics. J Hepatol 14: 381–384
12. Novati R, Thiers V, D'Arminio Monforte A, Maisonneuve P, Principi N, Conti M, Lazzarin A, Brechot C (1992) Mother-to-child transmission of hepatitis C virus detected by nested polymerase chain reaction. J Infect Dis 165: 720–723
13. Ochiya T, Tsurimoto T, Ueda K, Okubo K, Shiozawa M, Matsubara K (1989) An in vitro system for infection with hepatitis virus that uses primary human fetal hepatocytes. Proc Natl Acad Sci USA 86: 1875–1879
14. Salas-Prato M, Tanguay JF, Lefebvre Y, Wojciechowicz D, Liem HH, Barnes DW, Ouellette G, Muller-Eberhard U (1988) Attachment and multiplication, morphology and protein production of human fetal primary liver cells cultured in hormonally defined media. In Vitro Cell Dev Biol 24: 230–238
15. Tiollais P, Pourcel C, Dejean A (1985) The hepatitis B virus. Nature 317: 489–492
16. Van der Poel CL, Cuypers HTL, Reesink HW, Weiner AJ, Quan S, Di Nello R, Van Boyen JJP, Winkel I, Mulder-Folkerts D, Exel-Oehlers PJ, Schaasberg W, Leentvaar-Kuypers A, Polito A, Houghton M, Lelie PN (1991) Confirmation of hepatitis C virus infection by new four-antigen recombinant immunoblot assay. Lancet 337: 317–319

Authors' address: Dr. G. Carloni, IMS CNR, Viale Marx 15, I-00137 Rome, Italy.

II. Molecular biology of hepatitis B virus

Arch Virol (1993) [Suppl] 8: 43–52

© Springer-Verlag 1993

Molecular basis of the diversity of hepatitis B virus core-gene products

H.-J. Schlicht, G. Wasenauer, and **J. Köck**

Department of Virology, University of Ulm, Ulm, Federal Republic of Germany

Summary. All hepatitis B viruses examined to date code for at least two different core-gene products which are referred to as the c- and the e-protein. In the case of the human hepatitis B virus, they are known as the HBcAg and the HBeAg. Although these proteins share most of their primary amino acid sequence, they exhibit quite distinct properties. The e-protein is located in the cytoplasm and the nucleus of infected cells and very efficiently assembles into nucleocapsids. By contrast, the e-protein does not form particles. It enters the secretory pathway and is actively secreted by the cells. Here we describe the biosynthetic pathways by which the c- and e-proteins are expressed and summarize recent data from our laboratory showing that the antigenic and biophysical properties which distinguish the HBeAg from the HBcAg are primarily due to the 10 amino acid long portion of the HBeAg leader sequence that remains attached to the HBeAg after cleavage.

Introduction

In 1972, Magnius and Espmark described a new serological marker in the serum of HBV infected patients which they called the e-antigen [17]. The term e-antigen, which often is misinterpreted as "envelope antigen", was chosen because the letters a–d had already been assigned to other virus markers. Determination of the e-antigen in the serum of HBV infected patients later proved to be of immense practical value since its presence correlated very well with viremia and therefore with infectivity. Furthermore, in the vast majority of cases elimination of the HBeAg and subsequent appearance of the corresponding antibody, antiHBe, indicates that the virus is being eliminated and that the infection will either resolve or enter a usually mild chronic course which is characterized by low or undetectable virus levels.

Biochemical and serological studies later demonstrated that the HBeAg was related to one of the major structural viral proteins, the core-protein or HBcAg, which is the building block of the viral capsid [9, 18, 34, 35]. In fact, since the HBeAg was found to be slightly smaller than the HBcAg and since denaturation by boiling or protease treatment serologically converted the HBcAg into HBeAg, it was generally assumed that the HBeAg was a degradation product of the HBcAg. However, several recent reports convincingly demonstrated that the secretory core-gene product of hepadnaviruses is by no means derived from the capsid protein but rather represents a second independently expressed protein which, by an elaborate biosynthetic mechanism, acquires a structure which endows it with properties which are quite distinct from those of the HBcAg [4, 15, 23, 27, 33]. The description of this remarkable expression strategy which results in the production of two distinct proteins from one gene is the subject of this article.

Biophysical properties of HBcAg and HBeAg

As was mentioned above, the HBcAg, which in the following will be referred to as the HBc-protein, if the protein rather than its antigenicity is being discussed, represents the building block of the virus capsid. This capsid has an icosahedral structure and consists of 180 HBc-molecules which appear to be identical [12]. There is no evidence that other proteins besides the HBc-protein are included.

In line with its major function to form the capsid is the most prominent feature of the HBc-protein, its exceptionally strong tendency to form ordered aggregates. In fact, this protein forms nucleocapsids which are morphologically identical to those found in infected tissue in every cell type tested to date, e.g. *E. coli* [7], yeast [20], frog [41], insect [16] or mammalian cells [32]. In infected patients, the HBc-protein can mainly be detected in the cytoplasm and the nuclei of liver cells and is only released as part of a virus particle packaged into an envelope.

One consequence of HBc-assembly is the formation of a conformational epitope, the HBc-epitope, which is the serologic hallmark of assembled HBc-protein [8, 10, 24, 26, 39]. Antibodies which specifically bind to this epitope occur very early but are of no relevance for the course of the infection. Upon denaturation of the HBc-particles, HBc-antigenicity completely disappears [18, 34].

Compared to the HBc-protein, the features of the HBe-protein are quite distinct. First of all, there is no convincing evidence that this protein can form ordered aggregates under physiological conditions. Consequently, the HBe-protein is completely devoid of HBc-antigenicity.

Instead, it exhibits two non cross-reactive HBe-specific epitopes which are referred to as HBe1 and HBe2 or HBe-alpha and HBe-beta [9, 14, 24, 25]. Antibodies which bind to these epitopes occur only late during infection and correlate with virus elimination.

Moreover, in contrast to the HBc-protein, the HBe-protein is mainly located in the serum of infected patients, suggesting that it is released by the infected cells. This property together with the correlation between HBe-elimination and cessation of virus production has lead to the speculation that a function of the HBe-protein might be the induction of a virus specific immunosuppressed state [19]. However, evidence supporting this hypothesis is scarce.

Evidence from tissue culture experiments suggests that the e-protein of both the human and the duck hepatitis B virus is not only secreted but can also be incorporated into the cell membrane [29, 31]. Since e-protein specific antibodies can bind to this protein, it represents a potential target for an antibody mediated elimination of infected cells [30]. However, the significance of these findings is still unclear.

The HBe- and the HBc-protein are different proteins which are encoded by the same gene

Serological studies first suggested that the HBe-protein was a degradation product of the HBc-protein. This view was first challenged when Ou et al. demonstrated that HBe-secretion depended upon the expression of a small open reading frame, the so called preC region, which precedes the core-gene proper [23]. Further analyses which were also extended to the duck hepatitis B virus (DHBV) then lead to the discovery of an elaborate biosynthetic strategy by which two independent core-gene products are expressed from one gene [15, 23, 27, 33].

Figure 1 shows a schematic representation of the mechanism of c- and e-biosynthesis using HBV as an example. The expression strategy is essentially the same for both the duck and the human virus. As is shown in Fig. 1, there are two closely spaced in-phase initiation codons located at the 5' end of the preC/C-gene. From this gene, two different mRNAs are transcribed. The first mRNA, which is also referred to as the pregenome because it also serves as the replication template, is slightly smaller and starts between these two ATGs. Consequently, the first AUG on this mRNA, which is used for translation initiation, is the C-AUG. Translation of this mRNA results in the biosynthesis of the c-protein. This protein is located primarily in the cytoplasm of the cells where it assembles into nucleocapsids which are serologically defined as HBcAg. During the assembly process, the c-protein interacts with the

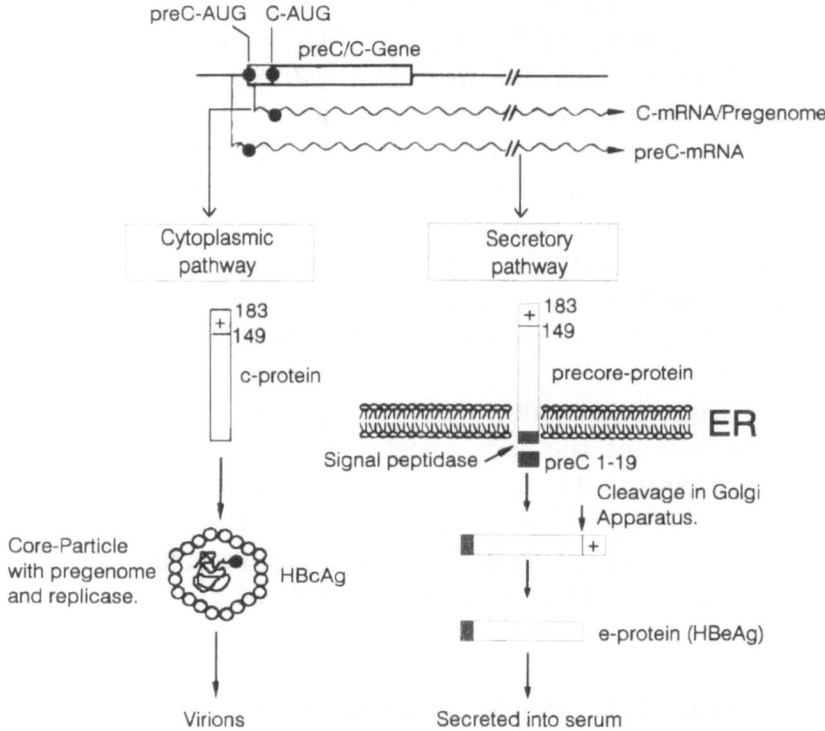

Fig. 1. Schematic representation of the two different biosynthetic pathways which are entered by the hepatitis B virus core gene products. Details are given in the text

primary replication complex consisting of the pregenomic RNA and the replicase, which is then incorporated into the nascent nucleocapsid. A new viral genome is then generated by reverse transcription of this RNA. Most recently it has been shown that neither the pregenomic RNA nor the polymerase alone can be encapsidated [1]. Thus, it is reasonable to assume that it is the polymerase/pregenome complex which interacts with the aggregating c-protein. In this context it should be stressed that nucleocapsid formation is independent from genome encapsidation and that perfect capsids are produced even if the c-protein alone is expressed in the cell.

While biosynthesis of the c-protein is a rather straightforward process, the biosynthetic mechanism by which the e-protein is produced is more complicated. This protein is encoded by the second mRNA, the preC-mRNA, which initiates upstream from the most 5'ATG. Translation of this mRNA therefore starts at the preC-AUG and results in the production of a preC/C or precore-protein which, besides the core-sequences, contains at its N-terminus the small peptide encoded by the preC-region. This small extra piece functions as a signal sequence for secretion and mediates the translocation of the precore-protein across the membrane of the endoplasmic reticulum. During intracellular transport, the precore protein is then subjected to two proteolytic processing steps. Firstly, 19 of the 29 preC amino acids are cleaved. Secondly, a strongly basic C-terminal sequence (amino acids 149–183 in case of HBV) which has the properties of a nucleic acid binding domain and which is essential for the formation of replication-competent nucleocapsids [13, 21, 28, 40] is removed [27, 34]. This processing step occurs late during HBe- and DHBe-synthesis, most likely during passage of the protein through the Golgi apparatus [31, 36]. Thus, with respect to its primary sequence the e-protein differs from the c-protein in that it contains a short extra sequence at its N-terminus and lacks the basic domain at the C-terminus.

With respect to these findings, three possible explanations for the serological and structural differences between the HBc- and the HBe-protein are obvious: either it is the different cellular compartment (cytoplasmic vs. secretory) which enforces the structural differences or it are the differences in the primary sequence or a combination of both. In fact, the third explanation turned out to be correct.

The non-cleaved portion of the preC-peptide determines the differences between the HBc- and the HBe-protein

One possible explanation for the distinct aggregational behaviors of the c- and the e-proteins, the differences in the C-terminus, could be ruled out since an HBc-protein which lacks this region still assembles into nucleocapsids ([2, 11]; see also Figs. 2A,B). In this context it should also be mentioned that the four cysteines within the HBc-protein seem to be unimportant for capsid-assembly and are also not required for genome replication [22, 42]. We therefore focused our work on the possible influence of the preC sequence on HBe-structure formation.

In case of the human hepatitis B virus, the only hepadnavirus for which such an analysis has been performed to date, the preC sequence consists of 29 amino acids. Of these, 19 are cleaved by the signal peptidase during the first processing step, leaving 10 preC amino acids at the mature HBe-protein [33]. That this short sequence is crucial for HBe-structure was first suggested by studies in which the preC region was replaced by a different, completely unrelated sequence, the signal peptide of an influenza hemagglutinin [32]. When such a recombinant HBe-protein was expressed, a secretory core-gene product was produced which not only assembled into particles but which also exhibited both HBe- and HBc-antigenicity.

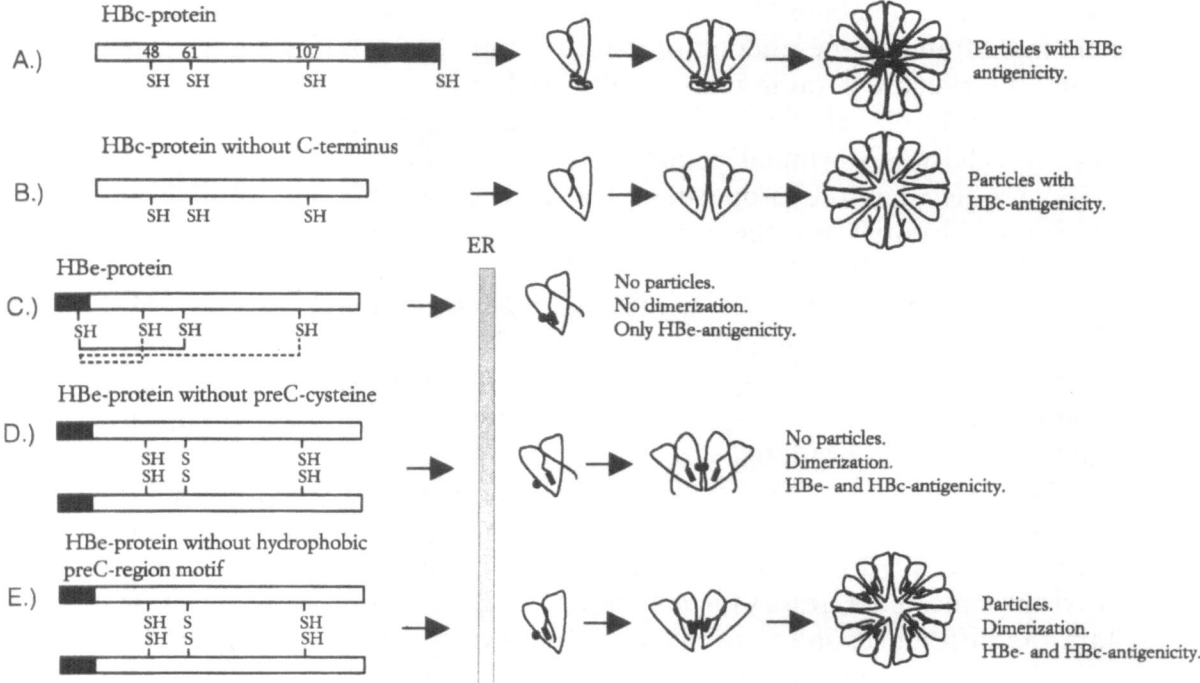

Fig. 2. Summary of the features of the different hepatitis B virus core gene products discussed in this review

Further analysis of this phenomenon revealed that at least two elements which are essential for proper HBe-biosynthesis reside in the 10 non-cleaved preC amino acids [37, 38]. One of these elements is a cysteine residue. In the wild-type protein, this cysteine forms an intramolecular disulfide bridge with one of the three internal cysteines (Fig. 2C). This protein does not aggregate, it occurs mostly in monomeric form and exhibits only HBe-antigenicity. The internal cysteine at aminoacid position 61 (see Fig. 2A for numbering) appears to be the most important reaction partner, but in principle any of the three internal cysteines can do. This intramolecular disulfide bridge is important for proper antigenicity of the HBeAg. If it does not form, a protein is produced which has a much reduced HBe-antigenicity.

If the cysteine in the preC region is mutated, a dramatic shift in quaternary structure and antigenicity occurs. Now the internal cysteine at position 61 forms an intermolecular disulfide bridge with the same cysteine in a second molecule (Fig. 2D). This HBe-dimer exhibits both HBe- and HBc-antigenicity. This shows that the cysteine in the preC region is essential for the production of an HBe-molecule with only e-antigenicity. However, this HBe-dimer does not assemble into particles.

Aggregation of the HBe-protein is rather inhibited by a second element which appears to be a strongly hydrophobic tripeptide (sequence WLW in one letter code). If this tripeptide is mutated, the HBeAg assembles into particles which have the same morphology as normal nucleocapsids (Fig. 2E). Assembly is more efficient if the preC-cysteine is also mutated, but small amounts of particulate HBeAg are produced also in the presence of the cysteine.

Although not experimentally proven, it can be assumed that the influence of the preC sequence on disulfide bond formation requires the environment of the endoplasmic reticulum. The signal peptidase which cleaves the preC region is confined to this cellular compartment. Furthermore, it is the oxidizing redox potential of the ER which permits the efficient formation of disulfide bridges. In the cytoplasm, such bonds form only very rarely because the redox potential is strongly reducing.

Conclusions

Recent studies have shown that the preC sequence of the human hepatitis B virus is an unusual signal sequence which not only mediates the translocation of the HBe-precursor across the membrane of the endoplasmic reticulum but also determines the structural and serologic properties of the mature HBeAg. A cysteine located within this region determines the quaternary structure and the antigenicity of the HBeAg via formation of an intramolecular disulfide bridge. If this cysteine is lacking, the HBeAg, which is predominantly a monomer with only HBe-antigenicity, is expressed as a disulfide linked homodimer showing both HBe- and HBc-antigenicity.

Why it might be important for the virus to express a secretory core-gene product which is serologically distinct from the core protein can only be speculated. To date, the biological relevance of the HBe-protein is still unclear. Since production of such a secretory core-gene product is a feature which is conserved between the most distantly related hepadnaviruses, the human and the duck hepatitis B virus, it is generally assumed that, although not essential for establishment of an infection [3, 5, 6, 27], it is of some benefit for the virus. For instance, it has been suggested that this protein may have a tolerogenic effect on the immune system [19]. If so, production of a secretory core-protein which lacks HBc-antigenicity makes sense since otherwise this protein would be eliminated by the HBc-specific antibodies which almost invariably already arise very early during HBV infection. In general, any possible function of the HBeAg which would require its continuous presence in the circulation would be void if it possessed HBc-antigenicity.

The finding that HBeAg assembles into particles if the preC sequence is replaced or mutated strongly suggests that the default folding pathway of HBV core-gene products leads to proteins which are bound to self-assemble into particles. How the hydrophobic motif in the preC region can interfere with the assembly process is under investigation. It appears possible that this sequence interacts with hydrophobic patches in the vicinity of the internal cysteines. If so, it is conceivable that through this interaction the N-terminus masks areas of the protein which serve as contact sites for the capsid subunits.

Prevention of HBe-aggregation could fulfil the same purpose as the inhibition of dimerization: to allow the synthesis of a secretory core-gene product with a certain antigenicity. Alternatively, it might also be possible that the function of the HBe-protein requires that it be in a non-aggregated state. What this function might be, however, is still one of the most fascinating questions in hepatitis B virus molecular biology.

References

1. Bartenschlager R, Schaller H (1992) Hepadnaviral assembly is initiated by the polymerase binding to the encapsidation signal in the viral RNA genome. EMBO J 11: 3413–3420
2. Birnbaum F, Nassal M (1990) Hepatitis B virus nucleocapsid assembly: primary structure requirements in the core protein. J Virol 64: 3319–3330
3. Bonino F, Rizzetto M, Will H (1991) Hepatitis B virus unable to secrete e antigen. Gastroenterology 100: 1138–1141
4. Bruss V, Gerlich W (1988) Formation of transmembraneous hepatitis B e-antigen by cotranslational in vitro processing of the viral precore protein. Virology 163: 268–275
5. Carman W, Jacyna M, Hadziyannis S, Karayiannis P, McGarvey M, Makris A, Thomas H (1989) Mutation preventing formation of hepatitis B e antigen in patients with chronic hepatitis B infection. Lancet 8663: 588–591
6. Chang C, Enders G, Sprengel R, Peters N, Varmus HE, Ganem D (1987) Expression of the precore region of an avian hepatitis B virus is not required for viral replication. J Virol 61: 3322–3325
7. Cohen BJ, Richmond JE (1982) Electron microscopy of hepatitis B core antigen synthesized in *E. coli*. Nature 296: 677–679
8. Colucci G, Beazer Y, Cantaluppi C, Tackney C (1988) Identification of a major hepatitis B core antigen (HBcAg) determinant by using synthetic peptides and monoclonal antibodies. J Immunol 141: 4376–4380
9. Ferns RB, Tedder RS (1984) Monoclonal antibodies to hepatitis B e antigen (HBeAg) derived from hepatitis B core antigen (HBcAg): their use in characterization and detection of HBeAg. J Gen Virol 65: 899–908
10. Ferns RB, Tedder RS (1986) Human and monoclonal antibodies to hepatitis B core antigen recognize a single immunodominant epitope. J Med Virol 19: 193–203
11. Gallina A, Bonelli F, Zentilin L, Rindi G, Muttini M, Milanesi G (1989) A recombinant hepatitis B core antigen polypeptide with the protamine-like domain

deleted self-assembles into capsid particles but fails to bind nucleic acids. J Virol 63: 4645–4652

12. Ganem D, Varmus HE (1987) The molecular biology of the hepatitis-B viruses. Annu Rev Biochem 56: 651–693

13. Hatton T, Zhou S, Standring D (1992) RNA- and DNA-binding activities in hepatitis B virus capsid protein: a model for their roles in viral replication. J Virol 66: 5232–5241

14. Imai M, Nomura M, Gotanda T, Sano T, Tachibana K, Miyamoto H, Takahashi K, Toyama S, Miyakawa Y, Mayumi M (1982) Demonstration of two distinct antigenic determinants on hepatitis B e antigen by monoclonal antibodies. J Immunol 128: 69–72

15. Junker M, Galle P, Schaller H (1987) Expression and replication of the hepatitis B virus genome under foreign promoter control. Nucleic Acids Res 15: 10117–10132

16. Lanford RE, Notvall L (1990) Expression of hepatitis B virus core and precore antigens in insect cells and characterization of a core-associated kinase activity. Virology 176: 222–233

17. Magnius L, Espmark J (1972) New specificities in Australia antigen positive sera distinct from the Le Bouvier determinants. J Immunol 109: 1017–1021

18. McKay P, Lees J, Murray K (1981) The conversion of hepatitis B core antigen synthesized in E. coli into e antigen. J Med Virol 8: 237–243

19. Milich DR, Jones JE, Hughes JL, Price J, Raney AK, McLachlan A (1990) Is a function of the secreted hepatitis B e antigen to induce immunologic tolerance in utero? Proc Natl Acad Sci USA 87: 6599–6603

20. Miyanohara A, Imamura T, Araki M, Sugawara K, Ohtomo N, Matsubara K (1986) Expression of hepatitis B virus core antigen gene in Saccharomyces cerevisiae: synthesis of two polypeptides translated from different initiation codons. J Virol 59: 176–180

21. Nassal M (1992) The arginine-rich domain of the hepatitis B virus core protein is required for pregenome encapsidation and productive viral positive-strand DNA synthesis but not for virus assembly. J Virol 66: 4107–4116

22. Nassal M (1992) Conserved cysteines of the hepatitis B virus core protein are not required for assembly of replication-competent core particles nor for their envelopment. Virology 190: 499–505

23. Ou J, Laub O, Rutter W (1986) Hepatitis B virus gene function: the precore region targets the core antigen to cellular membranes and causes the secretion of the e antigen. Proc Natl Acad Sci USA 83: 1578–1582

24. Salfeld J, Pfaff E, Noah M, Schaller H (1989) Antigenic determinants and functional domains in core antigen and e antigen from hepatitis B virus. J Virol 63: 798–808

25. Sallberg M, Ruden U, Wahren B, Noah M, Magnius LO (1991) Human and murine B-cells recognize the HBeAg/beta (or HBe2) epitope as a linear determinant. Mol Immunol 28: 719–726

26. Sallberg M, Ruden U, Magnius LO, Harthus HP, Noah M, Wahren B (1991) Characterisation of a linear binding site for a monoclonal antibody to hepatitis B core antigen. J Med Virol 33: 248–252

27. Schlicht HJ, Salfeld J, Schaller H (1987) The pre-C region of the duck hepatitis B virus encodes a signal sequence which is essential for the synthesis and secretion of processed core proteins but not for virus formation. J Virol 61: 3701–3709

28. Schlicht HJ, Bartenschlager R, Schaller H (1989) The duck hepatitis B virus core protein contains a highly phosphorylated C-terminus essential for replication but not for RNA packaging. J Virol 63: 2995–3000

29. Schlicht HJ, Schaller H (1989) The secretory core protein of human hepatitis B virus is expressed on the cell surface. J Virol 63: 5399–5404
30. Schlicht HJ, von Brunn A, Theilmann L (1990) Antibodies in anti-HBe positive patient sera bind to a HBe protein expressed on the cell surface of human hepatoma cells: implications for virus clearance. Hepatology 13: 57–61
31. Schlicht HJ (1991) Biosynthesis of the secretory core protein of duck hepatitis B virus: intracellular transport, proteolytic processing, and membrane expression of the precore protein. J Virol 65: 3489–3495
32. Schlicht HJ, Wasenauer G (1991) The quaternary structure, antigenicity and aggregational behavior of the secretory core-protein (HBe-protein) of human hepatitis B virus is determined by its signal sequence. J Virol 65: 6817–6825
33. Standring D, Ou J, Masiarz F, Rutter W (1988) A signal peptide encoded within the precore region of hepatitis B virus directs the secretion of a heterogeneous population of e antigens in Xenopus oocytes. Proc Natl Acad Sci USA 85: 8405–8409
34. Takahashi K, Imai M, Gotanda T, Sano T, Oinuma A, Mishiro S, Miyakawa Y, Mayumi M (1980) Hepatitis B e antigen polypeptides isolated from sera of individuals infected with hepatitis B virus: comparison with HBeAg polypeptide derived from Dane particles. J Gen Virol 50: 49–57
35. Takahashi K, Machida A, Funatsu G, Nomura M, Usuda S, Aoyagi S, Tachibana K, Miyamoto H, Imai M, Nakamura T, Miyakawa Y, Mayumi M (1983) Immunochemical structure of hepatitis B e-antigen in the serum. J Immunol 130: 2903–2907
36. Wang J, Lee A, Ou J (1991) Proteolytic conversion of hepatitis B virus e antigen precursor to end product occurs in a postendoplasmic reticulum compartment. J Virol 65: 5080–5083
37. Wasenauer G, Köck J, Schlicht HJ (1992) A cysteine and a hydrophobic sequence in the non-cleaved portion of the preC leader peptide determine the biophysical and immunological properties of the secretory core-protein (HBe-protein) of human hepatitis B virus. J Virol 66: 5338–5346
38. Wasenauer G, Köck J, Schlicht HJ (1993) Relevance of cysteine residues for biosynthesis and antigenicity of human hepatitis B virus e protein. J Virol 67: 1315–1321
39. Waters JA, Jowett TP, Thomas HC (1986) Identification of a dominant immunogenic epitope of the nucleocapsid (HBc) of the hepatitis B virus. J Med Virol 19: 79–86
40. Yu M, Summers J (1991) A domain of the hepadnavirus capsid protein is specifically required for DNA maturation and virus assembly. J Virol 65: 2511–2517
41. Zhou S, Standring D (1991) Production of hepatitis B virus nucleocapsidlike core particles in Xenopus oocytes: assembly occurs mainly in the cytoplasm and does not require the nucleus. J Virol 65: 5457–5464
42. Zhou S, Standring D (1992) Cys residues of the hepatitis B virus capsid protein are not essential for the assembly of viral core particles but can influence their stability. J Virol 66: 5393–5398

Authors' address: Dr. H.-J. Schlicht, Department of Virology, University of Ulm, Albert Einstein Allee 11, D-89081 Ulm, Federal Republic of Germany.

Arch Virol (1993) [Suppl] 8: 53–62

Characterization of the endogenous protein kinase activity of the hepatitis B virus

M. Kann[1,2], **R. Thomssen**[1], **H.G. Köchel**[1], and **W.H. Gerlich**[1,2]

[1] Department of Medical Microbiology, University of Göttingen, Göttingen
[2] Institute of Medical Virology, University of Giessen, Giessen,
Federal Republic of Germany

Summary. During the assembly of the nucleocapsid of the hepatitis B virus a protein kinase, probably of cellular origin, is encapsidated. This enzyme phosphorylates serine residue(s) localized within the lumen of the particle. By using purified, liver-derived core particles, we characterized the protein kinase activity in the presence of different ions and inhibitors. Controls were performed with cAMP-dependent protein kinase (PKA) and protein kinase C (PKC) and recombinant core particles. We showed that the endogenous protein kinase of the core particles was not inhibited by H89, a specific inhibitor of PKA. Staurosporine, a selective inhibitor of PKC inhibited the endogenous kinase activity only within the first minutes of the reaction. In contrast, quercetine, a selective inhibitor of the protein kinase M (PKM) did not inhibit during the first minutes but inhibited efficiently during later phases of incubation. PKM represents an enzymatically active proteolytic fragment of PKC. These results suggest that PKC is encapsidated into human core particles and is converted to PKM during the in vitro reaction. This conclusion implies the association of a protease activity localized with the HBV nucleocapsid inside liver-derived core particles.

Introduction

The nucleocapsid of the hepatitis B virus has been proposed to consist of 180 subunits of core proteins [18] which assemble in the form of disulfide-linked dimers [22]. During this assembly, the pregenomic RNA, the viral polymerase and a protein kinase are encapsidated in the lumen of the particle. The protein kinase phosphorylates serine residue(s) [6] of core proteins which are present in the liver of hepatitis-B-infected individuals [6] as well as the nucleocapsids of serum-derived virions [1]. On the molecular level, phosphorylation occurs in the carboxyterminal portion of the core molecules, localized between arginine clusters [14]. These positively charged clusters are involved in the nucleic acid binding activity of the core proteins [2]. They seem to face the lumen of the particles so that neither the nucleic acids [8, 15] nor the phosphate groups can be enzymatically removed from the assembled particles [6].

Since the phosphorylation also occurs in nucleocapsids which have been expressed in absence of any other viral protein in eucaryotic cells [13], but not in procaryotically expressed nucleocapsids, a eukaryotic, cellular origin of the protein kinase is most probable. Machida et al. [14] identified a phosphorylated sequence of 12 amino acids in the core protein from Dane particles using monoclonal antibodies. By comparing this sequence with the target sequences of different protein kinases [9] it can be concluded that protein kinase A or protein kinase C are most likely responsible for the phosphorylation, although other kinases cannot be excluded. To obtain a better understanding of the life cycle of the hepatitis B virus, the identification and characterization of the encapsidated protein kinase seems to be of great importance.

Direct identification of the kinase turned out to be difficult, probably because only one molecule may be present in core particles. Therefore, we tried to characterize the hepatitis-B-related protein kinase enzymatically. Since the rigorous treatment for disassembly of the disulfide-linked protein subunits might destroy the phosphorylation activity of the encapsidated protein kinase we studied the phosphorylation kinetics using inhibitors of low molecular weight. We validated the experimental design by control reactions, using PKA and PKC as enzymes and recombinant core particles as substrate.

Materials and methods

Liver-derived core particles (lHBc) were purified as described by Gerlich et al. [6], so that the purified core protein appeared as a single 22 kDa band in silver staining after separation on a 15% SDS-PAGE (Fig. 1). The final concentration of core particles was 5–10 ng/μl.

Recombinant core particles (rHBc) were produced and purified according to Uy et al. [21]. The purified core particles consisted of disulfide-bond linked core subunits, migrating as a double band in SDS-PAGE. The upper band of 22 kDa represents the full-length product, the lower one of 20 kDa an N-terminal deletion variant which lacks the first 12 amino acids due to an additional translation start in *E. coli* (K.H. Heermann, pers. comm.).

The assays for the protein kinases were carried out in a reaction mixture containing 10 mM sodium phosphate buffer pH 7.0, 10 mM $MgCl_2$, 10 mM DTT, 0.2 mM $CaCl_2$,

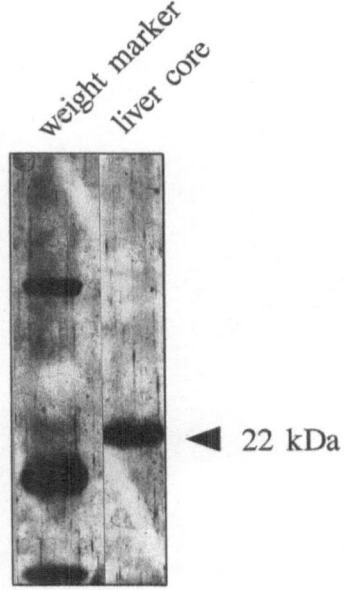

◀ 22 kDa

silver
staining

Fig. 1. Silver-staining of purified lHBc protein after SDS electrophoresis on 15% SDS-PAGE

10 µg/ml phosphatidyl serine (Sigma), 0.1 µCi/µl [gamma^{32}P]ATP (Amersham) and 0.25% bovine serum albumin (BSA) as carrier protein. When assays with purified protein kinases were performed, ATP was added to a final concentration of 3.6 µM. Protein kinase C (Boehringer) or cAMP-dependent protein kinase (catalytic subunit) were added to a concentration of 0.32 picomolar units/µl, complete protein kinase A of 4 picomolar units/µl, with 40 ng/µl of heat-denaturated rHBc protein as substrate. In the endogenous kinase assays, the final concentration of lHBc was 0.25–0.5 ng/µl.

The inhibitors used in our study were staurosporine (Boehringer) at a concentration of 20 nM, quercetine (Sigma) at 5 µM and H89 [N-[2-((3-(4-bromophenyl)-2-propenyl)-amino)-ethyl]-5-isoquinolinesulfonamide, di HCl] (Calbiochem) at 0.5 µM. The reactions were carried out at 30°C. Incorporation of radioactivity was determined by precipitation with trichloracetic acid (TCA) (2 × 10 min 10% TCA, 2 × 5 min 5% TCA) of an 2.5 µl aliquot on GF/C filters (Whatman). The precipitated radioactivity was measured in a liquid scintillation counter.

To ensure the specific phosphorylation of the core-substrate, the reactions were validated by autoradiography of the proteins after they were separated on a 15% SDS-PAGE.

Results

Purity of the lHBc particles

The absence of other, non-encapsidated protein kinases in the liver-core preparation was checked by a kinase reaction in the presence of

M. Kann et al.

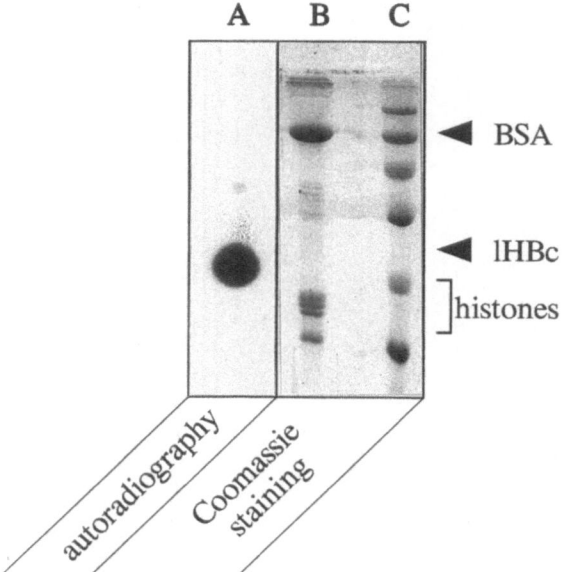

Fig. 2. Detection of ^{32}P-labeled lHBc protein by autoradiography (*A*) and Coomassie-staining of proteins after endogenous protein kinase reaction in presence of histones (*B*); molecular weight marker (*C*)

[gamma[32]P]ATP and 400 ng/µl of a histone mixture (Boehringer) (Fig. 2). After 30 min of incubation, no phosphorylated product other than lHBc was observed in SDS gels, indicating the encapsidation of a protein kinase activity within the lHBc particles and absence of free protein kinases.

Phosphorylation in the presence of different ions and activators

For the endogenous kinase reaction, no effect of cAMP, Ca^{++} or EGTA was observed after an incubation period of 30 min (Fig. 3). In contrast, the addition of 20 mM EDTA or 0.5 mM Zn^{++} led to a complete loss of activity. In the control experiments, cAMP had a strong stimulating effect on the complete cAMP-dependent protein kinase and no effect on PKC or on the catalytic subunit of PKA as shown in the same figure. Ca^{++} led to the expected stimulation of PKC. EDTA was able to in-activate all of the tested enzymes, indicating the necessity of divalent ions for all of the tested protein kinases. In contrast, addition of Zn^{++} decreased the activity of PKA but led to complete loss of the activity of PKC. Since no clear coordination of the core-related protein kinase to one of the tested kinases was possible by these results, kinetics of the phosphorylation in the presence of different inhibitors were performed.

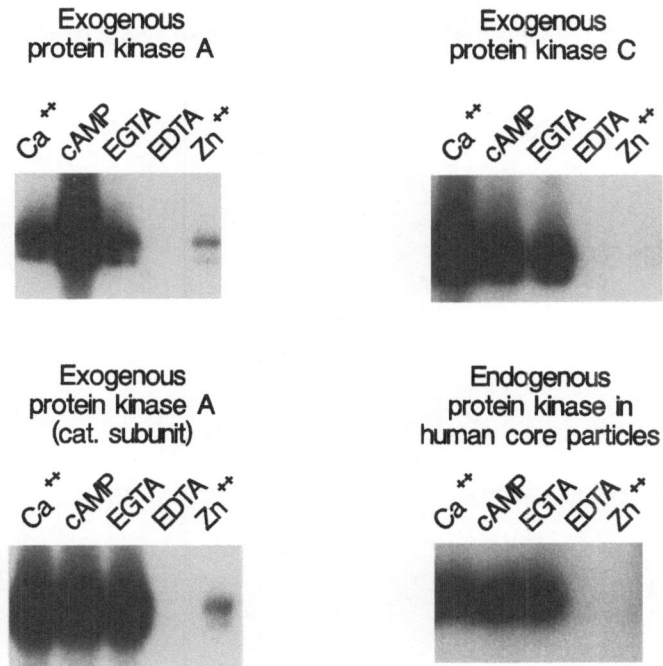

Fig. 3. Autoradiography of ^{32}P-rHBc, phosphorylated by PKA$_{(complete)}$, PKA$_{(catalytic}$ $_{subunit)}$, PKC or of ^{32}P-lHBc in the presence of different ions and activators after running on SDS electrophoresis. Reaction conditions: 10 mM sodium phosphate buffer pH 7.0; 10 mM MgCl$_2$; 10 mM DTT; 10 µg/ml phosphatidyl serine; 0.1 µCi/µl [gamma^{32}P] ATP; 0.25% BSA and Ca^{++} 0.2 mM; cAMP 50 µM; EGTA 1 mM; EDTA 20 mM; Zn^{++} 0.5 mM. Incubation for 30 min at 30°C

Kinetics of the phosphorylation in the presence of inhibitors

The PKA-specific inhibitor H89 [3] at a concentration of 0.5 µM led to 65% inhibition for the catalytic subunit on the PKA (Fig. 4c). In contrast, this inhibitor was not able to inhibit the PKC (Fig. 4a) or the lHBc protein kinase (Fig. 4e). A different pattern of inhibition was observed for the PKC specific inhibitor staurosporine [4]. This substance reduced the activity of PKC almost to the background levels (Fig. 4b) but had no effect on PKA (Fig. 4d). During the phosphorylation kinetics of the lHBc, an inhibition was only detectable within the first 4 min (Fig. 4f). After this period the rate of phosphorylation increased again to the level in the non-inhibited control. The third inhibitor used in our study was quercetine, known as an inhibitor of the protein kinase M, which represents an enzymatically active fragment of the PKC. In vivo, this fragment is generated by a proteolytic cleaveage of the PKC in its V3 region [10], which separates the regulatory subunit from the catalytic one. This inhibitor is specific for PKM and does not inhibit PKC. In

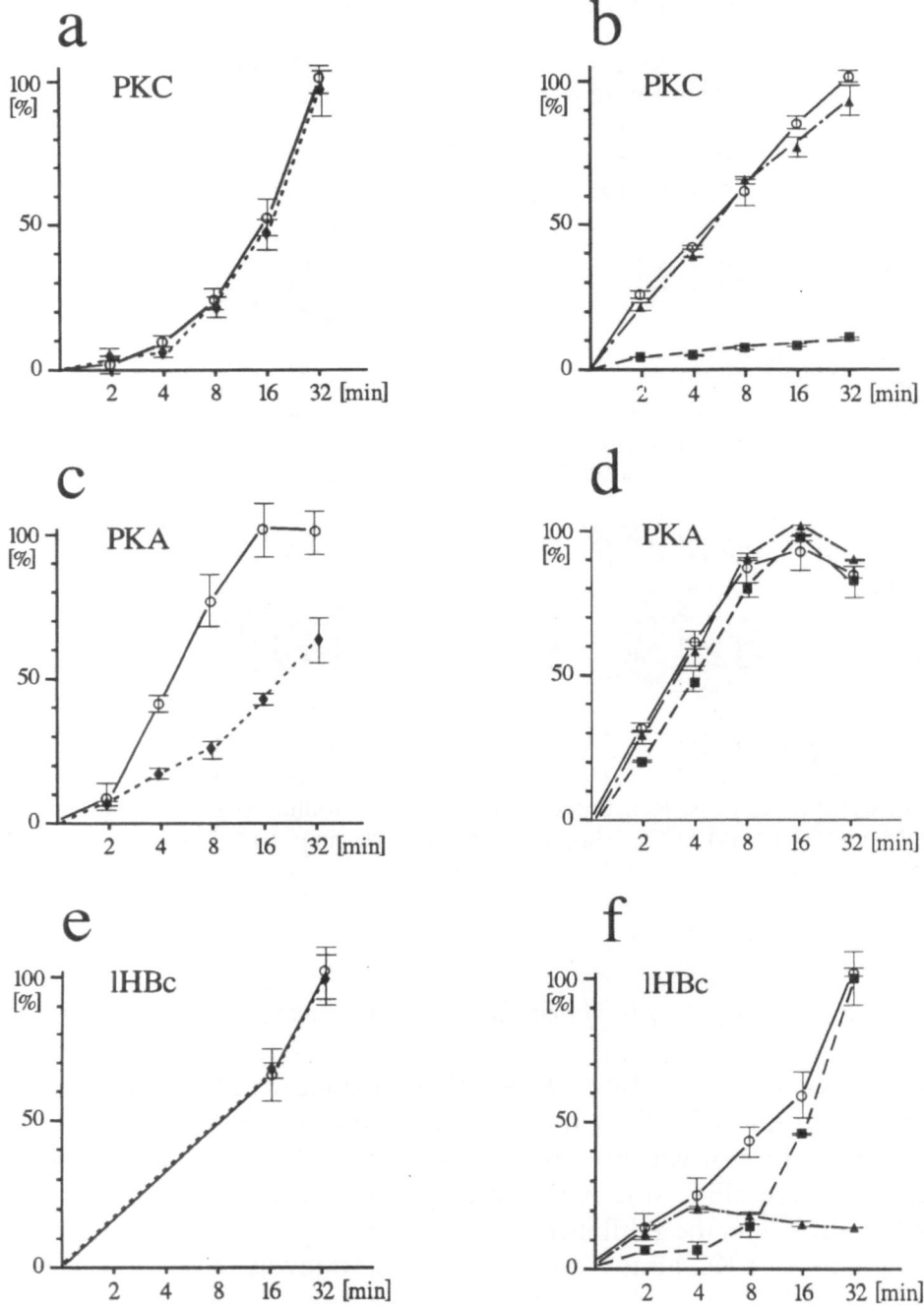

Fig. 4. TCA precipitates of phosphorylated rHBc by PKA$_{(complete)}$, PKA$_{(catalytic\ subunit)}$, PKC and of lHBc phosphorylated by the endogenous protein kinase kinase in the presence of different inhibitors; the points represent the mean values of three reactions. The degree of phosphorylation is expressed as percent of specific maximal labeling (M). The background activity (B) was subtracted from all values (the absolute data were: **a** B 26979 cpm, M 34541 cpm; **b** BZ5978 cpm, M 109259 cpm; **c** B 6970 cpm, M 104283 cpm; **d** B 1319 cpm, M 47518 cpm; **e** B 2037 cpm, M 12790 cpm; **f** B 1913, M 5718 cpm). Range I; without inhibitor ◯——◯; 0.5 µM H89 ◆----◆; 5 µM quercetine ▲—-▲; 20 nM staurosporine ■——■

agreement with Junco et al. [7], no inhibition could be observed for PKC (Fig. 4b). Furthermore, quercetine did not inhibit PKA. In contrast, the lHBc-associated protein kinase was inhibited, but only after more than 4 min incubation (Fig. 4f). This time course is diametrically opposed to the phosphorylation kinetics in the presence of staurosporine.

Discussion

In order to characterize the HBV-associated protein kinase, we investigated the influence of different inhibitors known to be specific for different cellular protein kinases. Previous studies [6] showed that this protein kinase phosphorylates (a) serine residue(s) within the carboxy-terminal portion of the core molecules, and that the phosphate group of the newly generated phosphoserine cannot be removed from the lHBc by phosphatases [6]. Thus, a luminal location of the phosphorylation site and of the core-associated protein kinase seems to be most probable. Because of the tight linkage of the core subunits, the kinase could not phosphorylate proteins outside the particle. This assumption could be verified in our experiments by addition of histones to the phosphorylation reaction mixture.

The protein kinase is believed to be of cellular origin. From the target sequences of the known cellular protein kinases it can be concluded that either the cAMP-dependent protein kinase or the protein kinase C might be the core-associated kinase, but other kinases such as cdc kinases could not be excluded. By our experiments a complete inhibition of the endogenous protein kinase was observed in the presence of Zn^{++}. As this ion is known to be essential for cdc kinases [20], an encapsidation of this enzyme in the HBV core particles probably does not occur. The presence of the complete protein kinase A consisting of two regulatory and two catalytic subunits could be ruled out by the missing stimulating effect of cAMP in the endogenous phosphorylation. Either the catalytic subunit of the PKA or the PKC were therefore most probably responsible for the endogenous kinase activity. For the PKC, a stimulating effect of Ca^{++} in the presence of phosphatidyl serine should be observed as demonstrated in the control experiment. However, after 30 min of incubation no stimulating effect was observed in the HBV core reaction, a result similar to the reaction of the catalytic subunit of PKA with recombinant cores as substrate. In contrast to the PKA, however, the HBV endogenous protein kinase was completely inhibited by Zn^{++}. This observation was similar to the phosphorylation activity of PKC. Therefore, we examined the endogenous protein kinase by comparing its activity in the presence of different inhibitors with the activity of the catalytic subunit of PKA and the activity of PKC. The missing inhibition by H89 indicates that the PKA was not present in the lHBc particles. Interestingly, the clearly detectable inhibition of staurosporine in the endogenous reaction lasted only for a few minutes. An instability of the inhibitor could be ruled out as a reason for this phenomenon because the inhibition in the control experiment with PKC was complete for more than half an hour. A possible explanation for the strange inhibition pattern generated by staurosporine was found by studying the effects of quercetine. This flavonoid shows an inhibitory effect on the protein kinase M (PKM) [7], which represents a proteolytic fragment of PKC generated either by Ca^{++}-dependent neutral proteases (calpain) or by serine proteases [11, 16, 19]. The PKM contains the catalytic centre of PKC and retains the phosphorylation activity, although it is independent of Ca^{++} and fatty acids [7]. In vivo, the generation of the PKM may be an alternative pathway of activation of the PKC [16].

In our inhibition kinetics we observed that quercetine was able to inhibit the endogenous kinase reaction at that time point at which the inhibitory effect of staurosporine began to decrease. Therefore, a conversion of PKC to PKM might have occurred during the incubation period. At the same time this result might explain the missing stimulatory effect of Ca^{++} in the reaction shown in Fig. 2 for the endogenous reaction. The inhibition pattern of the lHBc kinase was not due to freezing and thawing, since pure PKC did not show a PKM-like pattern after multiple freezing and thawing cycles (data not shown).

Our results, consequently, suggest that during the assembly of human core particles the PKC is encapsidated in the nucleocapsid of the hepatitis B virus. This kinase seems to be proteolytically cleaved, generating the PKM. This assumption is supported by observations of others [12, 17] who observed a stimulated proteolysis of PKC in the presence of basic polypeptides as they are present in the arginine clusters of the HBc subunits. In vivo, this cleavage would allow a phosphorylation of the core particles independent of Ca^{++} or membranes which seem to be prerequisites for the phosphorylation activity of PKC [5]. The characterization and identification of the putative HBc-associated proteolytic activity will be a subject of further investigations.

Acknowledgements

We thank B. Boschek for critically reading the manuscript. This work was supported by a grant of the Deutsche Forschungsgemeinschaft to H.G.K. (KO 1102/1-1).

References

1. Albin C, Robinson WH (1980) Protein kinase activity in hepatitis B virus. J Virol 34: 297–302
2. Birnbaum F, Nassal M (1990) Hepatitis B virus nucleocapsid assembly: primary structure requirements in the core protein. J Virol 64: 3319–3330
3. Chijiwa T, Mishima A, Hagiwara M, Sano M, Hayashi K, Inoue T, Naito K, Toshioka T, Hidaka H (1990) Inhibition of forskolin-induced neurite outgrowth and protein phosphorylation by a newly synthesized selective inhibitor of cyclic AMP-dependent protein kinase, N-[2-(p-bromocinamylamino)ethyl]-5-isoquinoline-sulfonamide (H89), of PC12D pheochromocytoma cell. J Biol Chem 265: 5267–5272
4. Davis PD, Hill CH, Keech E, Lawton G, Nixon JS, Sedgwick AD, Wadsworth J, Westmacott D, Wilkinson SE (1989) Potent selective inhibitors of protein kinase C. FEBS Lett 259: 61–63
5. Edashige K, Utsumi T, Sato E, Ide A, Kasai M, Utsumi K (1992) Requirement of protein association with membranes for phosphorylation by protein kinase C. Arch Biochem Biophys 296: 296–301
6. Gerlich WH, Goldmann U, Müller R, Stibbe W, Wolff W (1982) Specificity and localization of the hepatitis B virus-associated protein kinase. J Virol 42: 761–766
7. Junco M, Diaz-Guerra MJM, Boscá L (1990) Substrate-dependent inhibition of protein kinase C by specific inhibitors. FEBS Lett 263: 169–171
8. Kaplan PM, Greenman RL, Gerin JL, Purcell RH, Robinson WS (1973) DNA polymerase associated with human hepatitis B antigen. J Virol 12: 996–1005
9. Kemp BE, Pearson RB (1990) Protein kinase recognition sequence motifs. Trends Biochem Sci 15: 342–346
10. Kikkawa U, Kishimoto A, Nishisuka Y (1989) The protein kinase C family: heterogeneity and its implications. Annu Rev Biochem 58: 31–44
11. Kishimoto A, Mikawa K, Hashimoto K, Yasuda I, Tanaka S, Tominaga M, Kuroda T, Nishizuka Y (1989) Limited proteolysis of protein kinase C subspecies by calcium-dependent neutral proteases (Calpain). J Biol Chem 264: 4088–4092
12. Kuroda T, Mikawa K, Mishima H, Kishimoto A (1991) H1 histone stimulates limited proteolysis of protein kinase C subspecies by calpain II. J Biochem 110: 354–368
13. Lanford RE, Notvall L (1990) Expression of hepatitis B virus core and precore antigens in insect cells and characterization of a core-associated kinase activity. Virology 176: 222–233
14. Machida A, Ohnuma H, Tsuda F, Yoshikawa A, Hoshi Y, Tanaka T, Kishimoto S, Akahane Y, Miyakawa Y, Mayumi M (1991) Phosphorylation of the carboxy-terminal domain of the capsid protein of hepatitis B virus: Evaluation with a monoclonal antibody. J Virol 65: 6024–6030
15. Melegari M, Bruss V, Gerlich WH (1991) The arginine-rich carboxy-terminal domain is necessary for RNA packaging by hepatitis B core protein. In: Hollinger FB, Lemon SM, Margolis HS (eds) Viral hepatitis and liver disease. Williams & Wilkins, Baltimore, pp 164–168
16. Melloni E, Pontremoli S, Michetti M, Sacco O, Sparatore B, Horecker BL (1986) The involvement of calpain in the activation of protein kinase C in neutrophils stimulated by phorbolmyristic acid. J Biol Chem 261: 4101–4105
17. Mikawa K, Maekawa N, Goto R, Yaku H, Obara H, Kishimoto A, Kusunoki M (1991) Limited proteolysis of protein kinase C subspecies by calpain: stimulation by basic polypeptides. Ital J Biochem 40: 133–142

18. Onodera S, Ohori H, Yamaki M, Ishida N (1982) Electron microscopy of human hepatitis B virus cores by negative staining-carbon film technique. J Med Virol 10: 147–155
19. Pelaez F, de Herreros AG, de Haro C (1987) Purification and properties of protein kinase C from rabbit reticulocyte lysates. Arch Biochem Biophys 257: 328–338
20. Reed S, Hadwiger J, Lörincz AT (1985) Protein kinase activity associated with the product of the yeast cell division cycle gene CDC28. Proc Natl Acad Sci USA 82: 4055–4059
21. Uy A, Bruss V, Gerlich WH, Köchel HG, Thomssen R (1986) Precore sequence of hepatitis B virus inducing e antigen and membrane association of the viral core protein. Virology 155: 89–96
22. Zhou S, Standring DN (1992) Hepatitis B virus capsid particles are assembled from core-protein dimer precursors. Proc Natl Acad Sci USA 89: 10046–10050

Authors' address: Dr. M. Kann, Institute of Medical Virology, University of Giessen, Frankfurter Strasse 107, D-35392 Giessen, Federal Republic of Germany.

Arch Virol (1993) [Suppl] 8: 63–71

The hepatitis B virus X gene product transactivates the HIV-LTR in vivo

Clara Balsano[1,2], O. Billet[1], Myriam Bennoun[1], Catherine Cavard[1], A. Zider[1], Gisele Grimber[1], G. Natoli[2], P. Briand[1], and M. Levrero[2]

[1] Laboratoire de Genetique et Patologie Experimentale, INSERM Institut Cochin de Genetique Moleculaire, Paris, France
[2] Laboratorio di Espressione Genica, Fondazione A. Cesalpino e I Clinica Medica, Policlinico Umberto I, Rome, Italy

Summary. It has previously been shown that the hepatitis B virus (HBV) X gene product, HBx, transactivates homologous and heterologous transcriptional regulatory sequences of viruses, including the human immunodeficiency virus type 1 (HIV1) long terminal repeat (LTR), and various cellular genes in vitro. To evaluate the transactivating function of HBx in vivo, we generated transgenic mice carrying the X open reading frame under the control of the human antithrombin III (ATIII) gene regulatory sequences. These mice express the 16 Kd HBx protein in the liver, as demostrated by immunoprecipitation studies. Crossbreeding of HBx mice with transgenics carrying either the chloramphenicol acetyl transferase (CAT) bacterial or the lacZ reporter gene driven by the HIV1-LTR allowed us to demostrate, for the first time, the in vivo transactivating function of HBx protein.

Introduction

The hepatitis B virus (HBV) genome contains four recognized open reading frames (ORF). Three of these code for known viral structural proteins. The fourth open reading frame, the hepatitis B virus X gene, is conserved among all mammalian hepadnaviruses and encodes for a 16 Kd protein, HBx which is expressed in chronically infected patients [6]. HBx has been shown to transactivate transcription, regulated by several cis-acting sequences including autologous regulatory sequences [13, 17], heterologous viral (SV40, RSV, HTLV1, HIV1 and HIV2) and cellular (beta-interferon, class I and class II MHC, c-myc) regulatory sequences [1, 13, 15, 17]. The transactivating function is preserved if the carboxyterminal region of the X ORF is truncated as is often observed in integrated HBV DNA sequences [1, 13, 18]. HBx-containing NIH-3T3 cells have been reported to display an increased saturation density in culture and to form tumors in nude mice [16]. Moreover, depending upon the time and the level of expression of HBx in the liver, either no effect [12] or the sequential development of hepatoma and hepato-carcinoma have been described [11].

Most if not all the evidence provided so far on HBx-induced trans-activation have been obtained in conditions that are very far from the normal environment in which HBx is expressed and refer to experiments performed in immortalized or transformed cells transfected with rela-tively large amounts of HBx-expressing vectors. Thus, the obvious question is whether HBx is able to modulate gene expression in vivo and in normal hepatocytes carrying a single copy or a low number of copies of the X gene. We therefore generated transgenic mice carrying the X ORF under the contol of the human ATIII gene regulatory region. Crossbreeding of HBx mice with transgenics carrying either CAT or lacZ reporter genes driven by the HIV1-LTR allowed us to demostrate, for the first time, the in vivo transactivating properties of HBx.

Materials and methods

Production of the HIV1-CAT. HIV1-LacZ and ATIII-X transgenic mice

The HIV1-CAT transgene used to produce the HIV1-CAT transgenic mice (Cavard et al., in prep.) is described in Fig. 1. The HIV1-lacZ transgene and the resulting trans-genic mice have been previously described in detail ([3] and Fig. 1).

To obtain the plasmid pATIII-X, the 898-bp XhoI-AccI restriction fragment from plasmid pMLP-X was inserted between the SalI and AccI sites of plasmid pTZ-ATIII [4], a pTZ19R derivative containing the 700-bp 5' flanking region of the R allele of the human ATIII gene. The 1.6 kb insert, containing the human antithrombin III regu-latory sequences linked to the HBx-encoding sequence and the HBV polyadenylation signal, was excised from plasmid pATIII-X by EcoRI and SalI digestion (Fig. 1C). Transgenic mice were obtained by standard procedures following microinjection of C57B1/6 × DBA2 fertilized eggs with the chimeric ATIII-X gene. DNA and RNA analysis were performed using standard procedures [14]. Southern blot analysis of restriction enzyme digested DNA from 32 live-born progeny revealed that three of the animals (Tg16, Tg17 and Tg22) contained integrated copies of the ATIII-X hybrid gene in a head-to-tail tandem array organization (data not shown). Analysis of the progeny of the Tg16, Tg17 and Tg22 founders showed that the transgene was transmitted at a frequency of about 50%.

Expression patterns of the ATIII-X transgene were studied in animals that were heterozygous for the transgene by dot blot analysis of 20 µg of total RNA extracted by the guanidium isothyocianate-caesium chloride method. HBx transcripts were observed

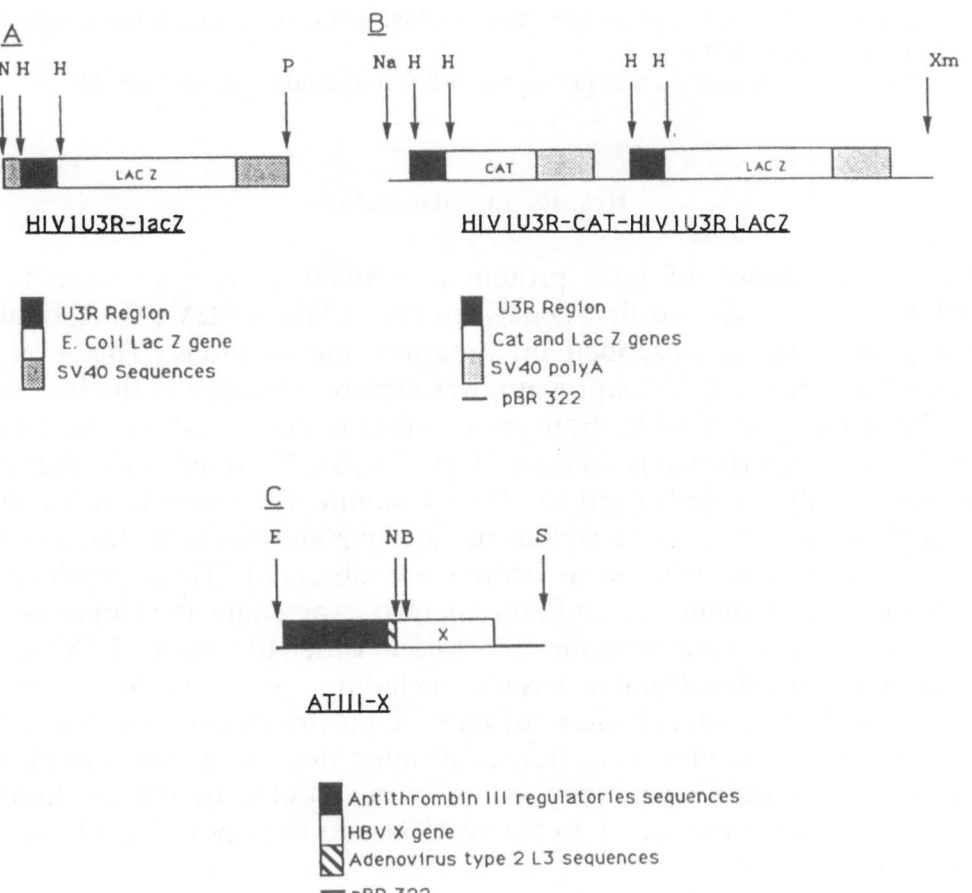

Fig. 1. Construction of the chimeric genes HIV-LacZ (**A**), HIV1-CAT (**B**), and ATIII-X (**C**)

in the livers, the kidneys, the lungs and the guts from two out of the three ATIII-X transgenic lines (Tg16 and Tg17), (data not shown). HBx protein was detected in the livers of transgenic animals by both indirect immunofluorescence and immunoprecipitation, performed as previously described [9] on 5 µm cryostat sections and liver extracts (10 mg) respectively. In both cases we used a 1/40 dilution of the specific rabbit polyclonal anti-X antiserum (a gift from L. Vitviski- Trépo, Lyon, France) as primary antibody.

The HIV1-CAT/ATIII-X and HIV1-lacZ/ATIII-X double transgenic mice were produced by mating the different transgenic homozygous animals.

Escherichia coli *beta-galactosidase activity assay and CAT assay*

Ten micrometer sections of liver specimens were immediately fixed for 10 min in 4% para-formaldehyde in 0.1 M PBS pH 7.2. Thereafter, slices were incubated over-night at 30°C in 0.1 M PBS containing 2 mM X-gal (5-bromo-4-chloro-3-indolyl-beta-Dgalactopyranoside), 4 mM potassium ferrocyanide, 4 mM potassium ferrocyanide and 4 mM magnesium chloride.

The CAT assay method was performed following standard procedures [8].

Results and discussion

To test the ability of HBx protein to activate gene expression from HIV1 LTR in vivo, we first generated the ATIII-X/HIV1-CAT double transgenic mice as described in Materials and methods. The level of HIV1-LTR driven CAT expression was clearly enhanced in the liver and in the kidney of double transgenic mice as compared to the single HIV1-CAT heterozygous animals (Figs. 2A,B). Maximal activation was detected in the animals aged less than 1 month. By contrast, in the liver of adult transgenic mice as well as in gut, lung and stomach, three of the HBx-expressing tissues, no activation was observed. These results suggest the requirement, in addition to HBx, for some developmentally regulated and/or tissue-specific cofactors in order to achieve HIV1-LTR activation. In HBx-negative tissues, including spleen, brain and heart (data not shown), no activation of HIV1-LTR driven expression of CAT reporter gene was observed, thus confirming that the increase of HIV1-LTR driven reporter gene expression in our double transgenic animals was due to the presence of an active HBx and does not result from the crossing procedures.

To determine whether the accumulation of the reporter-gene products occurs in the same cells that express HBx we generated ATIII-X/HIV1-lacZ double transgenic mice. LacZ expression in the liver of mice carrying both the HIV1-lacZ and the ATIII-X transgene was significantly higher than in mice carrying only the HIV1-lacZ transgene at 2, 4 and 8 days after birth (Fig. 3, Table 1). The accumulation of beta-galactosidase activity induced by HBx was observed in a limited number of hepatocytes, randomly distributed through the liver tissue of each tested animal (Figs. 4A,B). HBx expression and beta-galactosidase activity was detected in the same foci of hepatic cells, as illustrated on liver sections from 6-day-old double transgenic mice (Figs. 4C, D), although few HBx-expressing cells failed to demonstrate a beta-galactosidase activity. These results confirm the observations made in the ATIII-X/ HIV1-CAT double transgenics and strongly support the role of HBx in the HIV1-LTR activation phenomenon in vivo. The demonstration of in vivo and in vitro HIV1-LTR transactivation by HBx suggests that coinfection by HBV and HIV1, as already shown for cytomegalovirus and herpes simplex virus infections [5], might result in the stimulation of a latent HIV1 genome. Recently the productive HIV1 infection of CD4−

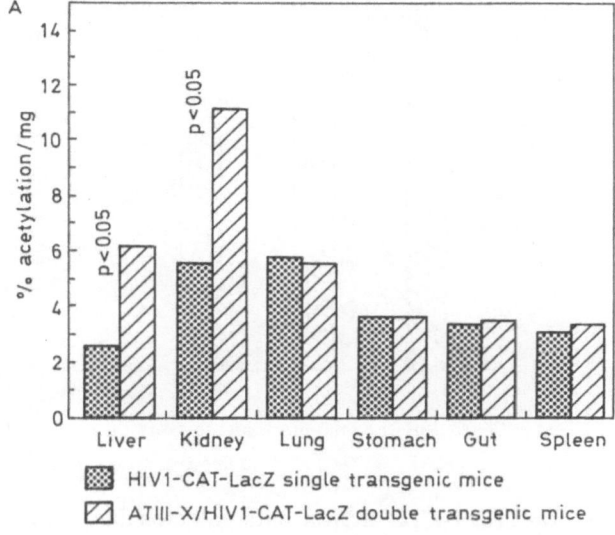

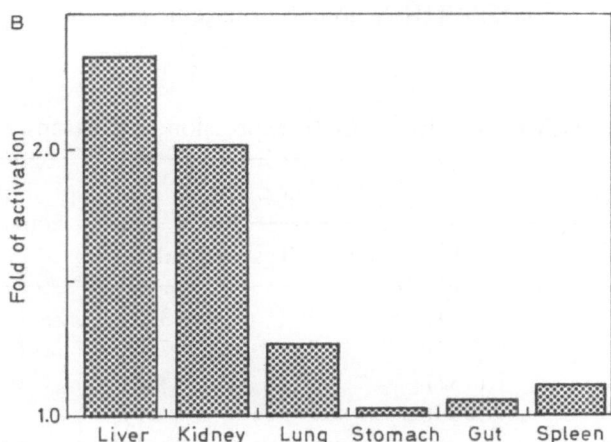

Fig. 2. Effect of HBx on HIV1-LTR driven CAT expression in vivo. **A** CAT assays were performed on liver, kidney, stomach, gut and spleen extracts obtained from animals aged less than 1 month. The results were expressed as percent of acetylation per mg of protein extract. Black bars correspond to HIV1-CAT single transgenics; stippled bars correspond to the ATIII-X/HIV1-CAT double transgenic animals. In **B** the fold activation is expressed as the ratio between the percent of acetylation in the ATIII-X/HIV1-CAT double transgenic animals as compared to the HIV1-CAT single transgenics

negative human hepatoma cell lines has also been reported [2], further suggesting that HIV1 tropism may not be limited to CD4+ cells. Although no direct evidence for HIV1 infection of human hepatocytes in vivo exists, these findings provide an explanation for the hepatic abnor-

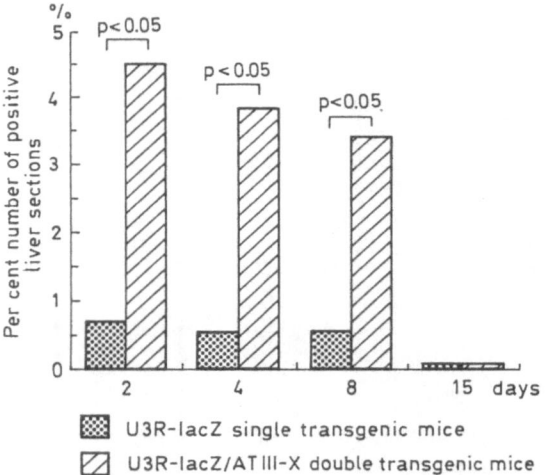

Fig. 3. Effect of pX on HIV1-LTR driven bata-galactosidase expression in vivo. This figure is a graphic representation of the experiment detailed in Table 1. Black bars correspond to HIV1-lacZ single transgenics; stippled bars correspond to the HIV1-lacZ/ATIII-X double transgenic animals

Table 1. HIV1-LTR driven LacZ expression in pX transgenic mice

Days	LacZ Tgs		X + LacZ Tgs		Fold activation	
	No. of Tgs	β-gal +	No. of Tgs	β-gal +		
2	4	1 (144)	2	17 (372)	6.5	p < 0.05
4	11	4 (759)	7	28 (724)	6.3	p < 0.05
8	4	1 (182)	4	13 (388)	6.2	p < 0.05
15	3	0 (500)	3	0 (364)		

The number of liver sections containing one or more foci of hepatocytes expressing beta-galactosidase in HIV1-lacZ transgenics (Tgs) was compared to that observed in HIV-lacZ/ATIII-X double transgenic animals. The total number of liver sections analyzed is indicated within brackets. To ensure that positive foci could not be counted several times in consecutive liver sections, the quantitative analysis was performed on one out of five consecutive slices. The fold activation is expressed as the ratio between the percent number of positive liver sections on HIV1-lacZ/ATIII-X double transgenics as compared to HIV1-lacZ single transgenics. Statistical analysis was performed using the chi-quare independence test

malities (hepatocellular necrosis and hepatic granuloma) described in association with both AIDS in man and SIV infection in Rhesus monkeys [7]. Since AIDS patients are often coinfected with HBV, the liver could represent one of the sites for a direct interaction between the two viruses,

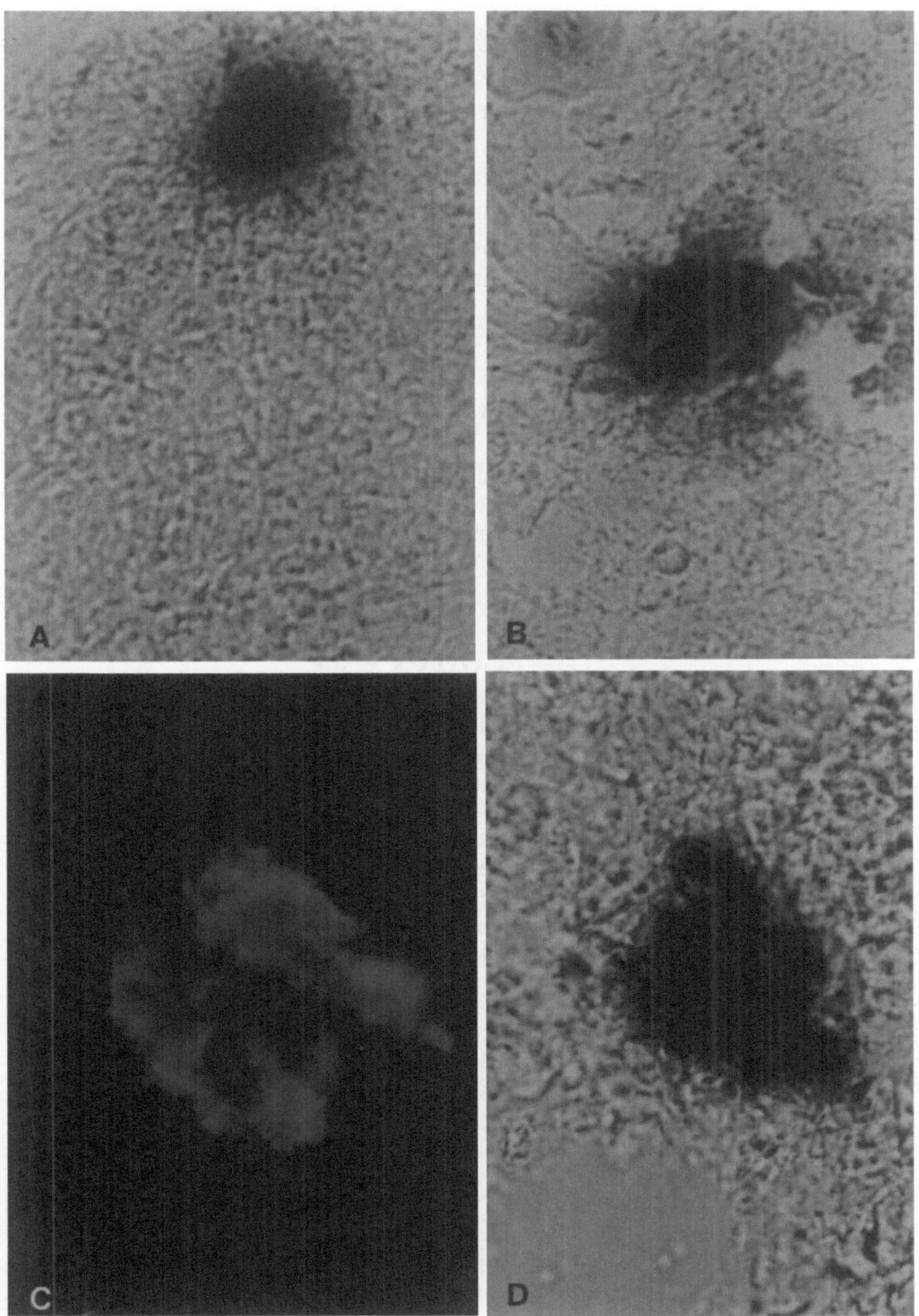

Fig. 4. In situ assay of bacterial beta-galactosidase activity in liver sections prepared from 6-day-old HIV1-lacZ/ATIII-X double transgenic mice (**A**) (×250) and from 8-day-old HIV1-lacZ/ATIII-X double transgenic mice (**B**) (×312). **C,D** Detection of HBx expression by indirect immunofluorescence (**C**) in hepatocytes positive for HIV1-LTR driven beta-galactosidase activity (**D**) in liver sections from 6-day-old HIV1-lacZ/ATIII-X double transgenic mice

especially in the presence of liver inflammation and regeneration, as observed in acute and chronic HBV infection. Our transgenics provide the first demonstration that such interaction might happen in vivo. On the other hand HBV has been shown to infect and replicate in lymphocytes [10] and the HBV transcriptional enhancer element can function in a variety of hepatic cells including B and T lymphocytes. Since the HBx protein has been shown to transactivate in vitro the HIV1-LTR in CD4+ lymphocytic cell lines, HBV could possibly influence HIV1 expression in the T-lymphocyte enviroments as well [15].

Acknowledgements

We thank C. Trépo, L. Vitviski-Trépo and M. Zakin for providing reagents and A. Kahn for discussion and advice. C.B. was supported by a short-term EMBO fellowship. This work was supported by the ARC, the Ligue Nationale contre le Cancer, the ARNS, the INSERM, the V AIDS Project, the A. Cesalpino Foundation.

References

1. Balsano C, Avantaggiati ML, Natoli G, De Marzio E, Will H, Perricaudet M, Levrero M (1991) Full-length and truncated versions of the hepatitis B virus (HBV) X-protein (pX) transactivate the c-myc protooncogene at the transcriptional level. Biochem Biophys Res Commun 176: 985–992
2. Cao Y, Friedman-Kien AE, Huang Y, Li XL, Mirabile M, Moudgil T, Zucker-Franklin D, Ho DD (1990) CD4–independent, productive human immunodeficiency virus type 1 infection of hepatoma cell lines in vitro. J Virol 64: 2553–2559
3. Cavard C, Zider A, Vernet M, Bennoun M, Saragosti S, Grimber G, Briand P (1990) In vivo activation by ultraviolet rays of the human immunodeficiency virus type 1 long terminal repeat. J Clin Invest 86: 1369–1374
4. Dubois N, Bennoun M, Allemand I, Molina T, Grimber G, Daudet-Monsac M, Abelanet R, Briand P (1991) Time course development of differentiated hepatocarcinoma and lung metastasis in transgenic mice. J Hepatol 13: 227–239
5. Ensoli B, Lusso P, Schachter F, Josephs SF, Rappaport J, Negro F, Gallo RC, Wong-Staal F (1989) Human herpes virus-6 increases HIV-I expression in co-infected T cells via nuclear factors binding to the HIV-I enhancer. EMBO J 8: 3019–3027
6. Feitelson MA, Clayton MM (1990) X antigen polypeptides in sera of hepatitis B virus-infected patients. Virology 177: 367–371
7. Gerber MA, Chen ML, Hu FS, Baskin GB, Petrovich L (1991) Liver disease in Rhesus monkeys infected with Simian Immunodeficiency Virus. Am J Pathol 139: 1081–1088
8. Gorman C, Moffat L, Howard B (1982) Recombinant genomes which express chloranphenicol acetyl transferase in mammalian cells. Mol Cell Biol 2: 1044–1051
9. Lane H, Lane D (1988) Antibodies: a laboratory manual. Cold Spring Harbor Laboratory, Cold Spring Harbor

10. Harrison TJ (1990) Hepatitis B virus DNA in peripheral blood leukocytes: a brief review. J Med Virol 31: 33–35
11. Kim CM, Koike K, Saito I, Miyamura T, Jay G (1991) HBx gene of hepatitis B virus induces liver cancer in transgenic mice. Nature 351: 317–320
12. Lee TH, Finegold MJ, Shen RF, Demayo JL, Woo SLC, Butel JS (1990) Hepatitis B virus tranactivatior X-protein is not tumorigenic in transgenic mice. J Virol 64: 5939–5947
13. Levrero M, Balsano C, Natoli G, Avantaggiati ML, Elfassi E (1990) Hepatitis B virus X-protein transactivates the long terminal repeats of human immunodeficiency virus type-1 and type-2. J Virol 64: 3082–3086
14. Sambrook J, Fritsch EF, Maniatis T (1989) Molecular cloning, a laboratory manual 2nd edn. Cold Spring Harbor Laboratory, Cold Spring Harbor
15. Seto E, Yen TSB, Peterlin BM, Ou JH (1988) Trans-activation of the human immunodeficiency virus long terminal repeat by the hepatitis B virus X protein. Proc Natl Acad Sci USA 85: 8286–8290
16. Shirakata Y, Kawada M, Fujiki Y, Sano H, Oda M, Yaginuma K, Kobayashi M, Koike, K (1989) The X gene of hepatitis B virus induced growth stimulation and tumorigenic transformation of mouse NIH-3T3 cells. Jpn J Cancer Res 80: 617–621
17. Spandau DF, Lee CH (1988) Trans-activation of viral enhancers by the hepatitis B virus X protein. J Virol 62: 427–434
18. Takada S, Koike K (1990) Trans-activation function of a 3' truncated X gene cell fusion product from integrated hepatitis-B virus DNA in chronic hepatitis tissues. Proc Natl Acad Sci USA 87: 5628–5632

Authors' address: Dr. M. Levrero, Istituto di I Clinica Medica, Policlinico Umberto I, Viale del Policlinico 155, I-00161 Rome, Italy.

Arch Virol (1993) [Suppl] 8: 73–79

Accumulation of a cellular protein bearing c-myc-like antigenicity in hepatic and non-hepatic delta antigen expressing cells

G. Tappero[1], G. Natoli[2], F. Negro[3], Antonina Smedile[3], F. Bonino[3], M. Rizzetto[4], and M. Levrero[2]

[1] Clinica Medica Generale, Osp. San Luigi Orbassano, Torino, [2] I Clinica Medica and Fondazione Andrea Cesalpino, Universita' La Sapienza, Rome, [3] Divisione di Gastroenterologia, Osp. Le Molinette, Torino, and [4] Cattedra di Gastroenterologia, Istituto di Medicina Interna, Universita di Torino, Torino, Italy

Summary. Patients with chronic but not acute hepatitis delta virus infection undergo a strong accumulation of a protein of cellular origin which specifically reacts with a panel of anti-c-myc antibodies and which is expressed in the same nuclei that express the delta antigen. In this paper we report on the in vitro characterization of this phenomenon. The delta antigen induced c-myc antigen accumulation can occur in vitro upon transfection of HBsAg positive and negative cell lines with HDAg expression vectors. Using recombinant vaccinia viruses expressing only p24 or p27 we demonstrate that structures common to the two isoforms of HDAg are responsible for the phenomenon, which is not restricted to cells of hepatic origin.

Introduction

Chronic hepatitis of viral origin are diseases in which necrosis, inflammation as well as liver cell regeneration are associated. Hepatocellular regeneration is a complex phenomenon involving the expression of a set of cellular genes known as immediate-early genes (IEG) which are required for the transition from quiescence to proliferation. The main group of IEG is constituted by proto-oncogenes encoding nuclear proteins involved in transcription regulation, such as c-fos, c-jun, c-myb, c-myc and others [6, 7].

Screening by indirect immunofluorescence the expression of some of these genes (c-fos, c-myc and c-myb) in liver biopsies from chronic hepatitis patients we found a strong and selective correlation between chronic hepatitis delta virus (HDV) infection and the nuclear accumulation of a protein bearing a c-myc-like antigenicity, which occurred in the same hepatocytes expressing the hepatitis delta antigen (HDAg) [8]. This phenomenon is specific and it is not the result of an immunological cross-reaction between HDAg and the anti-c-myc antibodies, since the staining of c-myc can be abolished by neutralizing the monoclonal antibody with a specific peptide [8].

To characterize this phenomenon better, we investigated its reproducibility in vitro, either by transfection of hepatic and non-hepatic cells with a HDAg expression vector or by infection of the same cells with recombinant vaccinia-viruses expressing only the small or the large form of HDAg (p24 and p27 respectively). Our results show that the phenomenon can also occur in vitro and that it is independent of the presence of hepatitis B virus products; both forms of HDAg are able to induce the c-myc accumulation, which can occur also in cells other than hepatocytes.

Materials and methods

Cell culture and transfection

HBsAg positive (PLC/PRF/5) and HBsAg-negative (HepG2) hepatocarcinoma cell lines as well as cervical carcinoma HeLa cells were maintained in D-MEM supplemented with 10% fetal calf serum and transfected following the standard calcium-phosphate coprecipitation method. Indirect immunofluorescence was performed 48 h post-transfection.

Plasmids

The plasmid pSVLD3 (a kind gift by J. Taylor) contains a tandem trimer of HDV-cDNA under the control of SV40 early promoter [5]. This plasmid directs the synthesis of the 24 Kd small form of hepatitis delta antigen from a poly-adenylated antigenomic mRNA [5].

Vaccinia virus stocks and infection of cultured cells

The wild type vaccinia virus and the recombinant vaccinia viruses expressing only p24 and p27 were kindly provided by M. Eckart and M. Houghton (Chiron Corporation, Emeryville, CA, U.S.A.). Virus stocks were grown and titered using Vero cells. HepG2 and HeLa cells were infected in serum-free medium using 10 plaque forming units per cell. After 1 h incubation the virus was removed and the cells were overlaid with D-MEM containing 5% serum. 16–18 h after infection the cells were analyzed by indirect immunofluorescence.

Indirect immunofluorescence

Staining of c-myc was performed using the sandwich immunofluorescence technique with the mouse monoclonal antibody from clone 9E10 of Evans [4] (Cambridge Research Biochemicals, U.K.) as first layer, and a fluorescein isothyocianate (FITC) labeled goat anti-mouse IgG antiserum as second layer. Staining of HDAg was performed using a tetramethylrodamine (TRITC) labeled preparation of human IgG against the HDAg. Transfected or infected cells were double-stained, first for the presence of c-myc and then for the presence of the HDAg.

Results and discussion

To confirm in vitro the correlation between the expression of c-myc and HDAg, HBsAg positive hepatocarcinoma cells PLC/PRF/5 were transfected with pSVLD3, a plasmid that allows the expression of the 24 Kd form of HDAg in transient transfection assays [5]. Spontaneous c-myc expression was found only in a very limited number of mock-transfected PLC/PRF/5 cells; on the other hand, a strong increase in the number of c-myc positive cells was observed after transfection of the HDAg expression vector. Double staining for c-myc and the delta antigen confirms, as occurs in vivo [8], the identity between c-myc-positive and HDAg-positive nuclei. Similar experiments were performed in HBV-negative hepatoblastoma HepG2 cells, and the same results were obtained (Figs. 1A,B): this suggests that the HDAg-induced accumulation of the c-myc-like antigenicity is entirely dependent on the delta antigen and does not require any HBV product. Moreover, since the phenomenon was reproducible also in non-hepatic cells (HeLa), it does not require liver-specific factors.

Two different isoforms of HDAg, with a relative molecular weight of 24 and 27 Kd respectively, exist in infected livers and sera [2]. This heterogeneity depends on the occurrence, during HDV replication, of a mutation at a single nucleotide in the termination codon for the small HDAg [10]. Thus, p24 and p27 share the 195 N-terminal aminoacids and differ in that p27 contains an additional 19 amino acids at its carboxy-terminus. To investigate the ability of the two HDAg isoforms to induce the c-myc accumulation, we made use of recombinant vaccinia-viruses expressing only p24 or p27. Infection of HeLa cells with these viruses leads to a strong expression of HDAg, with a pattern that is different for p24 and p27; in fact, p24 appeared localized almost exclusively in the nucleoli, while p27 displayed both a nucleolar and nucleoplasmic localization. Double staining for c-myc an HDAg of infected HeLa cells revealed that both forms of HDAg are able to induce the nuclear accumulation of the c-myc related protein. The accumulation of this

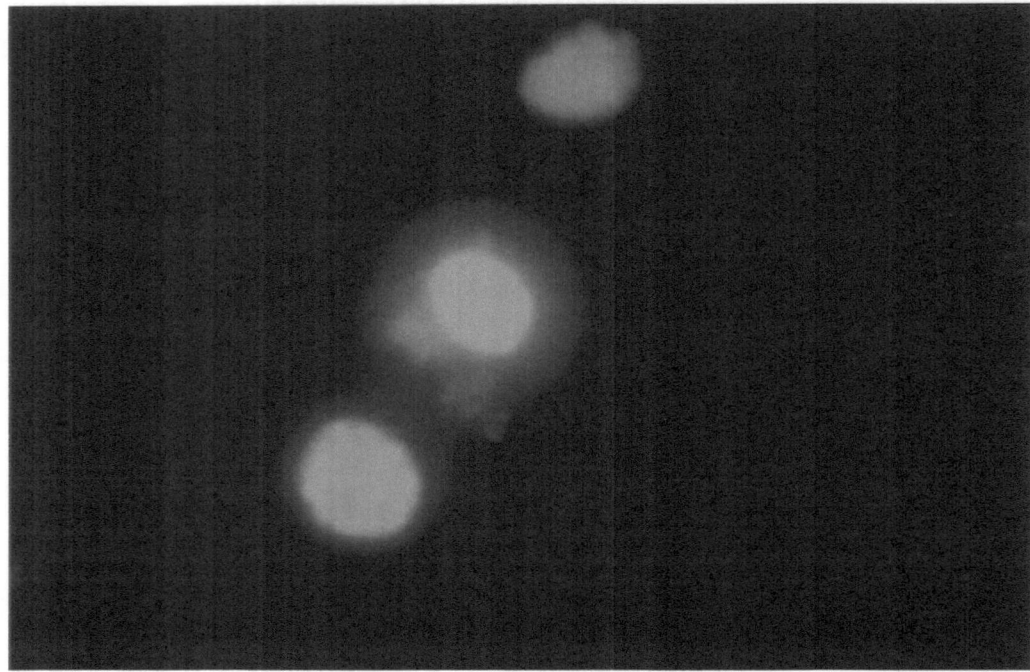

A

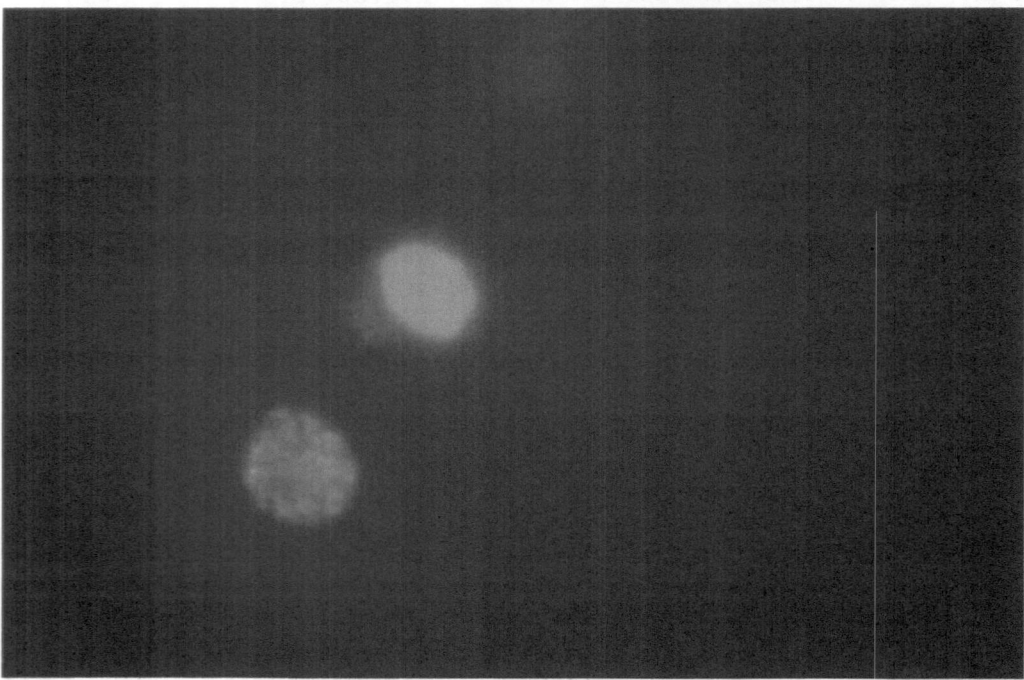

B

Fig. 1. Accumulation of a c-myc related protein in HepG2 cells transfected with pSVLD3. Double staining was performed with the sequential incubation of the cells with an ani-c-myc mouse monoclonal antibody (clone 9E10), a FITC conjugated goat anti-mouse serum and finally with a TRITC conjugated polyclonal anti-HDAg antiserum (**A** and **B** respectively)

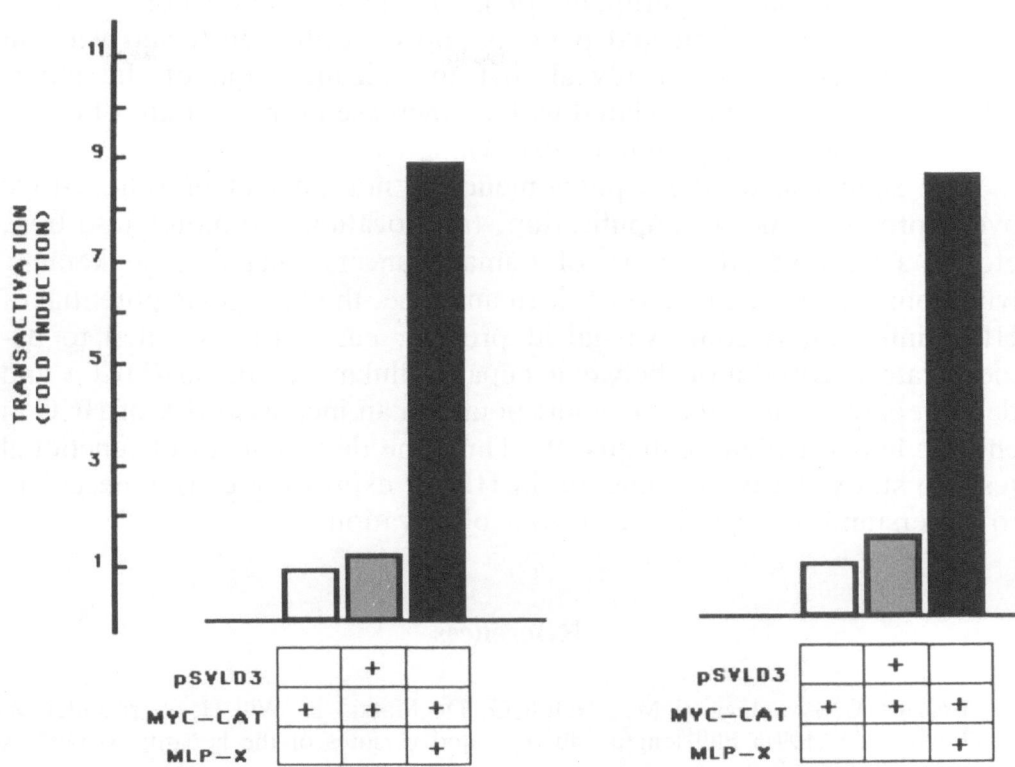

Fig. 2. Transactivation of the c-myc promoter by the HDAg. HepG2 and PLC/PRF5 cells were cotransfected with pSVLD3 and the c-MYC-CAT plasmid following the standard calcium phosphate procedure. CAT activity was measured 48 h after transfection. Black bars represent transactivation of the c-myc promoter by the plasmid pMLP-X, which encodes the X protein of the hepatitis B virus [1]

protein was under these conditions less pronounced than in vivo and in transfected cells, probably due to the negative effects of vaccinia virus infection on transcription of cellular genes. It must be stressed that the different pattern of nuclear expression of p24 and p27 exactly corresponded to the staining pattern of the c-myc related protein. The most reasonable explanation of this observation is the occurrence of a physical interaction between HDAg and the c-myc related protein; experiments of co-immunoprecipitation and cross-linking of the two proteins are in progress, in order to further evaluate this aspect of the phenomenon. Two different lines of evidence oppose the possibility that the HDAg induced c-myc accumulation depends on induction of c-myc transcription by HDAg. First, CAT assays on PLC/PRF/5 and HepG2 cells cotransfected with pSVLD3 and with a c-myc promoter/CAT construct reveal that HDAg is unable to transactivate the c-myc promoter (Fig. 2).

Second, preliminary experiments of in situ hybridization both on liver biopsies from HDV infected patients and on cells transfected with an HDAg expression vector reveal that the accumulation of the c-myc related protein is not associated with an increase in the amount of c-myc mRNAs (Negro et al., unpubl. results).

The significance of this phenomenon is not clear at present. c-Myc overexpression due to amplification, translocation, promoter insertion, etc. is a widespread feature of human cancers, including leukemias, lymphomas and breast cancer [3]. In any case, the oncogenic potential of HDV infection is controversial at present: early studies failed to demonstrate a correlation between hepatocellular carcinoma (HCC) and delta hepatitis, but a recent report points to an increased risk of HCC in chronic hepatitis delta patients [9]. Thus, the development of functional tests to study the c-myc function in HDAg expressing cells is necessary to understand the significance of our observation.

References

1. Balsano C, Avantaggiati ML, Natoli G, De Marzio E, Will H, Perricaudet M, Levrero M (1991) Full length and truncated versions of the hepatitis B virus X protein transactivate the c-myc proto-oncogene at the transcriptional level. Biochem Biophys Res Commun 176: 985–992
2. Bonino F, Hoyer B, Shih WK, Rizzetto M, Purcell RH, Gerin JL (1984) Delta hepatitis agent: structural and antigenic properties of the delta associated particles. Infect Immun 43: 1000–1005
3. Cole MD (1986) The myc oncogene: its role in transformation and differentiation. Annu Rev Genet 20: 361–364
4. Evan GI, Lewis GK, Ramsay G, Bishop JM (1985) Isolation of monoclonal antibodies specific for human c-myc proto-oncogene product. Mol Cell Biol 5: 3610–3616
5. Kuo MYP, Chao M, Taylor J (1989) Initiation of replication of the human hepatitis delta virus genome from cloned DNA: role of the delta antigen. J Virol 63: 1945–1950
6. McMahon SB, Monroe JG (1992) Role of primary response genes in generating cellular responses to growth factors. FASEB J 6: 2707–2715
7. Mohn KL, Laz TL, Hsu JC, Melby AE, Bravo R, Taub R (1991) The immediate-early growth response in regenerating liver and insulin-stimulated H-35 cells: comparison with serum-stimulated 3T3 cells and indentification of 41 novel immediate-early genes. Mol Cell Biol 11: 381–390
8. Tappero G, Natoli G, Anfossi G, Rosina F, Smedile A, Bonino F, Angeli A, Balsano F, Rizzetto M, Levrero M (1993) Expression of the c-myc proto-oncogene product in cells infected with the hepatitis delta virus. Hepatology (in press)
9. Verme G, Brunetto MR, Oliveri F, Baldi M, Forzani B, Piantino P, Ponzetto A, Bonino F (1991) Role of hepatitis delta virus infection in hepatocellular Carcinoma. Dig Dis Sci 36 1134–1136

10. Weiner AJ, Choo QL, Wang KS, Govindarajan S, Redeker AG, Gerin JL, Houghton M (1998) A single antigenomic open reading frame of the hepatitis delta virus encodes the epitope(s) of both hepatitis delta antigen polypeptides p24 and p27. J Virol 62: 594–599

Authors' address: Dr. M. Levrero, Istituto di I Clinica Medica, Policlinico Umberto I, Viale del Policlinico 155, I-00161 Rome, Italy.

Arch Virol (1993) [Suppl] 8: 81–87

Duck hepatitis B virus infection, aflatoxin B1 and liver cancer in ducks

Lucyna Cova[1], Agnes Duflot[1], M. Prave[2], and C. Trepo[1]

[1] INSERM U271, Lyon, and [2] Ecole Nationale Veterinaire de Lyon, Lyon, France

Summary. The association between chronic infection by hepadnaviruses isolated from human (HBV), woodchuck (WHV), ground squirrel (GSHV) and development of hepatocellular carcinoma (HCC) in their respective hosts is well established (reviewed in [11, 15, 17]). By contrast, the association of duck hepatitis B virus (DHBV) infection with HCC is less documented. Pekin ducks congenitally infected with DHBV and followed for several years throughout the world do not develop liver tumors: HCC has been found only in domestic ducks from a single area of China, Qidong. Several factors such as DHBV carrier rate, breed and age of ducks, subtype of DHBV and environmental carcinogens are suspected to contribute to this striking difference between the geographical repartition of liver cancer in DHBV-carrier ducks. In this brief review we will consider successively the role of these different factors in duck liver oncogenesis.

DHBV carrier rate

The high prevalence (50%) of DHBV infection in Qidong, which correlates with the high incidence of HCC in ducks from this area, was initially suspected to be a risk factor in duck liver oncogenesis [13]. However DHBV seems to be highly endemic in other parts of the world. DHBV infection was reported in 1 to 6% of ducks in France and in up to 60% of ducks from commercial flocks in US [1, 11]. Recently, the rate of DHBV infection has been identified within a similar range, i.e. 40–50%, in ducks from two areas of China with high (Qidong) and low (Shanghai) incidence of duck HCC [8]. In addition, in Australia, where a carrier rate of 70% in a naturally infected flock is the highest reported in the literature, neither advanced liver disease nor HCC were observed in

DHBV-infected ducks [7]. Altogether these data suggest that factors other than DHBV-carrier rate play a role in duck liver oncogenesis.

Breed of ducks

The host range of DHBV is less narrow than that of mammalian hepadnaviruses. DHBV was initially detected in the sera from Pekin ducks (*Anas domesticus*) in mainland China [26] and then in commercial flocks of Pekin ducks and in other duck breeds (Indian Runner and Khaki Campbell) in the USA, Australia and Europe ([1, 7, 12], reviewed in [16]). The Pekin duck, which is of Chinese origin, was introduced into Europe, America and Australia at the end of the 19th century. DHBV was also isolated from domestic geese (*Anser domesticus*), and from wild mallards (*Anas Platyrynchos*) as well as from maned ducks (*Chenonetta jubata*), another species of wild duck ([2, 6], reviewed in [16]). However, except for the Qidong area, HCC has, to date never been reported among domestic and wild duck that carry DHBV. Since in Qidong a higher ratio of liver disease and HCC was initially observed in domestic brown ducks, as compared to the Pekin ducks, the former were suspected to be more susceptible to liver disease [24]. However, a recent Chinese study has shown that if the incidence of HCC in brown ducks was high in Qidong, it was low in brown ducks in Shanghai, and this difference correlated with the incidence of human liver cancer [8]. This suggested that environmental factors rather that a particular breed of ducks might be involved in differences of HCC prevalence in human and duck between these two areas [8]. Whether brown ducks are more susceptible than Pekin ducks to liver disease has to be investigated by experimental DHBV transmission. This will not be easy to perform, however, since the brown ducks are bred only in China.

Age of ducks

Histological analysis of persistant infection in DHBV-infected ducks reveals milder hepatitic inflammation, as compared with the lesions observed in persistently infected woodchucks and ground squirrels, which range from no lesions in ducks congenitally infected with DHBV to the portal inflammation and necrosis in experimentally infected birds [9, 10, 13, 14, 20]. Accordingly, it has been suggested that HCC may take longer to develop in ducks than in ground squirrels and woodchucks. Indeed, the absence of HCC in DHBV carrier Pekin ducks in US and Europe might have been related, in the initial reports, to the limited time of observation (2–4 years) while the natural life span of ducks is of 10 years. Actually, after 10 years of investigation no HCC was reported in DHBV-carrier ducks outside China [11]. In addition, in China HCC has been observed in ducks which were no more than 3 years of age [24].

DHBV isolate

The difference in the oncogenic potential between Chinese and other DHBV isolates was also suspected to play a role in the geographical repartition of liver cancer in ducks. Uchida et al. [22] has shown that both DHBV genomes cloned from two Chinese isolates (DHBV-S5 and S31) were both 3027 bp in length and 6 bp longer than the American isolate of this virus. In addition, a new open reading frame was found in a complementary strand of each viral genome, although the product of this open reading frame has not yet been identified [22]. The Chinese isolates of DHBV were transmitted to Japanese ducklings and one of the isolates (DHBV-S5) seemed to induce more chronic persistent hepatitis than other DHBV strains [22]. In spite of this, neither advanced liver disease nor HCC was observed in these animals [14, 22]. Gu et al. [8] have recently described the presence of human HBV DNA sequences in some HCCs from Qidong ducks and suggested that an HBV variant which contains sequences coding for altered surface or core antigen might infect ducks. Using the same approach, i.e. Southern blot analysis of total duck liver DNA followed by hybridization with an HBV-specific probe, we have tested for but not detected human HBV DNA sequences in any of the 8 Qidong duck HCCs analyzed (Cova et al., unpubl.).

Aflatoxin B_1 exposure

HCC has been reported only in domestic ducks from Qidong, an area of China known for high human HCC incidence, in which both HBV and the aflatoxin B_1, AFB_1 are risk factors [19]. The high prevalence of human and duck liver cancer in Qidong may indicate the presence of common environmental risk factors. Several studies have therefore investigated the role AFB_1 exposure in the experimentally induced and naturally occurring duck liver tumors.

Experimental administration of AFB_1 to ducks

Ducks are highly susceptible to the cancerogenic effect of AFB_1 and were used by us [3] and by others [5, 21] as an experimental system to study the respective role of DHBV infection and AFB_1 exposure in the induction of liver tumors. In our study, AFB_1 was administered to DHBV infected and uninfected ducks at two doses (0.02 and 0.08 mg/kg) over a two year period; a reduced survival and higher incidence of liver tumors was observed in the DHBV infected ducks AFB_1 treated than in the uninfected controls [3]. Although these results suggest a synergic effect between DHBV and AFB_1 in the duck liver oncogenesis, they were not statistically significant. Two other studies which used higher doses and different schedules of AFB_1 administration to younger ducks have not reported significant interactions between DHBV and AFB_1 [5, 21]. In some AFB_1 induced tumors, in DHBV-infected ducks, we have observed an accumulation of viral DNA multimers [3], while in a similar study Cullen et al. [5] found integration of DHBV DNA.

One hypothesis on the mode of interaction between AFB_1 and hepadnavirus is that viral infection may alter aflatoxin metabolism. We have studied in vivo the metabolism of AFB_1 in adult ducks and demonstrated lower levels of AFB_1 binding to liver DNA and plasma protein in the DHBV infected as compared to noninfected animals [3]. More recently, we have investigated the metabolism of aflatoxin in newly hatched ducklings, however no influence of DHBV infection on AFB_1 adducts in the liver could be demonstrated [23]. This finding, different from that previously reported in adult ducks, suggests that age of ducks might be an important factor in these interactions.

AFB_1 exposure and DHBV infection in domestic Chinese ducks

There are limited data on the correlation between DHBV infection and liver disease occurring in Qidong ducks since, in three previous studies mostly paraffin embedded material was available and this considerably hampered the molecular analysis of these samples. These studies [10, 13, 24] describe a total of only 4 HCCs and all of them were well differentiated HCC of the trabecular type (Table 1). Recently, we analyzed a larger panel of 16 frozen duck liver samples selected for liver disease on routine pathologic examination of domestic Chinese brown ducks in local farms in Qidong 1988–1989 [4]. Out of these samples, we found 8 HCCs of different morphological types e.g. scirrhous, pseudoglandular and even undifferentiated. The presence of liver cirrhosis in Qidong ducks was observed by us and by Omata et al. [13] (Table 1). In addition we have found bile duct proliferation in these liver samples (Table 1). This biliary proliferation has not been reported as being associated with DHBV infection but has been observed in ducks experimentally exposed to AFB_1 [3, 21].

Table 1. Liver pathology observed in Chinese ducks

Pathology/sample origin	Omata et al. [13] Qidong (n = 24)	Marion et al. [10] Qidong (n = 14)	Yokosuka et al. [24] Qidong (n = 23)	Cova et al. [4] Qidong (n = 16)
Hepatocellular carcinoma (DHBV+)	1	1	1	4
Hepatocellular carcinoma (DHBV−)	0	1	0	4
Cirrhosis	7	0	0	1
Portal inflammation[a]	15	11	15	9
Biliary proliferation	0	0	0	4
No pathology[b]	2	1	7	2
AFB1 adducts	NT	NT	NT	1

[a] Portal tracts infiltrated (++ to ++++) mainly by mononuclear inflammatory cells

[b] Includes fatty change (steatosis)

NT Not tested

In two previous studies, liver disease and HCC in ducks from Qidong were not always associated with detectable virus [10, 13]. It has been suggested that a low level of DHBV replication might occur in some of these livers, although it was below the limit of sensitivity of a conventional dot blot assay. We have taken advantage of the high specificity and sensitivity of PCR to look for the presence of DHBV DNA in the Chinese duck HCCs. We have demonstrated that in 4 HCCs, PCR failed to show any DHBV DNA, indicating that liver tumors do occur in the ducks from Qidong in the absence of DHBV infection [4].

Unlike the HCCs of human and woodchucks, only a single case of Chinese duck HCC with integrated DHBV DNA has been reported to date [24]. We have found another case of DHBV DNA integration into cellular DNA in only one out of 4 DHBV positive HCCs, indicating that viral integration is not a prerequisite for tumor development [4].

The exposure of domestic ducks from Qidong to AFB_1 was suggested, although it has never been proven. We have taken advantage of the availability of frozen duck liver samples to search for the presence of AFB_1 adducts. We detected AFB_1-DNA adducts by hplc-immunoassay in one of 8 duck HCCs analyzed [4].

Together, these findings suggest a possible role of AFB_1 in duck liver oncogenesis. Since Qidong is the only area where liver cancer has been reported in ducks this raises the question of whether DHBV is an oncogenic virus. In addition, as summarized here, a direct link between DHBV infection and HCC has so far not been demonstrated. The differences in the oncogenicity between mammalian and avian hepadnaviruses might be related not only to the milder liver disease induced by DHBV in its host, but also to a direct effect of viral gene products such as the X gene which is lacking in DHBV and which can transactivate cellular transforming genes [15, 25].

References

1. Cova L, Hantz O, Arliaud-Gassin M, Chevallier A, Bertillon P, Boulay J, Jacquet C, Chomel B, Vitvitski L, Trépo C (1985) Comparative study of DHBV DNA levels and endogenous DNA polymerase in naturally infected ducklings in France. J Virol Methods 10: 251–260
2. Cova L, Lambert V, Chevallier A, Hantz O, Fourel I, Jacquet C, Pichoud C, Boulay J, Chomel B, Vitviski L, Trépo C (1986) Evidence for the presence of duck hepatitis B virus in the wild migrating ducks. J Gen Virol 67: 537–547
3. Cova L, Wild CP, Mehrotra R, Turusov V, Shirai T, Lambert V, Jacquet C, Tomatis L, Trépo C, Montesano R (1990) Contribution of aflatoxin B1 and hepatitis B virus infection in the induction of liver tumors in ducks. Cancer Res 50: 2156–2163
4. Cova L, Mehrotra R, Wild C, Chutimataewin S, Cao S, Duflot A, Yu S, Montesano R, Trepo C (1993) Duck hepatitis B virus infection and liver cancer in domestic Chinese ducks. Br J Cancer (in press)
5. Cullen JM, Marion PL, Sherman GJ, Hong X, Newbold JE (1990) Hepatic neoplasms in aflatoxin B1-treated, congenital duck hepatitis B virus-infected and virus-free Pekin ducks. Cancer Res 50: 4072–4080
6. Dixon RJ, Jones NF, Campbell M, Brooker A, Bowden S (1989) The Australian maned duck hepatitis B virus: a potential novel hepatitis B virus of ducks. In: Marion P, Schaller H (eds) Hepatitis B viruses, abstracts. Cold Spring Harbor Laboratory, Cold Spring Harbor, p 28
7. Freiman JS, Cossart YE (1986) Natural duck hepatitis B virus infection in Australia. Aust J Exp Biol Med Sci 64: 477–484
8. Gu JR (1992) Viral hepatitis B and primary hepatic cancer. In: Wen YN, Xu ZY, Melnick JL (eds) Viral hepatitis in China; problems and control strategies. Karger, Basel, pp 56–72 (Monographs in Virology, vol 19)
9. Lambert V, Cova L, Chevallier P, Mehrotra R, Trépo C (1991) Natural and experimental infection of wild mallard ducks with duck hepatitis B virus. J Gen Virol 72: 417–420
10. Marion PL, Knight SS, Ho BK, Guo YY, Robinson WS, Popper H (1984) Liver disease associated with duck hepatitis B virus infection of domestic ducks. Proc Natl Acad Sci USA 81: 898–902
11. Marion PL, Trepo C, Matsubara K, Price P (1991) Experimental models in hepadanavirus research: Report of a workshop. In: Hollinger FB, Lemon SM, Margolis HS (eds) Viral hepatitis and liver diseases. Williams and Wilkins, Baltimore, pp 866–879
12. Mason WS, Seal G, Summers J (1980) Virus of Pekin ducks with structural and biological relatedness to human hepatitis B virus. J Virol 36: 829–836

13. Omata M, Uchiumi K, lto Y, Yokosuka O, Mori J, Terao K, Wei-Fa Y, O'Connel AP, London WT, Okuda K (1983) Duck hepatitis B virus and liver disease. Gastroenterology 835: 260–267
14. Omata M, Yokosuka O, Imazeki F, Matsuyama Y, Uchiumi K, Ito Y, Mori J, Okuda K (1984) Transmission of duck hepatitis B virus from Chinese carrier ducks to Japanese ducklings: a study of viral DNA in serum and tissue. Hepatology 4: 603–607
15. Rogler CE (1991) Cellular and molecular mechanisms of hepatocarcinogenesis associated with hepadnavirus infection. In: Mason WS, Seeger C (eds) Current topics in microbiology and immunobiology, vol 168. Springer, Berlin Heidelberg New York Tokyo, pp 104–140
16. Schödel F, Weimer T, Fernholtz D, Schneider R, Sprengel R, Wildner G, Will H (1989) The biology of avian hepatitis B viruses. In: McLachlan A (ed) Molecular biology of the hepatitis B virus. CRC Press, Ann Arbor, pp 54–80
17. Sherker AH, Marion P (1991) Hepadnaviruses and hepatocellular carcinoma. Annu Rev Microbiol 45: 475–508
18. Sprengel R, Kuhn C, Will H, Schaller H (1985) Comparative sequence analysis of duck and human hepatitis B virus genomes. J Med Virol 15: 323–329
19. Sun T, Wu S, Wu Y, Chu Y (1986) Measurement of individual aflatoxin exposure among people having different risk to primary hepatocellular carcinoma. In: Hayashi Y (ed) Diet, nutrition and cancer. Jap Soc Press, Utrecht, pp 225–235
20. Uchida T, Suzuki K, Arii M, Shikata T, Fukuda R, Tao Y (1988) Geographical pathology of duck livers infected with duck hepatitis B virus from Chiba and Shimane in Japan and Shanghai in China. Cancer Res 48: 1319–1325
21. Uchida T, Suzuki K, Esumi M, Arii M, Shikata T (1988) Influence of aflatoxin B1 intoxication on duck livers with DHBV infection. Cancer Res 48: 1559–1565
22. Uchida M, Esumi M, Shikata T (1989) Molecular cloning and sequence analysis of duck hepatitis B virus genomes of a new variant isolated from Shanghai ducks. Virology 173: 600–606
23. Wild CP, Jansen LM, Coval L, Montesano R (1992) Molecular dosimetry of aflatoxin exposure: contribution to understanding the multifactorial aetiopathonegisis of primary hepatocellular carcinoma (PHC) with paticular reference to hepatitis B virus. Environ Health Perspect 99: 115–122
24. Yokosuka O, Omata M, Zhou YZ, Imazeki F, Okuda K (1985) Duck hepatitis B virus DNA in liver and serum of Chinese ducks: integration of viral DNA in a hepatocellular carcinoma. Proc Natl Acad Sci USA 82: 5180–5184
25. Zahm PH, Hofschneider PH, Koshy R (1988) The HBV X-ORF encodes a transactivator: a potential factor in viral hepatocarcinogenesis. Oncogene 3: 169–177
26. Zhou YZ, Kou PY, Shao LQ (1980) A virus possibly associated with hepatitis and hepatoma in ducks. Shanghai Med J 3: 1–3

Authors' address: Dr. Lucyna Cova, INSERM U271, 151 Cours A. Thomas, F-69003 Lyon, France.

III. Pathogenic and protective immune responses against hepatitis viruses

Arch Virol (1993) [Suppl] 8: 91–101

Cell mediated immune response to hepatitis B virus nucleocapsid antigen

C. Ferrari, Amalia Penna, A. Bertoletti, and F. Fiaccadori

Cattedra Malattie Infettive, Università di Parma, Parma, Italy

Summary. A coordinated and efficient development of humoral and cell-mediated immune responses is believed to be required for complete eradication of viral infections. During the course of hepatitis B virus (HBV) infection, the HLA class II and class I-restricted T cell responses to HBV nucleocapsid antigens are vigorous in patients with acute infection who succeed in clearing the virus but weak or totally absent in patients with chronic persistence of the virus. These findings suggest a role for these responses in the pathogenesis of hepatitis B and in HBV clearance. Molecular analysis of T cell recognition of the HBV nucleoprotein defines the presence of immunodominant core epitopes recognized by helper and cytotoxic T cells that may represent the starting point for the design of alternative strategies for prevention and treatment of HBV infection.

Introduction

During the last few years the molecular basis of T cell activation and T cell recognition of viral antigens has been defined in great detail, allowing the development of new experimental approaches for the study of the T cell response, based on the combined use of synthetic peptides and recombinant proteins. This has led to a better understanding of some immunopathogenetic mechanisms involved in virus clearance and cellular injury in hepatitis B virus (HBV) infection.

HLA class II restricted T cell response to HBV nucleocapsid antigens

While B cells recognize antigen molecules in their native conformation without the need for processing, T cells recognize antigens in the form of short peptides associated with histocompatibility leukocyte antigen (HLA) molecules. In particular, HLA class II restricted CD4+ T cells recognize peptide fragments associated with HLA class II molecules, usually derived from extracellular antigens which are proteolytically processed in acidified cellular vesicles (endosomes or lysosomes) after endocytosis by specialized antigen presenting cells [1]. Following synthesis in the endoplasmic reticulum (ER), transport of class II molecules to the proper peptide binding compartment (endosomes) and prevention of endogenous antigen binding in the ER (where cytosolic peptides derived from endogenous antigens are delivered for HLA class I binding) seem to be a function of the invariant chain [2].

Based on these principles, purified HBV nucleocapsid antigens from recombinant sources or from HBV infected subjects have been used as an exogenous source of antigen for in vitro stimulation of CD4+ T cells and for analysis of their function in different groups of HBV patients with acute or chronic HBV infection [3, 4]. The results of these studies show that HBV infected patients who develop a self-limited acute hepatitis express a vigorous HLA class II restricted peripheral blood T cell response to hepatitis B core antigen (HBcAg) and hepatitis B e antigen (HBeAg) (Fig. 1A). This response is significantly lower in patients with chronic HBV infection who do not succeed in clearing the virus [3, 4] (Fig. 1A). This low response to HBV nucleocapsid antigens seems to increase during acute exacerbations of chronic hepatitis [5]. Additional studies are now required to understand whether these defective peripheral blood responses are due to an actual lesion of the immune system in these patients, or are the result of specific viral strategies to evade immune surveillance (such as virus mutations), or whether they reflect a preferential intrahepatic sequestration of virus responsive T cells in the chronic stages of HBV infection, as previously reported [6, 7].

Interestingly, the development of a strong CD4-mediated response to nucleocapsid antigens during acute hepatitis B seems to be temporally associated with the clearance of HBV envelope antigens from the serum [3]. This finding, together with the evidence that this response is defective in the peripheral blood of patients who do not spontaneously clear the virus, suggests that the optimal development of a CD4-mediated response to HBV nucleocapsid antigens following HBV infection may be required for efficient virus elimination. Since the behavior of this response parallels that of HLA class I restricted cytotoxic T lymphocytes (CTL), the activity of which is similarly vigorous in acute and weak in chronic HBV infection, the apparent involvement of nucleocapsid-specific CD4+ T cells in the pathogenesis of HBV clearance may reflect the helper-dependent activation of nucleocapsid-specific CTL and the elimination of infected cells through their killer function. Alternatively, viral clearance could be facilitated by a direct cooperation between

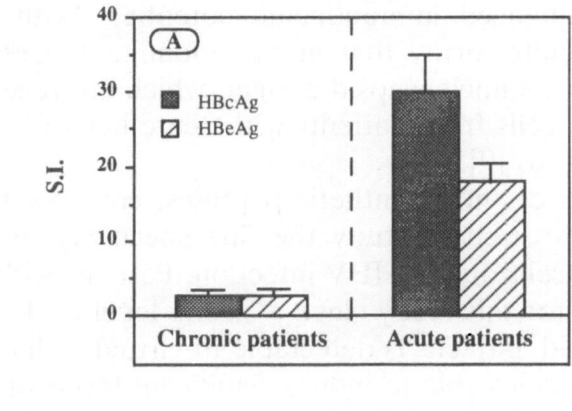

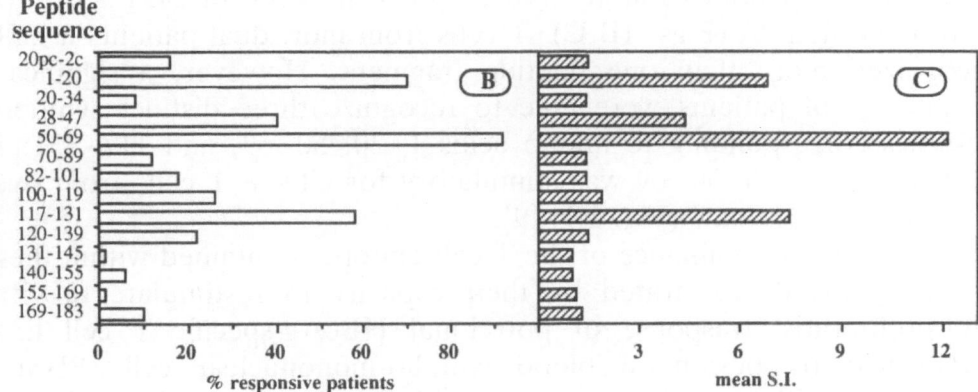

Fig. 1. A HLA class II restricted T cell proliferative response to HBcAg and HBeAg in patients with acute (21 cases) and chronic (42 cases) HBV infection. The bars represent the mean values of stimulation index in the two groups of patients. The stimulation index (*S.I.*) defines the level of T cell proliferative response and is the ratio between the ^{3}H-thymidine incorporated by T cells stimulated with antigen and the ^{3}H-thymidine incorporated by T cells cultured in the absence of antigen. **B,C** Fine specificity of the HLA class II restricted T cell response to HBV nucleocapsid antigens. PBL from 27 patients with acute hepatitis B were stimulated with different concentrations of the indicated synthetic peptides and their proliferative response measured after 7 days of culture. Results are expressed as percent of patients who gave a significant response to the individual core peptides (**B**) and as mean S.I. (**C**). The T cell response was HLA class II restricted because inhibitable by incubation with anti-HLA class II monoclonal antibodies (not shown)

HBcAg-specific helper T cells and HBV envelope-specific B cells, through a mechanism of "intermolecular" help [8], leading to amplification of the virus-neutralizing anti-envelope antibody response.

Fine specificity of the CD4-mediated response to HBV nucleoprotein

In view of its potential role in HBV clearance, the CD4-mediated response to HBcAg could be exploited for the development of anti-HB vaccines with enhanced immunogenic potential. With respect to this possibility, it is noteworthy that immunodominant peptides have been identified within the nucleocapsid antigen which are recognized by core-specific CD4+ T cells from patients with acute hepatitis B and different genetic backgrounds [9].

A large panel of short synthetic peptides, covering the entire HBV core region, were used to study the fine specificity of CD4+ nucleo-capsid-specific T cells during HBV infection. Patients with acute hepatitis B were selected for this study since a strong level of T cell response to HBV uncleocapsid antigens is detectable in virtually all of them. Several immunogenic peptides able to induce significant levels of CD4 activation were identified [9] (Figs. 1B,C). T cells from individual patients usually recognized more than one peptide fragment. However, a significant proportion of patients were able to recognize three distinct synthetic peptides corresponding to amino acids 1–20, 50–69, and 117–131: in particular, peptide 50–69 was stimulatory for CD4+ T cells from over 90% of acute patients (Fig. 1B) [9].

The immunodominance of the T cell epitopes contained within these peptides was demonstrated by their capacity to restimulate in vitro the proliferative response of polyclonal HBcAg-specific T cell lines established by peripheral blood lympho-mononuclear cell (PBMC) stimulation with the whole nucleocapsid protein [9]. In the reciprocal experiments, peptide-induced polyclonal T cell lines established by PBMC stimulation with synthetic peptides were able to recognize the native core molecule, demonstrating that the epitopes represented by the stimulatory peptides are actually generated by intracellular processing of the whole core protein [9].

Experiments are now needed to define precisely the HLA restriction elements for these peptides. Nonetheless, their "promiscuous" recognition by T cells of different HLA haplotypes is certainly encouraging for their use in a synthetic vaccine that might be effective in a large proportion of the general population.

CTL response against HBV nucleocapsid antigens

Identification by CTL of the infected cells that must be eliminated depends upon the recognition of short peptides associated with HLA class I molecules, derived from the intracellular processing of viral antigens synthesized endogenously within the infected cells [10–12]. Newly synthesized viral proteins are probably degraded by cytoplasmic proteases into peptides which are delivered to trans-membrane peptide transporters that transfer them accross the membrane of the ER from the cytoplasm to the ER lumen [12], where they bind to HLA class I molecules.

In several infections the viral nucleoprotein has been shown to serve as a major target for HLA class I restricted CTL [10, 13–17]. The potential importance of this antigen for the activation of an anti-viral CTL response in HBV infection was suggested by Mondelli et al. [18]. They showed that peripheral blood lymphocytes from patients with acute and chronic hepatitis B may kill autologous hepatocytes in vitro. This cytotoxicity was specifically inhibited following incubation of the target cells with anti-nucleocapsid but not with anti-envelope antibodies, suggesting that core antigen specific antibodies and T cells may recognize the same or overlapping epitopes.

Due to technical limitations inherent in those early studies, however, the reproducibility and specificity of the cytolytic activity, its HLA restriction elements and its cellular phenotype could not be established, and only patients undergoing diagnostic liver biopsies could be studied. Obviously, a different approach was necessary if definitive analysis of the CTL response to HBV encoded antigens was to be achieved.

Since HBV does not efficiently infect human cells in vitro, thereby precluding the possibility of producing stimulator/target systems by the conventional approach of human cell infection in vitro, a strategy involving the combined use of short synthetic peptides mimicking the processed antigen fragments and recombinant expression vectors inducing endogenous expression of HBV antigens was developed to analyze the HBV specific CTL response [19, 20] (Fig. 2). Following this approach, a panel of short overlapping synthetic peptides covering the entire amino acid sequence of the HBV core and e proteins was synthesized [19, 21]. In addition, two different eukaryotic expression systems were produced in F.V. Chisari's and H.-J. Schlicht's laboratories: recombinant vaccinia viruses that allow rapid infection of human cell lines in vitro followed by rapid and efficient, though transient, expression of HBV core and precore polypeptides [22] and Epstein Barr virus (EBV)-based plasmid vectors (EBO) that permit stable transfection of EBV-transformed human B cell lines with stable and efficient expression of HBV antigens [20].

HBV-specific CTL were never detected in freshly isolated PBMC without prior stimulation in vitro, suggesting that HBV-specific CTL are probably present in the peripheral blood compartment at very low frequency [19, 21]. In contrast, HBV-specific CTL activity was detectable following PBMC stimulation in vitro with synthetic peptides [19, 21]. This CTL activity was reproducibly induced in HLA-A2 positive patients with acute hepatitis B by a peptide corresponding to HBV core residues 11–27 (Fig. 3). This amino acid sequence is conserved among

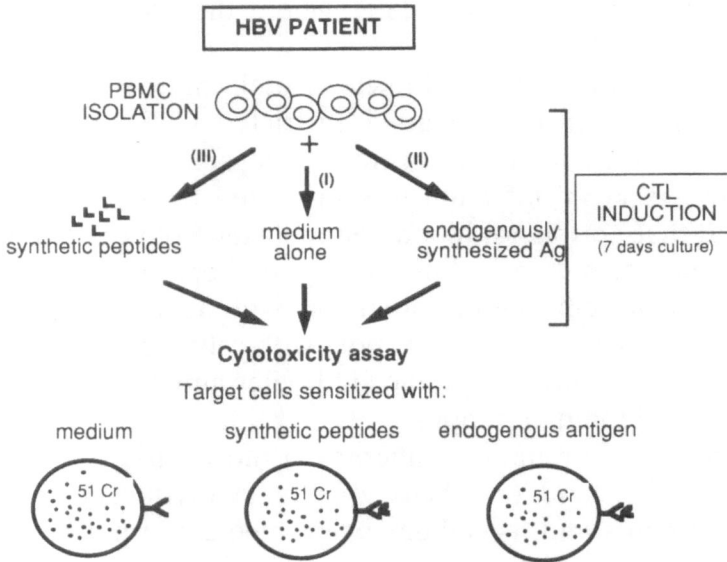

Fig. 2. Experimental strategy for the analysis of HBV-specific cytotoxic activity. PBL isolated from patients with acute and chronic HBV infection were tested for HBV-specific cytolytic activity either (*I*) immediately after isolation, without prior steps of in vitro stimulation, or (*II*) after 1 or 2 weeks of stimulation with autologous stimulator cells transfected or infected with recombinant HBV expression vectors (see below), or (*III*) after stimulation with HBV core synthetic peptides. Target cells were represented by autologous Epstein Barr virus transformed B-lymphoblastoid cell lines or autologous phytohaemagglutinin-stimulated T cell blasts either infected or transfected with HBV expression vectors inducing endogenous synthesis of HBV antigens, or preincubated with synthetic peptides. For the features of the two eukaryotic expression systems employed in these studies see [20, 22]

the major subtypes of HBV and is selectively presented to CTL by HLA-A2, one of the most common HLA class I alleles in the human population.

Importantly, peptide-induced CTL were able to recognize endogenous HBcAg and HBeAg encoded by recombinant vaccinia and EBO plasmid vectors in autologous HLA class I positive target cells [19] (Fig. 3), indicating that the CTL epitope represented by residues 11–27 of HBcAg is actually generated by the intracellular processing of endogenously synthesized core and precore proteins and that, therefore, it is likely available to CTL recognition in vivo on the surface of virus infected liver cells, at the site of antigen synthesis.

The repeated failure to detect nucleocapsid-specific cytotoxic activity in normal control subjects further confirms that these core-specific CTL must have encountered the antigen in vivo and that they have been primed by the natural HBV infection [19–21].

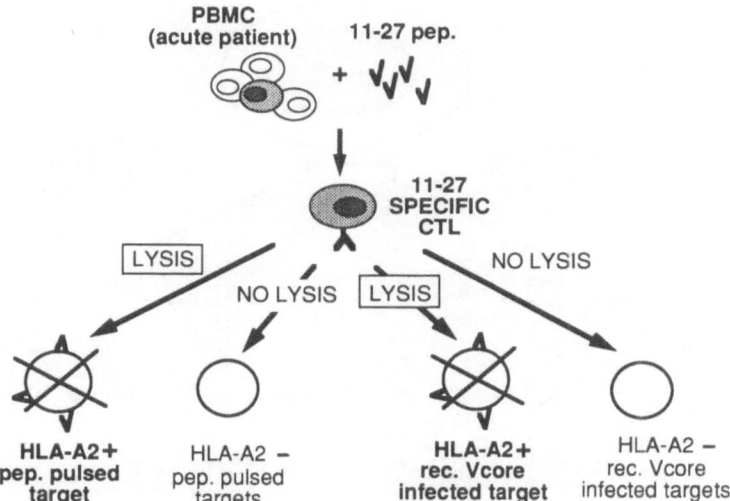

Fig. 3. HBcAg-specific cytotoxic T cells can be induced by stimulation of PBMC from HLA-A2+ patients with acute HBV infection with core peptide 11–27. These CTL selectively kill HLA-A2+ target cells either pre-pulsed with the synthetic peptide 11–27 or infected with recombinant vaccinia virus inducing endogenous synthesis of HBcAg [19, 21]

Analysis of the CTL response in patients at different stages of HBV infection shows that the CTL response against HBcAg 11–27 is strong in the peripheral blood of patients who recover from infection (i.e. acute patients) but is absent or very weak in patients who do not succeed in clearing the virus (i.e. chronic patients) [21]. Moreover, this CTL epitope is commonly recognized by HLA-A2 positive patients with acute self-limited hepatitis B (who are primed in vivo by the HBV nucleocapsid protein) since virtually all of them develop a vigorous CTL response against this core region peptide [21].

The different responses in patients with acute and chronic evolution of HBV infection suggest a pathogenic role for nucleocapsid-specific CTL in the clearance of HBV infection raising the possibility that a failure to mount a vigorous CTL response to nucleocapsid antigens may be a factor predisposing to virus persistence and chronic evolution of HBV infection.

Molecular analysis of the CTL response to HBV nucleocapsid antigens

Recent elution studies show that naturally processed peptides present in the groove of the HLA class I molecules are usually 8–10 amino acids long and that allele-specific binding motifs define peptides that are able

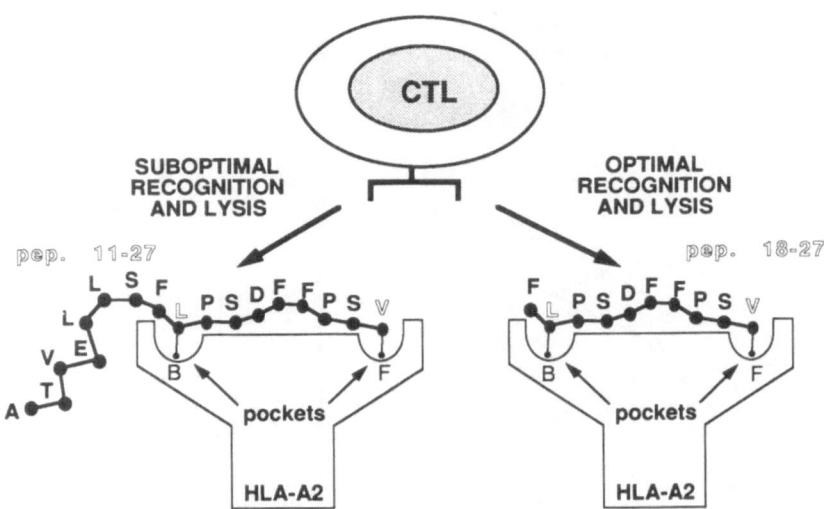

Fig. 4. Peptide 18–27 represents the minimal sequence optimally recognized by HLA-A2 restricted, HBcAg-specific CTL. Leucine at position 2 and valine at the C-terminus act as anchor residues (outlined letters), most likely by projecting their side chains into the polymorphic pockets B and F located in the floor of the HLA binding groove, and thereby stabilizing the interaction of the peptide with the HLA molecule [33]. The peptide is likely to be accommodated into the HLA groove in an extended conformation with the central residues protruding out of the cleft, that appears to be closed at both ends [27, 28]. The aminoterminal residues of peptide 11–27 appear likely to extend out of one end of the binding groove, thereby creating a less favorable interaction with the HLA molecule or the T cell receptor

to bind distinct HLA molecules [23–29]. For the HLA-A2 restricted T cell epitopes, the binding motif is represented by a leucine or isoleucine at position 2 and valine or a residue with aliphatic hydrocarbon side chains at the C-terminus [26, 30]. Moreover, the short natural peptides thus far isolated are more efficiently recognized by CTL than the longer synthetic analogues, because of a higher binding affinity to HLA class I molecules [31, 32]. This property makes them ideal candidates for the design of synthetic vaccines aimed at inducing a protective CTL response in vivo.

To precisely define the optimal amino acid sequences recognized by HBV nucleocapsid-specific CTL, a panel of truncated and overlapping synthetic peptides of different length covering the core region 11–27 were used [33]. The HLA-A2 restricted CTL epitope contained within the HBV core sequence 11–27 was mapped at residues 18–27 that define the optimally recognized peptide (Fig. 4). This 10-mer contains the predicted HLA-A2 binding motif with leucine in position 2 and valine at the C-terminus [33].

In view of the selective association of this nucleocapsid-specific CTL response with the acute stage of self-limited HBV infection and the wide recognition of peptide 18–27 in the HLA-A2 haplotype, this core sequence might represent a potential candidate for a specific immune therapy capable of inducing in vivo, an efficient anti-HBV specific CTL response in HLA-A2 positive patients.

Conclusion

In conclusion, much has been learned during the last few years about the immune mechanisms of virus clearance and liver cell injury in HBV infection thanks to several important discoveries that have clarified hitherto obscure aspects of the immune response. In particular, molecular cloning of the T cell receptor genes, resolution of the three-dimensional structure of human HLA class I molecules, sequence analysis of the antigen fragments bound to HLA molecules in vivo, fine molecular dissection of the intracellular pathways of antigen processing and presentation to T cells have provided a fundamental framework for studies designed to elucidate the complex network of specific immune responses involved in disease progression and viral clearance in HBV-infected individuals. The new pieces of information derived from these studies will hopefully represent the starting point for the design of alternative and more effective preventive and therapeutic strategies against HBV, directed to modulate specific stages of the HBV-specific immune response.

References

1. Germain RN (1991) The second class story. Nature 353: 605–606
2. Teyton L, O'Sullivan D, Dickson PW, Lotteau V, Sette A, Fink P, Peterson PA (1990) Invariant chain distinguishes between the exogenous and endogenous antigen presentation pathway. Nature 348: 39–44
3. Ferrari C, Penna A, Bertoletti A, Valli A, Degli Antoni A, Giuberti T, Cavalli A, Petit MA, Fiaccadori F (1990) Cellular immune response to hepatitis B virus-encoded antigens in acute and chronic hepatitis B virus infection. J Immunol 145: 3442–3449
4. Jung M-C, Spengler U, Schraut W, Hoffmann R, Zacoval R, Eisenburg J, Eichenlaub D, Riethmuller G, Paumgartner G, Ziegler-Heitbrock HWL, Will H, Pape GR (1991) Hepatitis B virus antigen-specific T-cell activation in patients with acute and chronic hepatitis B. J Hepatol 13: 310–316
5. Tsai SL, Chen MY, Yang PM, Sung JL, Huang JH, Hwang LH, Chang TH, Chen DS (1992) Acute exacerbations of chronic type B hepatitis are accompanied by increased T cell responses to hepatitis B core and e antigens. J Clin Invest 89: 87–93

6. Ferrari C, Penna A, Giuberti T, Tong MJ, Ribera E, Fiaccadori F, Chisari FV (1987) Intrahepatic, nucleocapsid antigen-specific T cells in chronic active hepatitis B. J Immunol 139: 2050–2058

7. Ferrari C, Mondelli MU, Penna A, Fiaccadori F, Chisari FV (1987) Functional characterization of cloned intrahepatic, hepatitis B virus nucleoprotein-specific helper T cell lines. J Immunol 139: 539–544

8. Milich DR, McLachlan A, Thornton GB, Hughes JL (1987) Antibody production to the nucleocapsid and envelope of the hepatitis B virus primed by a single synthetic T cell site. Nature 329: 547–549

9. Ferrari C, Bertoletti A, Penna A, Cavalli A, Valli A, Missale G, Pilli M, Fowler P, Giuberti T, Chisari FV, Fiaccadori F (1991) Identification of immunodominant T cell epitopes of the hepatitis B virus nucleocapsid antigen. J Clin Invest 88: 214–222

10. Townsend ARM, Rothbard J, Gotch FM, Bahadur G, Wraith D, McMichael AJ (1996) The epitopes of influenza nucleoprotein recognized by cytotoxic T lymphocytes can be defined with short synthetic peptides. Cell 44: 959–968

11. Gotch FM, Rothbard J, Howland K, Townsend A, McMichael AJ (1987) Cytotoxic T lymphocytes recognize a fragment of influenza virus matrix protein in association with HLA-A2. Nature 326: 881–882

12. Monaco JJ (1992) A molecular model of MHC class I restricted antigen processing. Imunol Today 13: 173–179

13. Townsend ARM, Gotch FM, Davey J (1985) Cytotoxic T cells recognize fragments of the influenza nucleoprotein. Cell 42: 457–467

14. Nixon DF, Townsend ARM, Elvin JG, Rizza CR, Gallway J, McMichael AJ (1988) HIV-1 gag-specific cytotoxic T lymphocytes defined with recombinant vaccinia virus and synthetic peptides. Nature 336: 484–487

15. Yedwell JW, Bennink JR, Mackett M, Lefrancois L, Lyles DS, Moss B (1986) Recognition of cloned vescicular stomatitis virus internal and external gene products by cytotoxic T lymphocytes. J Exp Med 163: 1529–1538

16. Holt CA, Osorio K, Lilly F (1986) Friend virus-specific cytotoxic T lymphocytes recognized both gag and env gene-encoded specificities. J Exp Med 164: 211–226

17. Bangham CRM, Openshaw PJM, Ball LA, King AMQ, Wertz GW, Askonas BA (1986) Human and murine cytotoxic T cells specific to respiratory syncytial virus recognize the viral nucleoprotein (N), but not the major glycoprotein (G) expressed by vaccinia virus recombinants. J Immunol 137: 3973–3977

18. Mondelli M, Mieli-Vergani G, Alberti A, Vergani D, Portman B, Eddleston ALWF, Williams R (1982) Specificity of T lymphocyte cytotoxicity to autologous hepatocytes in chronic hepatitis B virus infection: evidence that T cells are directed against HBV core antigen expressed on hepatocytes. J Immunol 129: 2773–2777

19. Bertoletti A, Ferrari C, Fiaccadori F, Penna A, Margolskee R, Schlicht HJ, Fowler P, Guilhot S, Chisari FV (1991) HLA class-I human cytotoxic T cells recognize endogenously synthesized hepatitis B virus nucleocapsid antigen. Proc Natl Acad Sci USA 88: 10445–10449

20. Guillhot S, Fowler P, Margolskee R, Ferrari C, Bertoletti A, Chisari FV (1992) Hepatitis B virus (HBV) specific cytolytic T cell response in man: production of target cells by stable expression of HBV encoded proteins in immortalized human B cell lines. J Virol 66: 2670–2675

21. Penna A, Chisari FV, Bertoletti A, Missale G, Fowler P, Giuberti T, Fiaccadori F, Ferrari C (1991) Cytotoxic T lymphocytes recognize an HLA-A2-restricted epitope within the hepatitis B virus nucleocapsid antigen. J Exp Med 174: 1565–1569

22. Schlicht HJ, Schaller H (1989) The secretory core protein of human hepatitis B virus is expressed on the cell surface. J Virol 63: 5399–5403
23. Rotzschke O, Falk K, Deres K, Schild H, Norda M, Metzger J, Jung G, Rammensee HG (1990) Isolation and analysis of naturally processed viral peptides as recognized by cytotoxic T cells. Nature 348: 252–254
24. Van Bleek GM, Nathenson SG (1990) Isolation of an endogenously processed immunodominant viral peptide from the class I H-2K^b molecule. Nature 348: 213–216
25. Jardetzky TS, Lane VS, Robinson RA, Madden DR, Wiley DC (1991) Identification of self peptides bound to purified HLA-B27. Nature 353: 326–328
26. Hunt DF, Henderson RA, Shabanowitz J, Sakaguchi K, Michel H, Sevilir N, Cox AL, Appella E, Engelhard VH (1992) Characterization of peptides bound to the class I MHC molecule HLA-A2.1 by mass spectrometry. Science 255: 1261–1263
27. Fremont DH, Matsumura M, Stura AE, Peterson PA, Wilson IA (1992) Crystal structures od two viral peptides in complex ith murine MHC class I H-2K^d. Science 257: 919–927
28. Matsumura M, Fremont DH, Peterson PA, Wilson IA (1992) Emerging principles for the recognition of peptide antigens by MHC class I molecules. Science 257: 927–934
29. Guo H-C, Jardetzky TS, Garrett TPJ, Lane WS, Strominger JL, Wiley DC (1992) Different length peptides bind to HLA-Aw68 similarly at their ends but bulge out in the middle. Nature 360: 264–366
30. Falk K, Rotzschke O, Stevanovic S, Jung G, Rammensee H (1991) Allele-specific motifs revealed by sequencing of self-peptides eluted from MHC molecules. Nature 351: 290–296
31. Cerundolo V, Elliott T, Elvin J, Bastin J, Ramenssee H-G, Townsend A (1991) The binding affinity and dissociation rates of peptides for class I major histocompatibility complex molecules. Eur J Immunol 21: 2069–2075
32. Christinck ER, Luscher MA, Barber BH, Williams DB (1991) Peptide binding to class I MHC on living cells and quantitation of complexes required for CTL lysis. Nature 352: 67–70
33. Bertoletti A, Chisari FV, Penna A, Guilhot S, Galati L, Missale G, Fowler P, Schlicht H-J, Vitiello A, Chesnut RC, Fiaccadori F, Ferrari C (1993) Definition of a minimal optimal cytotoxic T cell epitope within the hepatitis B virus nucleocapsid protein. J Virol 67: 2376–2380

Authors' address: Dr. C. Ferrari, Cattedra Malattie Infettive, Università di Parma, Via Gramsci 14, I-43100 Parma, Italy.

Arch Virol (1993) [Suppl] 8: 103–111

© Springer-Verlag 1993

Clinical and immunological aspects of hepatitis B virus infection in children receiving multidrug cancer chemotherapy

R. Repp[1], B.v. Hörsten[1], A. Csecke[1], J. Kreuder[1], A. Borkhardt[1], W.R. Willems[2], F. Lampert[1], and W.H. Gerlich[2]

Department of [1]Pediatrics and of [2]Medical Virology, Justus-Liebig-University, Giessen, Federal Republic of Germany

Summary. For two reasons hepatitis B virus infection is an important problem in patients with cancer. First, multidrug cancer chemotherapy may reactivate or worsen a previously benign chronic HBV infection. Second, patients undergoing cancer chemotherapy are at an increased risk of acquiring and spreading HBV which may result in an endemic infection. HBV reactivation may precipitate into a severe acute disease including fulminant hepatitis. In contrast, the acquisition of HBV during cancer chemotherapy commonly takes a mild clinical course but frequently leads to persistently high viremia. This state of immunotolerance to viral antigens allows viral replication without any sign of liver cell destruction. Withdrawal of chemotherapy does not cause significant changes if infection occurred during cytotoxic chemotherapy. Infection with HBV during cancer chemotherapy, therefore, may be considered as a model of an induced antigen-specific immunotolerance. In agreement with this hypothesis, vaccination against HBV during cancer chemotherapy does not prevent spread of HBV in oncology wards as it does not produce significant anti-HBs titers. Furthermore, vaccination even suppresses the immune response to later booster doses after chemotherapy has been withdrawn.

Introduction

The prevention of hepatitis B virus (HBV) infection requires specific measures for protecting medical staff and hospitalized patients. Long-term hospitalized patients undergoing hemodialysis or cancer chemotherapy are at particulary high risk of acquiring HBV infections [10, 24, 34]. Leukemic patients were among the first in whom hepatitis B surface antigen (HBsAg) was detected [5]. Immunosuppressive side effects of cancer chemotherapy in these patients are of special interest because they allow the study of HBV infection in the absence of the host's immune response. Currently, it is believed that the antiviral immune reaction is the major cause of liver cell destruction in HBV infection [3, 8, 30, 41]. Because of this, cancer chemotherapy seems to be somehow beneficial by suppression of immune-mediated liver cell destruction. However, immunological interactions between a virus and its host are complex and require a more differentiated consideration. In particular, there is an important difference between an acquisition of HBV while multidrug chemotherapy is being carried out and a pre-existing HBV carriage before chemotherapy was initiated.

Reactivation of preexisting or latent HBV infection

Chronic HBV carriage may be considered as a kind of equilibrium between viral replication and elimination of infected liver cells. Clinically silent, persistent HBV infection may occur in patients with a good immunological defense where only minor residual viral activity is present. Chronic aggressive hepatitis occurs in the presence of a cytotoxic immune response without sufficient neutralization of the virus. This leads to continous reinfection and destruction of hepatocytes [1, 7, 36, 37]. Multidrug cancer chemotherapy alters this balance by a severe suppression of the immune system. Indeed, there are several reports on reactivation of previously silent HBV infection by cancer chemotherapy and other immunosuppressive agents [11, 12, 21, 31, 38, 42]. AntiHBe positive HBsAg carriers with low viremia may convert to HBeAg positivity [19, 20, 28]. AntiHBc positive people may all of a sudden again become HBsAg positive. In such patients memory cells against HBV antigens are present. Only the suppression or deletion of proliferating cells allows the virus-infected cells to escape elimination. As soon as the immunosuppressive therapy is withdrawn a massive rebound effect may precipitate into acute or even fulminant hepatitis B [4, 9, 17, 27, 35]. The situation is further complicated by the fact that HBe-antigen negative variants of HBV or superinfection with hepatitis D virus may occur, both of which are reported to aggravate the course of hepatitis B [19, 20, 28, 42]. Upon diagnosis of the malignant disease, therefore, all patients should be tested for seromarkers of HBV carriage and viral replication prior to the onset of the multidrug chemotherapy. In case of positive results, they should be monitored carefully during chemotherapy and follow up [42]. Quantitative PCR might be a powerful tool to detect

HBV reactivation very early but this technique is still rather problematic and is not usually available in a routine laboratory. A fatal outcome due to HBV reactivation seems to be a rare event in children with malignancies, but HBV carriage, indeed, may undermine the results of cancer therapy [28].

Immunotolerance after acquisition of HBV during cancer chemotherapy

Several reports show an increased prevalence of HBV seromarkers in patients with malignancies [2, 15, 32, 34, 40]. Multiple venous punctures, transfusion of blood products, and destruction of mucous membranes by multidrug cancer chemotherapy allow parenterally transmissible pathogens such as HBV to easily enter the host. Particularly, in pediatric oncology units there seems to be a high risk of patient to patient transmission of the virus [3, 13, 29]. In Germany, at least 3 pediatric oncology therapy centres were afflicted by endemic HBV infections within the last 10 years [3] (pers. comm.). However, the prevalence of endemic HBV transmission may even be underestimated. Many centres have not screened all patients for HBV seromarkers routinely, because HBV infection usually is very rare in this country.

At the Pediatric Oncology unit of the University of Giessen, Germany, 74 children were inadvertently infected in an endemic HBV outbreak while under multidrug cancer chemotherapy from 1984–1986 [3]. At the time of admission all of these patients had been negative for HBsAg. Most of the patients suffered from an acute lymphoblastic leukemia. Until 1989, 20 children had died because of the primary malignant disease but an adverse prognostic influence of the HBV infection on the malignant disease was not detected [16]. More than 90% of the patients who survived (49 out of 54) developed an immunotolerant HBV carrier state without any clinical, serological, or histological signs of liver cell destruction. Their sera are still positive for HBsAg, HBeAg and HBV-DNA, even though anti-cancer chemotherapy has already been withdrawn for more than 5 years. In all those carriers HBV-DNA titers were higher than 10^8 genomes/ml serum. Twenty percent of these patients have not even developed anti-HBc antibodies although HBcAg is very immunogenic. Liver biopsies were taken from 36 patients in 1989. These revealed a normal histology or at most a minimal hepatitis with some ground-glass hepatocytes but no signs of inflammation. Furthermore, clinical signs of hepatitis or an elevation of transaminase values (ALT, AST) above 200 U/l were never detected in these patients, even though they were examined carefully in accordance to the guide-lines of the cancer therapy protocols. Temporary, slight elevations of liver enzymes are a common side-effect of some anti-cancer drugs and these values did not differ significantly from non-infected patients during therapy, and after therapy had been withdrawn they were always normal. In contrast, 9 of the patients' relatives presented with an acute hepatitis B. Only those repeated cases of acute hepatitis B among the patient's relatives suggested that the children at the pediatric oncology unit were infected with HBV. When monitoring for HBsAg was carried out, in addition to one test at time of admission, the very high number of subclinical HBV infections among these patients was diagnosed. Probably, the virus had been brought to the ward by one of the patients relatives, an adult HBV carrier from northern Africa.

Persistence of HBV infection in more than 90% of these patients is more frequent than reported in other studies on HBV infection during treatment for childhood malignancies [13, 15]. However, these studies did not test patients for HBsAg and anti-HBc prior to chemotherapy and, therefore, a reactivation of a previous HBV infection could not be ruled out. For this reason, the lower carrier rate might be due to a mixture of primary infections and cases of HBV reactivation. In this connection, two further aspects need to be considered. First, multidrug cancer chemotherapy is not equally severe over an extended period. Usually, it consists of several intensive cycles followed by periods of recovery when the immune system and other tissues affected may reconstitute. It is conceivable that the outcome of an HBV infection may vary depending upon the different intensities of cancer chemotherapy at the time of HBV acquisition. Second, a different pathogenicity of HBV strains might also alter the course of infection as might be the case in the natural infection [22, 23]. However, little is known about the role of this variable in the outcome after HBV infection during cancer chemotherapy. In our patients the genome of the virus was classified as genotype "D" corresponding to the HBsAg subtype "ayw" [25] but most other reports from this field do not consider this point and, therefore, no comparison can be drawn.

A high incidence of a chronic viral carrier state similar to HBV infection has already been found in some animals after infection with other low cytopathogenic viruses, such as duck hepatitis B virus (DHBV) and lymphocytic choriomeningitis virus (LCM), if an immunosuppressive therapy with cyclosporine had been started prior to infection and had been carried out during the incubation period [6, 14, 33]. In view of these findings, HBV infection during multidrug cancer chemotherapy can be considered as another model of an induced selective long-term immunotolerance. It may be worthwhile to study this further if multidrug cancer chemotherapy should become a possible approach for transplantation medicine.

Therapy and prevention of HBV infection in children with cancer

Interferon has been used to treat children who acquired HBV during cancer chemotherapy but no beneficial effect, not even an enduring reduction of viral replication, could be achieved [13]. This is not surprising, since patients with very high HBV-DNA titers in their blood usually do not respond well to interferon treatment [26]. Furthermore, the findings presented here provide further evidence against interferon treatment. First, these patients are not at a high risk of developing cirrhosis because liver cell destruction was not found in liver biopsies. Second, there may even be an increased risk of chronic active hepatitis after interferon treatment because interferon could disturb the stable state of immunotolerance. Third, there is no evidence for an adverse effect of the HBV infection on the primary malignant disease from which these patients suffered. Despite the absence of liver cell destruction, patients who acquired HBV during cancer chemotherapy might still be at an increased risk of developing primary liver cell carcinoma in later life and they remain a potential source of infection [2]. Most attention, therefore, should be paid to the prevention of HBV infection in children with cancer. Since most HBV infections during chemotherapy take an asymptomatic course, repeated screening for HBsAg or HBV-DNA seems to be the only way to detect potential endemic HBV transmissions very early. If HBV infection occurs during anti-cancer therapy patients are very likely to become highly infectious HBV carriers. In addition to a generally high susceptibility to persistent HBV infection, this further increases the risk of endemic HBV transmission in an oncology unit. In children with cancer, shedding of HBV in saliva might be an important route of virus transmission [13]. Destruction of mucous membranes, a major side effect of many anti-cancer drugs, may lead to an enhanced contamination of saliva with blood and causes an HBV contamination of toys and other objects which children frequently place into their mouths. For psychological and social reasons it is very difficult to isolate infectious children over the whole period of anti-cancer therapy. Therefore, we now passively immunize all the patients attending our Pediatric Oncology unit with anti-HBs hyperimmunoglobulins. Even though this is very expensive, it appears to be the most effective way to stop endemic HBV transmission if many patients have already become HBV carriers. In addition, the staff of a pediatric oncology unit should be vaccinated against HBV and any patients' relatives who attend the ward regularly should be tested for seromarkers of HBV infectivity.

Tolerance induction by active immunization

Active immunization against HBV does not seem to have a beneficial effect in patients receiving chemotherapy. A vaccination program was begun in 1986 in an attempt to stop the endemic HBV transmissions in our ward, but it was discontinued as none of the patients had developed significant anti-HBs ($>10\,U/l$) levels while being on cancer chemotherapy. Similar results have been obtained in heart transplant recipients receiving immunosuppressive therapy with cyclosporine [39]. Surprisingly, there was even a poor response to later booster doses in our patients who had been vaccinated during chemotherapy. To prove the significance of these findings, two groups of patients were compared retrospectively. The study group consisted of 10 patients who had received a full course of 3 doses HB-vaccine during cancer chemotherapy. In addition, these patients were revaccinated with 3 or more additional doses of vaccine after chemotherapy had been withdrawn for at least 3 months. A control group consisted of 6 patients who had been treated with cancer chemotherapy but who had only received 3 vaccine doses after its withdrawal. Even though the number of vaccine doses had been much higher in the study group, patient's geometric mean anti-HBs titers were significantly lower ($p < 0.05$). Similar to the outcome after HBV infection during multidrug chemotherapy these results suggest a long-term tolerogenic effect of antigens which are present in a person during cancer chemotherapy. An extended beneficial effect of multidrug cancer chemotherapy has already been reported in a kidney transplant recipient [18]. This patient had developed a secondary lymphoma possibly due to cyclosporine therapy during recurrent episodes of severe graft rejection. After successful treatment of the lymphoma with multidrug chemotherapy, no further episodes of graft rejection occurred. However, prior to a possible application of multidrug chemotherapy for long-term immunosuppression, e.g. in transplantation medicine, prospective studies with different antigens will be required.

Acknowledgements

This work was supported by the "Forschungshilfe Station Peiper" and by the "Deutsche Forschungsgemeinschaft" with a "Habilitandenstipendium" to R. Repp (Re 841/1-1).

References

1. Baker JR Jr (1992) The immune response in chronic hepatitis B virus infection: the "core" of the problem. Hepatology 16: 498–500
2. Beasley RP, Hwang LY, Lin CC, Chien CS (1981) Heaptocellular carcinomas and hepatitis B virus. A prospective study of 22, 707 men in Taiwan. Lancet ii: 1129–1133
3. Bertram U, Repp R, Fischer HP, Willems WR, Lampert F (1990) Hepatitis-B-Endemie bei zytostatisch behandelten Kindern. Dtsch Med Wochenschr 115: 1253–1254
4. Bird GLA, Smith H, Portmann B, Alexander GJM, Williams R (1989) Acute liver decompensation on withdrawal of cytotoxic chemotherapy and immunosuppressive therapy in hepatitis B carriers. Q J Med 73: 895–902
5. Blumberg BS, Alter HJ, Visnich S (1965) A "new" antigen in leukemia sera. JAMA 191: 101–106
6. Cote PJ, Korba BE, Baldwin B, Hornbuckle WE, Tennant BC, Gerin JL (1992) Immunosuppression with cyclosporine during the incubation period of experimental woodchuck hepatitis virus infection increases the frequency of chronic infection in adult woodchucks. J Infect Dis 166: 628–631
7. Eddleston ALWF, Mondelli M (1986) Immunopathological mechanisms of liver injury in chronic hepatitis B virus infection. J Hepatol 3 [Suppl 2]: 17–23
8. Ferrari C, Penna A, Bertoletti A, Valli A, Antoni AD, Giuberti T, Cavalli A, Petit M-A, Fiaccadori F (1990) Cellular immune response to hepatitis B virus-encoded antigens in acute and chronic hepatitis B virus infection. J Immunol 145: 3442–3449
9. Flowers MA, Heathcote J, Wanless IR, Sherman M, Reynolds WJ, Cameron RG, Levy GA, Inman RD (1990) Fulminant hepatitis as a consequence of reactivation of hepatitis B virus infection after discontinuation of lowdose methotrexate therapy. Ann Intern Med 112: 381–382
10. Franco E, Olivadese A, Valeri M, Albertoni F, Petrosillo N (1992) Control of hepatitis B virus infection in dialysis units in Latium, Italy. Nephron 61: 329–330
11. Grunmayer ER, Panzer S, Ferenci P, Gander H (1989) Recurrence of hepatitis B in children with serological evidence of past B virus infection undergoing anti-leukemic chemotherapy. J Hepatol 8: 232–235
12. Hoofnagle JH, Dusheiko GM, Schafer DF, Jones EA, Micetich KC, Young RC, Costa J (1982) Reactivation of chronic hepatitis B virus infection by cancer chemotherapy. Ann Intern Med 96: 447–449
13. Hovi L, Saarinen UM, Jalanko H, Pohjanpelto P, Siimes MA (1991) Characteristics and outcome of acute infection with hepatitis B virus in children with cancer. Pediatr Infect Dis J 10: 809–812
14. Karayiannis P, Goldin R, Luther S, Carman WF, Monjardino J, Thomas HC (1992) Effect of cyclosporin-A in woodchucks with chronic hepatitis delta virus infection. J Med Virol 36: 316–321
15. Kumar A, Misra PK, Rana GS, Mehrotra R (1992) Infection with hepatitis A, B, Delta, and human immunodeficiency viruses in children receiving cycled cancer chemotherapy. J Med Virol 37: 83–86
16. Lampert F, Willems WR, Bertram U, Berthold F (1987) No adverse prognostic influence of hepatitis B virus infection in acute childhood lymphoblastic leukemia. Blut 55: 115–120

17. Lau JYN, Lai CL, Lin HJ, Lok ASF, Liang RHS, Wu PC, Chan TK, Todd D (1989) Fatal reactivation of chronic hepatitis B virus infection following withdrawal of chemotherapy in lymphoma patients. Q J Med 73: 911–917

18. Lien Y-HH, Schröter GPJ, Weil R III, Robinson WA (1991) Complete remission and possible immune tolerance after multidrug combination chemotherapy for cyclosporine-related lymphoma in a renal transplant recipient with acute pancreatitis. Transplantation 52: 739–742

19. Locasciulli A, Alberti A, Rossetti F, Santamaria M, Santoro N, Madon E, Miniero R, Lo Curto M, Tamaro P, Paolucci P, Casale F, Nespoli L, Tucci F, Masera G (1985) Acute and chronic hepatitis in childhood leukemia: a multicentric study from the Italian Cooperative Group for Therapy of Acute Leukemia (AIL-AIEOP). Med Pediatr Oncol 13: 203–206

20. Locasciulli A, Santamaria M, Masera G, Schiavon E, Alberti A, Realdi G (1985) Hepatitis B virus markers in children with acute leukemia: the effect of chemotherapy. J Med Virol 15: 29–33

21. Lok ASF, Liang RHS, Chiu EKW, Wong K-L, Chan T-K, Todd D (1991) Reactivation of hepatitis B virus replication in patients receiving cytotoxic therapy. Report of a prospective study. Gastroenterology 100: 182–188

22. Milich DR, Hughes JL, McLachlan A, Langley KE, Thornton GB, Jones JE (1990) Importance of subtype in the immune response to the pre-S(2) region of the hepatitis B surface antigen. I. T cell fine specificity. J Immunol 144: 3535–3543

23. Milich DR, Jones JE, McLachlan A, Bitter G, Moriarty A, Hughes JL (1990) Importance of subtype in the immune response to the pre-S(2) region of the hepatitis B surface antigen. II. Synthetic pre-S(2) immunogen. J Immunol 144: 3544–3551

24. Oguchi H, Miyasaka M, Tokunaga S, Hora K, Ichikawa S, Ochi T, Yamada K, Nagasawa M, Kanno Y, Aizawa T, Watanabe H, Yoshizawa S, Sato K, Terashima M, Yoshie T, Oguchi S, Tanaka E, Kiyosawa K, Furuta S (1992) Hepatitis virus infection (HBV and HCV) in eleven Japanese hemodialysis units. Clin Nephrol 38: 36–43

25. Okamoto H, Tsuda F, Sakugawa H, Sastrosoewingnjo R, Imai M, Miyakawa Y, Mayumi M (1988) Typing hepatitis B virus by homology in nucleotide sequence: comparison of surface antigen subtypes. J Gen Virol 69: 2575–2583

26. Perrillo RP (1990) Factors influencing response to interferon in chronic hepatitis B: Implications for Asian and Western populations. Hepatology 12: 1433–1435

27. Pinto PC, Hu E, Bernstein-Singer M, Pinter-Brown L, Govindarajan S (1990) Acute hepatic injury after the withdrawal of immunosuppressive chemotherapy in patients with hepatitis B. Cancer 65: 878–884

28. Ratner L, Peylan-Ramu N, Wesley R, Poplack DG (1986) Adverse prognostic influence of hepatitis B virus infection in acute lymphoblastic leukemia. Cancer 58: 1096–1100

29. Repp R, Mance A, Bertram U, Niemann H, Gerlich WH, Lampert F (1991) Persistent hepatitis B virus replication in mononuclear blood cells as a source of reinfection of liver transplants. Transplantation 52: 935

30. Repp R, von Hörsten B, Müller A, Bertram U, Kreuder J, Netz H, Lampert F (1991) Persistent immunotolerance to hepatitis B virus antigens after infection during polychemotherapeutic treatment for malignancies. In: Touraine JL, Traeger J, Bétuel H, Dubernard JM, Revillard JP, Dupuy C (eds) Transplantation and clinical immunology vol 23: virus and transplantation. Excerpta Medica, Amsterdam, p 384

31. Rostoker G, Rosenbaum J, Ben Maadi A, Nedelec G, Deforge L, Vidaud M, Lang P, Lagrue G, Goossens M, Weil B (1990) Reactivation of hepatitis B virus by corticosteroids in a case of idiopathic nephrotic syndrome. Nephron 56: 224
32. Steinberg SC, Alter HJ, Leventhal BG (1975) The risk of hepatitis transmission to family contacts of leukemic patients. J Pediatr 87: 753–756
33. Stitz L, Soeder D, Deschl U, Frese K, Rott R (1989) Inhibition of immunemediated meningoencephalitis in persistently borna disease virus-infected rats by cyclosporine A. J Immunol 143: 4250–4256
34. Tabor E, Gerety RJ, Mott M, Welburg J (1978) Prevalence of hepatitis B in high risk setting. A serologic study of patients and staff in a pediatric oncology unit. Pediatrics 61: 711–715
35. Thomas HC (1989) Acute liver decompensation on withdrawal of cytotoxic chemotherapy and immunosuppressive therapy in hepatitis B carriers: implications for the treatment of chronic HBV carriers. Q J Med 73: 873–874
36. Thomas HG (1990) The hepatitis B virus and the host response. J Hepatol 11 [Suppl 1]: S83–S89
37. Thomas HC (1990) Hepatitis B virus-induced liver disease. Springer Semin Immunopathol 12: 1–3
38. Vandercam B, Cornu C, Gala JL, Geubel A, Cahill M, Lamy ME (1990) Reactivation of hepatitis B virus in a previously immune patient with human immunodeficiency virus infection. Eur J Clin Microbiol Infect Dis 9: 701–702
39. Wagner D, Wagenbreth I, Stachan-Kunstyr R, Flik J (1992) Failure of vaccination against hepatitis B with Gen H-B- Vax-D in immunosuppressed heart transplant recipients. Klin Wochenschr 70: 585–587
40. Wands JR, Chura CM, Roll FJ, Maddrey WC (1975) Serial studies of hepatitis associated antigen and antibody in patients recieving anti tumor chemotherapy for myeloproliverative and lymphoproliverative disorders. Gastroenterology 68: 105–111
41. Wen Y-M, Xu Y-Y (1991) Pathogenesis and immune responses in hepatitis B patients and carriers. Monogr Virol 19: 31–39
42. Yoshiba M, Sekiyama K, Sugata F, Okamoto H, Yamamoto K, Yotsumoto S (1992) Reactivation of precore mutant hepatitis B virus leading to fulminant hepatic failure following cytotoxic treatment. Dig Dis Sci 37: 1253–1259

Authors' address: Dr. R. Repp, Universitäts-Kinderklinik, Abteilung Hämatologie/ Onkologie, Justus-Liebig-Universität, Feulgenstrasse 12, D-35392 Giessen, Federal Republic of Germany.

Arch Virol (1993) [Suppl] 8: 113–121

Hepatitis C virus infection in type II essential mixed cryoglobulinemias

M. Ballaré, G. Airoldi, M.R. Brunetto, P. Manzini, G. Bordin, A. Touscoz, F. Bonino, and A. Monteverde

Division of General Medicine II, Ospedale Maggiore, Novara, and of Gastroenterology, Molinette Hospital, Torino, Italy

Summary. The possible relationship between essential mixed cryoglobulinemias (EMCs) and hepatitis C virus (HCV) has been investigated in eight patients with type II EMCs and biochemical signs of liver damage, whose serum tested positive in the ELISA for anti-HCV. Sera were tested using the 2nd generation RIBA assay, while serum HCV-RNA was measured semiquantitatively by a RT-PCR in whole serum, cryoprecipitates and supernatants. In all patients a percutaneous liver biopsy and a bone marrow biopsy were performed. At liver biopsy, chronic active hepatitis and/or cirrhosis were present in 6 patients; in the remaining two, a lymphoplasmacytoid infiltration of elements positive for kappa light chains was found. In all patients a bone marrow biopsy showed a paratrabecular infiltration of monoclonal lymphoplasmacytoid elements similar to those found in the liver of the two patients described above. Antibodies against structural and non-structural HCV proteins were detectable in the serum of all patients. HCV-RNA was amplified from the whole sera, cryoprecipitates and supernatants: significantly higher concentrations were found in cryoprecipitates than in supernatants. Our results confirm the high prevalence of HCV infection and ongoing viral replication in patients with type II EMC and suggest the possible implication of HCV in EMC pathogenesis.

Introduction

Type II essential mixed cryoglobulinemias (EMC) are characterized by reversible, temperature-dependent precipitation from serum of a monoclonal IgM rheumatoid factor (RF) complexed with a polyclonal IgG. Although the etiology of EMC is unclear, it has been hypothesized that infection by hepatotropic viruses may play a role in the pathogenesis of this disorder because of the high incidence of coexisting liver-function abnormalities [3]. Earlier studies suggested a possible etiological role for hepatitis B virus [13, 18], but this hypothesis has not been substantiated [11]. Hepatitis C virus (HCV) has been implicated in the pathogenesis of EMC for the high prevalence of serum anti-HCV antibodies [4, 6, 8–10, 16, 17, 24]; however the specificity of these reactivities was questioned for the possibility of false positive results due to the presence of serum rheumatoid factors [12, 25].

The aim of this study was to evaluate the true incidence of HCV infection in a selected group of patients with type II EMC, biochemical signs of liver disease and positivity of serum in the assay for anti-HCV.

Materials and methods

Eight patients with type II EMC, biochemical signs of liver damage and a positive test in 2nd generation ELISA for anti-HCV entered the study. Their mean age was 64 years, ranging from 53–75 years; the mean follow-up was 9 years, ranging from 6 to 14 years (Table 1). All patients had repeated episodes of the classic cryoglobulinemic syndrome, with fatigue, arthralgias and purpura of the lower extremities.

Cryoglobulins were measured according to Meltzer and Franklin [21, 22]. Blood samples were allowed to clot at 37°C for longer than 2h; the serum obtained was centrifuged at 37°C and then stored at 4°C in glass cuvettes for at least 7 days. After multiple washing and redissolving, cryocrit was measured as protein content and expressed as a percentage of the whole serum.

Serum anti-HCV was tested by the 2nd generation RIBA assay (Ortho Diagnostic Systems). Serum HCV-RNA was determined from whole serum, cryoprecipitates and

Table 1. Clinical features of patients with EMC type II

Pt.	Age (years)	Follow-up (years)	ASAT	ALAT	Liver histology
			(times normal)		
1	60	9	2	1.5	CAH
2	58	9	3	4	CAH
3	53	12	2	3	Cirrhosis
4	72	14	2	1.5	Lymph.inf.
5	65	7	2	2	Lymph.inf.
6	57	7	3	3	CAH
7	75	6	2	3	Cirrhosis
8	66	6	4	5	CAH

ASAT Aspartate aminotransferase; *ALAT* alanine aminotransferase; *CAH* chronic active hepatitis; *Lymph.inf.* lymphoblasmacytoid infiltration of portal tracts supernatants (100 μl). Briefly, 10 μl of RNA extracted from serum (or cryoprecipitates or supernatants) according to Chomczynski and Sacchi [5] were reverse transcribed in a 25 μl cDNA reaction. Thirty-five cycles of PCR were then performed using primers encompassing the 5′ non-coding region of HCV (5′-AGTCTTGCGGCCGCAGCGCC-AAATC-3′ and 5′-TTCGCGGCCGCACTCCACCATGAATCACTCCCC-3′). Hepatitis C virus specific products were detected after electrophoresis on an ethidium bromide stained gel and on nitrocellulose membrane by hybridization with ^{32}P labeled probe [1, 15].

In all patients a percutaneous liver biopsy with a Menghini's needle and a bone marrow biopsy with Jamshidi's needle were performed. Bone marrow biopsies were fixed in 10% unbuffered formaline for 24 h, decalcified in the bisodium salt of ethylen-diamino-tetra-acetic acid (EDTA) for 2 h and embedded in Paraplast (Oxford/Lancer) at 57°C, while liver biopsies were divided in two parts: one of these was used for routine histology (formaline fixation, Paraplast embedding), and the other was frozen in liquid nitrogen and stored at −80°C. Sections were cut at 3 μm from routinely processed tissue blocks, and stained according to the following methods: hematoxylin and eosin, Giemsa, periodic acid Schiff, and Gomori's silver impregnation. Sections of 6 μm were cut from frozen blocks and air-dried overnight at room temperature. After a sequential treatment in acetone and chloroform (for 10 min at room temperature in both cases), they were incubated in normal horse serum (Vectastain TM ABC Kit, PK 4002) for 30 min at room temperature. After removing the excess of normal serum, the following monoclonal antibodies were applied for 30 min: anti-IgA, anti-IgG, anti-κ, anti-λ (all purchased from Becton-Dickinson), anti-IgM, pan-B (CD22), T1 (CD5), T3 (CD3), T4 (CD4), T8 (CD8), T11 (CD2) (all obtained from Dako). After washing, the sections were incubated in biotinylated horse anti-mouse serum (Vectastain TM ABC Kit) preadsorbed with normal human serum, for 30 min. After a further washing, they were treated with avidin-biotin-peroxidase-complexes (Vectastain TM ABC Kit) for 60 min. After repeated washings, peroxidase activity was demonstrated with 3-amino-9-ethyl-carbazol (Sigma, no. A5754) applied for 40 min. After a further wash in distilled water, the sections were counterstained with Mayer's hematoxylin and mounted in Kaiser's glycerin. All washes and dilutions were performed with 0.05 M Tris buffer saline (pH 7.6) [20].

Results

Serum cryoglobulins were present at 10–80 volume %, and rheumatoid factor activity was positive in all patients.

On liver biopsy, chronic active hepatitis and/or cirrhosis were present in 6 patients. Histological signs of lymphoplasmacytoid infiltration were found in the remaining two: immunohistochemical staining of these lymphocytes was positive for kappa light chains, showing the monoclonal nature of these infiltratres (Fig. 1).

In all patients the bone marrow biopsy showed a paratrabecular infiltration of monoclonal lymphoplasmacytoid elements, phenotypically similar to that found in the liver of the two patients described above (Fig. 2).

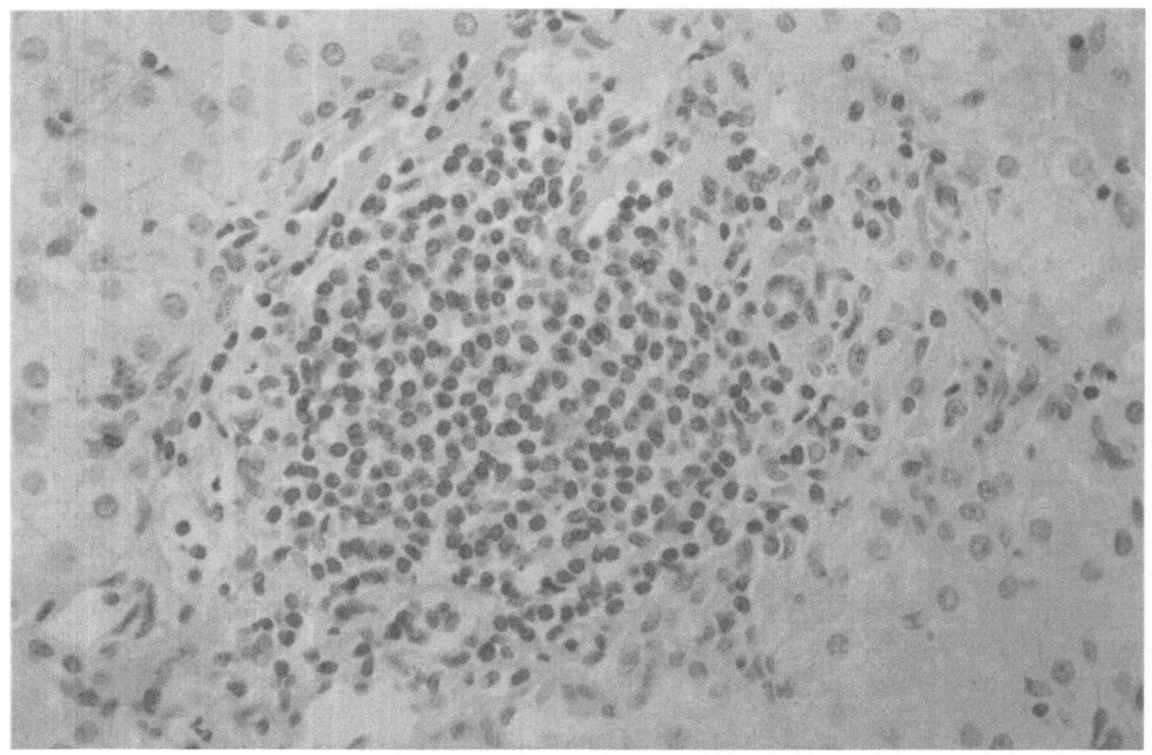

A

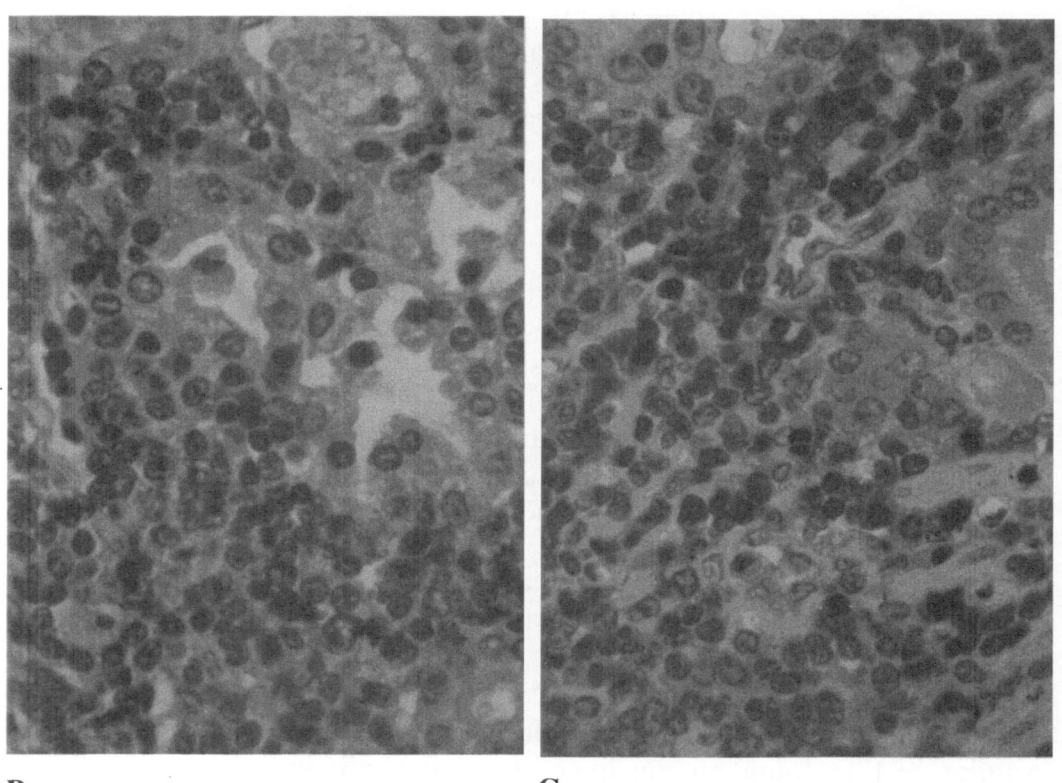

B   **C**

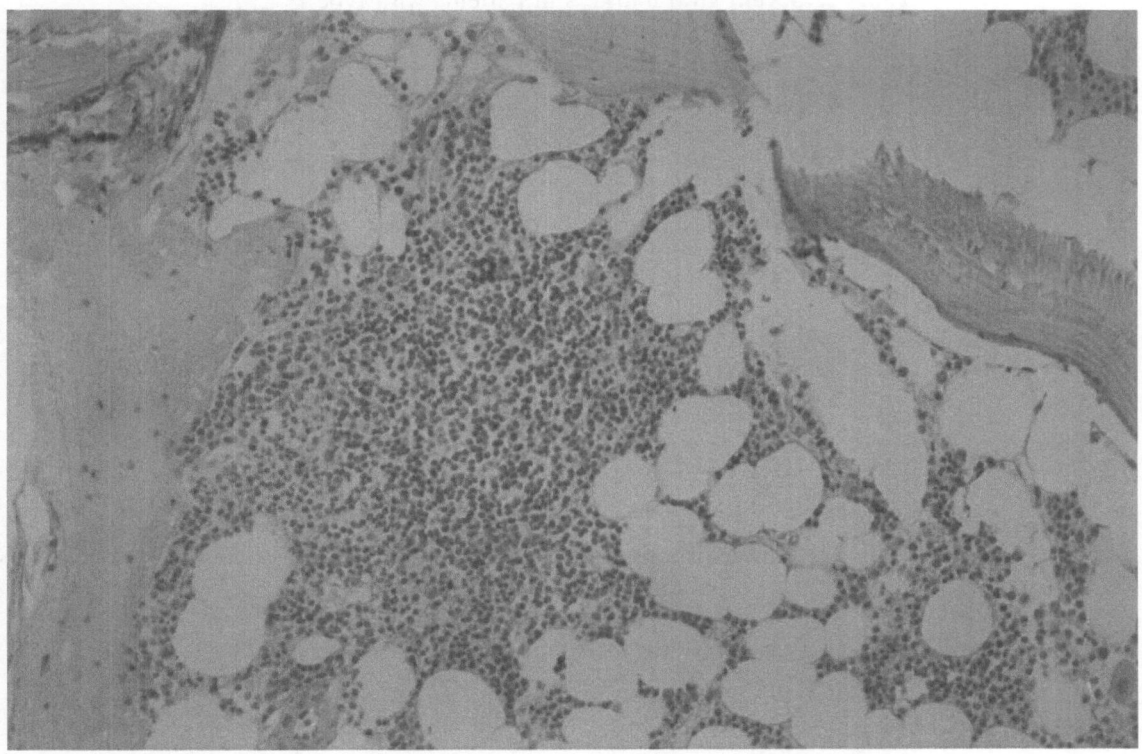

Fig. 2. Paratrabecular lymphoplasmacytoid infiltration of the bone marrow of a patient with type II EMC. Giemsa stain (×300)

Antibodies against structural and non structural HCV proteins (2nd gen. RIBA assay) were detectable in the serum of all patients (Table 2).

HCV-RNA was amplified from the whole sera, cryoprecipitates and supernatants. A significantly higher ethidium bromide signal was observed in cryoprecipitates than in supernatants (Table 3). This difference was confirmed by the results of end-point dilutions of the two fractions: HCV-RNA was still amplified from 2 log higher dilutions of cryoprecipitates than those of supernatants.

←——

Fig. 1. Liver biopsy of a patient with type II EMC, showing a lymphoplasmacytoid infiltration of portal tracts (**A**). Lymphoid elements express kappa light chains (**B**), while immunohistochemical determination of lambda light chains (**C**) is negative. Avidin-biotin complex method on frozen sections. Amino-ethyl-carbazol and Mayer's hematoxylin (×200)

Table 2. Serum viral markers in patients with type II EMC

Pt.	Anti-HCV RIBA				HBsAg	anti-HBc
	5.1.1	c100.3	c33c	c22.3		
1	−	−	+	++	−	+
2	+++	+++	+++	+++	−	−
3	+	−	+++	+++	−	−
4	−	+	+	+	−	−
5	−	−	+	+	−	−
6	−	−	+	++	−	+
7	−	−	+	++	−	−
8	++	+	+++	+++	−	+

Table 3. Semiquantitative analysis of serum HCV-RNA in patients with type II EMC

Pt.	Whole serum	Cryoprecipitate	Supernatants
1	2	2	1
2	2	2	1
3	1	1	1
4	2	2	1
5	1	1	1
6	2	2	1
7	2	2	1
8	2	2	1

Score system [1]: *2* more than 10000 genome equivalents/ml g.e./ml; *1* between 1000 and 10000 g.e./ml; *0* less than 1000 g.e./ml

Discussion

Our results confirm the high prevalence of HCV infection and ongoing viral replication in patients with type II EMC [4, 6, 8–10, 16, 17, 24]. The possible implication of HCV in EMC pathogenesis is also suggested by the finding of similar lymphocytic infiltration in the liver and bone marrow of our patients, with monoclonal populations of B-lymphocytes responsible for the production of the IgM rheumatoid factor. Evidence of mononuclear cells infection, particularly of B-lymphocytes, by HCV, as found by other authors [2, 23, 26], may be the link between EMC and the virus; the pathogenic mechanism of the disease, and factors determining its evolution in a form so different from the great part of liver diseases found in chronic HCV infection remains to be clarified.

We have previously reported the consistent finding of monoclonal lymphopasmacytoid infiltates in the bone marrow and/or liver of patients with type II EMC [20]. A complete review of 67 type II EMC patients followed in our center since 1985 confirms that bone marrow infiltrations, mostly with a nodular, paratrabecular pattern, virtually identical morphologically and immunochemically to a lymphoplasmacytoid immunocytoma (Kiel classification), were present in 62 patients in at least one biopsy. At liver biopsy, 40 patients had periportal infiltrates without piecemeal necrosis, a picture resembling that of chronic persistent hepatitis. An immunochemical study on frozen sections showed a sharp prevalence of monotypic CD22+ B cells expressing surface IgMκ in 29 cases and surface IgMλ in 1 case. These features are especially prominent in early disease, changing over the long term into chronic active hepatitis and cirrhosis, with prevalence of T lymphocytes. These results shows that our patients, at least once in the course of their disease, manifested histological evidence of a monoclonal lymphoproliferation; thus, EMCs could be considered as clinical syndromes superimposed on a low-grade lymphoproliferation with extranodal involvement, mainly confined to the bone marrow and the liver. Whereas only a minor percentage of these atypical proliferations (9% in our series) shows evidence of spreading and frank lymphomatous evolution, a similar behaviour can be found with Sjogren's syndrome (a relentless anatomic progression and clinical deterioration, lymphomatous evolution), in which a high incidence of HCV infection has recently been reported [7, 14].

Thus, the pathogenetical link between HCV and EMCs could be the induction of lymphoproliferation, perhaps through a direct localization of the virus in B lymphocytes. In the majority of patients only a polyclonal activation of B lymphocytes takes place; in only a minority of patients, the lymphoproliferation evolves into a monoclonal population, with the appearance of EMC.

References

1. Abate ML, Saracco G, Negro F, Brunetto MR, Bonino F (1992) Quantitative detection of hepatitis C viraemia by HCV-RNA RT-PCR: clinical applications. Ital J Gastroenterol 24: 251
2. Artini M, Natoli G, Avantaggiati ML, Balsano C, Chirillo P, Bonavita MS, Levrero M (1992) HCV replicates in the peripheral blood mononuclear cells from chronic HCV carriers. Proceedings of the first annual meeting on hepatitis C virus and related viruses, Venice 1992, p 32

3. Brouet JC, Clauvel JP, Danon F, Klein M, Seligmann M (1974) Biological and clinical significance of cryoglobulins. A report of 86 cases. Am J Med 57: 775–778

4. Casato M, Taliani G, Pucillo LP, Goffredo G, Laganà B, Bonomo L (1991) Cryoglobulinemia and hepatitis C virus. Lancet 337: 1047

5. Chomzczynski A, Sacchi F (1987) Single step method of RNA isolation by acid guanidinium, thiocyanate-phenol-chloroform extraction. Anal Biochem 162: 156–159

6. Chung RT, Agnello V, Weiner NJ, Dienstag JL, Kaplan LM (1992) A role for hepatitis C virus infection in the pathogenesis of essential mixed cryoglobulinemia: selective concentration of HCV antigen and RNA in cryoprecipitates. Gastroenterology 104: 1092

7. deBandt M (1992) Role du virus de l'hépatite C dans les cryoglobulinémies mixtes "essentielles" et le syndrome de Gourgerot-Sjogren. Presse Med 21: 1750–1752

8. Disdier P, Harlé JR, Weiller PJ (1991) Cryoglobulinemia and hepatitis C infection. Lancet 338: 1151

9. Ferri C, Greco F, Longombardo G, Bombardieri S (1991) Antibodies to hepatitis C virus in patients with mixed cryoglobulinemias. Arthritis Rheum 34: 1606–1610

10. Ferri C, Greco F, Longombardo G, Bombardieri S (1991) Antibodies against hepatitis C virus in mixed cryoglobulinemia patients. Infection 19: 417–420

11. Galli M, and the GISC (1991) Cryoglobulinemia and serological markers of hepatitis B virus. Lancet 338: 758

12. Galli G, Monti G, Monteverde A, Invernizzi F, Pietrogrande M, Di Girolamo M, Mazzaro C, Migliaresi S, Mussini C, Ossi E, Renoldi P (1992) Hepatitis C virus and mixed cryoglobulinemias. Lancet 339: 989

13. Gorevic PD, Kassab HJ, Levo Y, Kohn R, Meltzer M, Prose P, Franklin EC (1980) Mixed cryoglobulinemia. Clinical aspects and long-term follow-up of 40 patients. Am J Med 69: 287–290

14. Haddad J, Deny P, Munz-Gotheil C, Harley D (1992) Lymphocytic sialadenitis of Sjogren syndrome associated with chronic hepatitis C virus liver disease. Lancet 339: 321–324

15. Han JH, Shymala V, Richman KH, Greenmount A (1991) Characterization of the terminal region of hepatitis C virus RNA: identification of conserved sequences in the 5' untranslated region and poly(A) tails at the 3' end. PNAS 88: 1711–1715

16. Harlé JR, Disdier P, Durand JM, Weiller PJ (1991) Mixed cryoglobulinemia in hepatitis C infection: ten cases. Presse Med 20: 1233

17. Houlf S, Romain P, Szymanski L, Bonkovski H (1992) Hepatitis C: An important cause of mixed cryoglobulinemia. Gastroenterology 102: 105

18. Levo Y, Gorevich PD, Kassab HJ, Zucker-Franklin D, Franklin EC (1977) The association between hepatitis B virus and essential mixed cryoglobulinemia. N Engl J Med 296: 1501–1504

19. Lunel F, Musset L, Cacoub P, Perrin M, Frangeul JC, Piette O, Bousquet P, Godeau D, Valla Y, Le Charpentier P, Opolon P, Haraux JM (1992) Prevalence of mixed cryoglobulinemia in 112 patients with viral and non viral chronic hepatitis or cirrhosis. Proceedings of the first annual meeting on hepatitis C virus and related viruses, Venice 1992, p 179

20. Monteverde A, Rivano MT, Allegra GC, Monteverde AI, Zigrossi P, Baglioni B, Falini B, Bordin G, Pileri S (1988) Essential mixed cryoglobulinemia, type II: a manifestation of a low-grade malignant lymphoma? Acta Haematol 79: 20–25

21. Meltzer M, Franklin EC (1966) Cryoglobulinemia. A study of 29 patients. I. IgG and IgM cryoglobulins and factors affecting cryprecipitability. Am J Med 40: 828–836

22. Meltzer M, Franklin EC, Elias K, McCluskey RT, Cooper N (1966) Cryoglo-
    bulinemia. A clinical and laboratory study. II. Cryoglobulins with rheumatoid
    factor activity. Am J Med 40: 837–856
23. Muller HM, Kallinovski B, Solbach C, Goeser T, Theilmann L, Pfaff E (1992)
    Peripheral blood mononuclear leukocytes as a potential site for extrahepatic
    replication of hepatitis C virus: predominant role of B-lymphocytes in viral pro-
    pagation. Proceedings of the first annual meeting on hepatitis C virus and related
    viruses, Venice 1992, p 31
24. Pascual M, Perrin L, Giostra E, Schifferli JA (1990) Hepatitis C virus in patients
    with cryoglobulinemia type II. J Infect Dis 335: 559–560
25. Theilmann L, Blazek M, Goeser T, Gmelin K, Kommerel B, Fiehn W (1990) False
    positive anti-HCV tests in rheumatoid arthritis. Lancet 335: 1346
26. Zignego AL, Macchia D, Monti M, Thiers V, Romagnani S, Gontallini P, Brechot
    C (1992) Hepatitis C virus infection of fresh and cultured peripheral blood mono-
    nuclear cells. Ital J Gastroenterol 24: 310

Authors' address: Dr. A. Monteverde, Via Scavini 4A I-28100 Novara, Italy.

Arch Virol (1993) [Suppl] 8: 123–131

Detection of hepatitis B virus DNA by polymerase chain reaction in vaccinated and non-vaccinated Senegalese children

M. Chabaud[1], N. Depril[1], P. Le Cann[1], D. Leboulleux[2], R. Nandi[1], A-M. Coll-Seck[2], and P. Coursaget[1]

[1] Institut de Virologie de Tours, Tours Cedex, France
[2] Service des Maladies Infectieuses, Faculté de Médecine et Pharmacie, Dakar Sénégal

Summary. An hepatitis B immunization programme was initiated in Senegal in 1978, and infants included in this controlled study have been followed for a period of 2–12 years after immunization. During this period HBV infections have been observed both in vaccinated and non-vaccinated infants. The polymerase chain reaction was used to search for HBV DNA sequences in the sera of 153 children with evidence of serum markers of past or present HBV replication. Amplified HBV DNA sequences were detected in 93% of the HBsAg positive individuals, in 58% of those only positive for antiHBc antibodies and in 7.8% of antiHBs and antiHBc positive infants. The results confirm the high efficiency and long-lasting effectiveness of HB vaccine.

Introduction

Hepatitis B virus (HBV) infections represent a major public health problem because of the ability of HBV to cause chronic liver disease. Even though many chronic carriers remain asymptomatic for a long period of time, they subsequently develop cirrhosis and/or primary hepatocellular carcinoma [1].

The development of a vaccine against HBV offers the potential for reduction of morbidity and mortality associated with chronic sequels of HBV infection including liver cancer. During a long-term follow-up study of hepatitis B vaccine in Senegalese infants [6, 8], we observed HBV infections in vaccine and in control groups, as detected by HBsAg and/or antiHBc antibodies. With time, the level of protective antibodies decrease, and HBV events have been observed 5 to 12 years after immunization in 10–20% of the recipients. However, the HBsAg chronic carrier state has not been observed in HB vaccine responders except in one Senegalese infant who had received only a set of two primary doses at one month interval.

On the other hand, the detection of HBV DNA sequences by the polymerase chain reaction (PCR) in liver and serum of patients has been reported, although HBsAg was not identified in the serum [4, 7, 13, 14, 16, 19, 21, 24]. These results suggested that a persistent HBV infection may develop in individuals despite an apparent clearance of the virus based on the serological profile. Thus, to further confirm the protective efficacy of HB vaccine, it is necessary to check for the absence of persistence of HBV DNA in vaccinated subjects who experienced HBV infection. In this study, we have used PCR to investigate the presence of HBV DNA in sera of vaccinated and non-vaccinated Senegalese children with evidence of past or present HBV infection.

Materials and methods

An hepatitis B immunization programme was initiated in Senegal in 1978 [17, 18]. The infants included in this study have been followed for a period of 2 to 12 years after immunization. Infants were 3 to 24 months old at the time that they were included in the study and were immunized according to a schedule comprising 3 or 4 doses of plasma-derived vaccine. A control group of infants, who did not receive hepatitis B vaccine, was followed in order to determine the protective efficacy of the vaccine.

A total of 166 of these children were included in this study: 41 HBsAg carriers (HBsAg and antiHBc positive), 16 children HBsAg positive but antiHBc negative, 77 subjects who were seroconverted to anti-HBs and antiHBc, 19 children who were only antiHBc positive children, and 13 immunized children (antiHBs positive) without markers of HBV infection used as negative control. The HB vaccine group included 62 children with an incomplete protocol (1–3 primary doses), and 42 vaccinated children who had 3–4 vaccine doses.

Sera were analyzed for the presence of HBsAg, antiHBs and antiHBc using commercially available radioimmunoassays (Abbott Diagnostika, Wiesbaden-Delkenheim, Federal Republic of Germany).

HBV-DNA was purified from serum according to the procedure described by Thiers et al. [22]. Amplification of HBV DNA was carried out by the PCR. In each case, PCR amplification was performed at least twice and a negative control using water as template was included in each PCR run. The S region of HBV DNA was amplified using primers [22] encompassing the S region 5'CATCTTCTTGTTGGTTCTTCTG3' (position 429–450) and 5'TTAGGGTTTAAATGTATACCC3' (position 844–824). Amplification was carried out in a final volume of 50 μl containing purified DNA, 0.4 μmol/l of each primer, 200 μmol/l of each deoxyribonucleoside triphosphate, 1.5 mmol/l magnesium chloride, 10 mmol/l Tris-HCl buffer (pH 8.3), 0.01% gelatin and 2.5 units of Taq polymerase (Perkin-Elmer-Cetus, Norwalk, CT, U.S.A.). The samples were subjected to 30 cycles of amplification using a DNA Thermal Cycler (Perkin-Elmer-Cetus, Norwalk, CT, U.S.A.). For each cycle, denaturation was at 94°C for

30 sec, annealing of primers at 50°C for 30 sec, and elongation at 75°C for 1 min. After the last cycle the mixture was incubated at 72°C for 3 min.

Amplified HBV-DNA was visualized in agarose gel by ethidium bromide staining (Fig. 1) and subsequently by dot blot hybridization onto a nylon membrane using a digoxigenin-labelled DNA probe (Fig. 2). The probe was obtained by random primed incorporation of digoxigenin-labelled deoxyuridine-triphosphate (dUTP) during amplification of HBV-DNA purified from the serum of an HBsAg chronic carrier. After molecular hybridization, the hybrids were detected by enzyme-linked immunoassay using an anti-digoxigenin alkaline phosphatase conjugate and subsequent enzyme-catalyzed color reaction with 5-bromo 4-chloro 3-indolyl phosphate and nitroblue tetrazolium salt (DNA labelling and detection kit, Boehringer Mannheim, Mannheim, Federal Republic of Germany).

Results

Thirteen immunized infants that were positive for antiHBs, but without any serologic markers of HBV infection were used as negative control. No HBV DNA was found in sera from these subjects (Table 1). HBV DNA was detected in 92.7% of 41 HBsAg and antiHBc positive infants, and in 81.2% of 16 HBsAg positive and antiHBc negative children. In all, HBV DNA was detected in 89.5% of HBsAg positive children.

HBV-DNA sequences were detected by PCR in 57.9% of sera from Senegalese children positive only for antiHBc antibodies (Table 2). HBV-DNA was detected in 4 out of 5 children from the control group, in 5 out of 8 of those who were fully immunized and in 2 out of 6 infants with an incomplete immunization. It was possible to determine the time of HBV infection in 7 children. HBsAg was detected 6–11 years before HBV-DNA testing in 4 of them. In the 3 others antiHBc seroconversion without detection of HBsAg was observed 7–10 years before HBV-DNA testing. It must be noted that two of these subjects developed antiHBs antibodies after HBV infection and 7 years later were found positive only for antiHBc. HBV-DNA was not detected in these two samples that were only positive for antiHBc (for example see Fig. 3).

Among the 77 infants with evidence of past HBV infection (antiHBs and antiHBc positive), HBV DNA was detected in 6 (7.8%). Three were from the control group and the three others had received an incomplete HB vaccination schedule, one of which tested HBsAg positive during vaccination 11 years before HBV-DNA testing. For the two others, HBV infection took place between 4 and 11 years after immunization.

No significant difference in HBV-DNA detection was observed between children from the control or vaccine groups that were positive for antiHBc alone or positive for antiHBs and antiHBc (p = 0.52, p = 0.78, respectively).

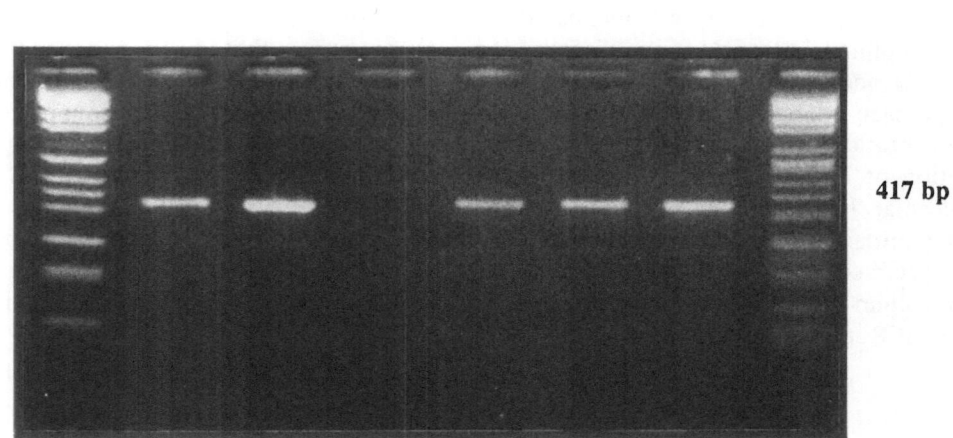

Fig. 1. A representative ethidium-bromide-stained agarose gel of amplified HBV DNA. *1* and *8* Molecular weight size markers; *4* negative control; *2, 3, 5–7* HBsAg positive Senegalese children

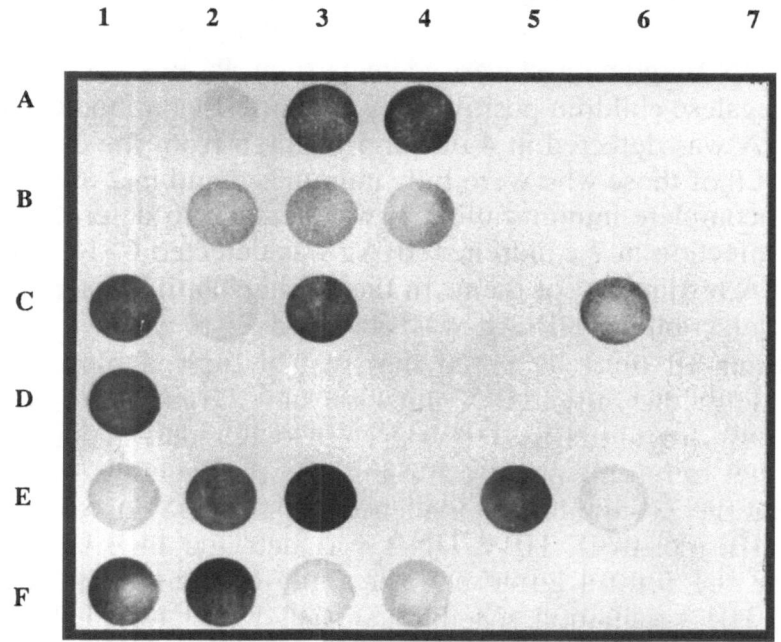

Fig. 2. Dot blot analysis of amplified HBV DNA using digoxigenin labelled probe. *D3, D4* Negative controls; *E3* positive control

Table 1. HBV-DNA detection by PCR in Senegalese children with an HBsAg positive infection or without HBV infections (antiHBs alone due to HB immunization)

	No. tested	HBV DNA
Children with active HBV infection		
HBsAg and antiHBc positive	41	38 (92.7%)
HBsAg positive alone	16	13 (81.2%)
Control group		
AntiHBs positive alone	13	0 (0.0%)

Table 2. HBV-DNA detection by PCR in HB vaccinated and non-vaccinated Senegalese children with serological evidence of past HBV infection

	No. tested	HBV DNA
AntiHBs and antiHBc pos.		
vaccine group		
incomplete immunization	35	3 (8.6%)
fully immunized	14	0 (0.0%)
control group	28	3 (10.0%)
total	77	6 (7.8%)
AntiHBc alone		
vaccine group	14	7 (50.0%)
control group	5	4 (80.0%)
total	19	11 (57.9%)

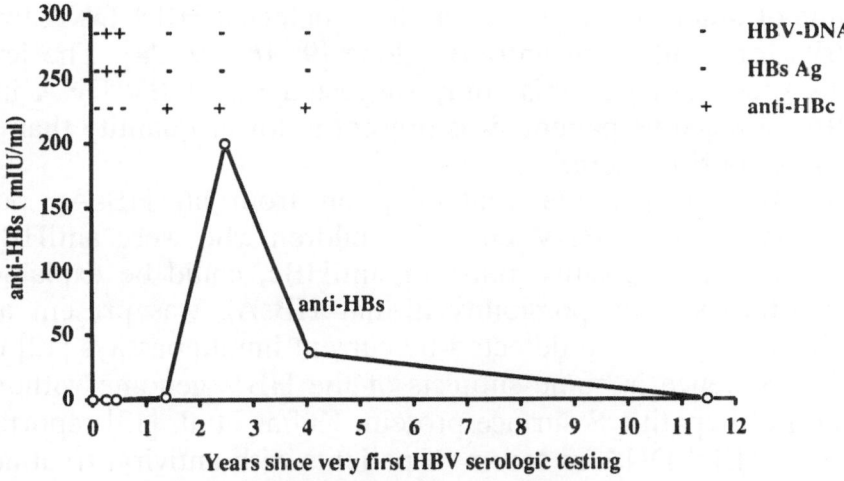

Fig. 3. HBV serologic markers and HBV-DNA in a Senegalese infant from the control group with an HBsAg positive infection

Among all immunized infants tested, HBV-DNA was detected in HBsAg negative samples from 3 nonresponders to the vaccination, in 3 children with evidence of HBV infection during immunization, in 5 children incompletely immunized and in two vaccine responders who showed evidence of HBV infection during the period 4–11 years after the very first vaccine dose.

In all, HBV-DNA was detected by dot-blot hybridization in 38 HBsAg and antiHBc positive children, in 13 children only positive for HBsAg, in 12 only positive for antiHBc and in 6 positive for both antiHBc and antiHBs. Among them, HBV DNA was detected by ethidium bromide staining in 31 (82%) of the infants with evidence of HBsAg and anti-HBc positive infection, in 6 (46%) of the children that were HBsAg positive and antiHBc negative, in two (17%) of the children of the only antiHBc group and in 2 (33%) of the antiHBs and antiHBc positive children.

Discussion

HBV DNA was detected in a small proportion (7.8%) of Senegalese children that were negative for HBsAg but positive for antiHBc and antiHBs. This was similar to the results of Wang et al. [24] and Shih et al. [21] who detected HBV DNA in 3.6–7.4% of the antiHBs and antiHBc positive healthy subjects or blood donors from Taiwan. The detection of HBV-DNA in antiHBs positive individuals has also been reported by others [3, 5, 20].

In the group of infants that positive only for antiHBc antibodies, HBV-DNA sequences were detected in 58% of the cases. This confirms the results of other investigators who have detected HBV DNA by PCR in 20–65% of adults with antiHBc alone [9, 16, 19, 24]. The level of HBV-DNA estimated in this study suggested that HBV-DNA in sera from HBsAg-negative patients was present in lower quantity than DNA from HBsAg positive sera.

In addition to possible contamination from an HBsAg positive tube, the detection of HBV DNA in children who were antiHBs and antiHBc positive or positive only for antiHBc, could be explained by other hypotheses. One possibility is that HBsAg was present at low levels which could not be detected by current immunoassays [12] or the possible persistence in some subjects of the HBV genome without the production of hepatitis B surface protein. Kuhns et al. [13] reported the detection of HBV DNA in serum of patients with antiviral treatment, 6 to 67 months after disappearance of HBsAg. These authors did not detect HBsAg mRNA by in situ hybridization, suggesting either the cessation of active transcription of the residual HBV DNA, or the production of HBV antigens at very low levels. This was confirmed by Blum et al. [2] who demonstrated HBV DNA persistence in the liver after serological recovery from HBV infection. Another possibility is that the HBsAg was masked by HBsAg-antiHBs immune complexes [2, 23]. Finally, there remains the possibility of the presence of HBV variants with critical mutations or deletions in the viral genome, making the virus inefficient in replication or undetectable by current immunoassays. This hypothesis is supported by the finding of HBV mutants in patients with HBV-DNA but negative for HBsAg [10, 15, 22].

Luo et al. [16] has followed, for 4–6 years, subjects with only antiHBc and positive for HBV-DNA and showed that the great majority of them are chronic carriers of the virus and did not develop antiHBs. Our data confirm these results since virus persistence was observed in 3 out of 5 cases that were only antiHBc positive followed for a period ranging from 6–11 years.

From the data obtained, there is no evidence of chronic carriage of HBV by the detection of its genome in infants who had responded to the vaccination. However, 3 cases were detected in infants who were nonresponders to the vaccination, and in 3 cases of HBV infection before completion of vaccination. These results are in confirmation with the detection of a small number of chronic HBsAg carriers observed in vaccination studies due either to nonresponse to vaccination or to HBV infection during vaccination [7, 11, 17]. These results confirmed the high efficiency and long-lasting effectiveness of HB vaccine.

References

1. Beasley RP (1982) Hepatitis B virus as the etiologic agent in hepatocellular carcinoma-epidemiologic considerations. Hepatology 2: 21–26
2. Blum HE, Liang TJ, Galun E, Wands JR (1991) Persistence of hepatitis B viral DNA after serological recovery from hepatitis B virus infection. Hepatology 14: 56–62
3. Bréchot C, Degos F, Lugassy C, Thiers V, Zafrani S, Franco D, Bismuth H, Trépo C, Benhamou JP, Wands J, Isselbacher K, Tiollais P, Berthelot P (1985) Hepatitis B virus DNA in patients with chronic liver disease and negative tests for hepatitis B surface antigen. N Engl J Med 321: 270–276
4. Bréchot C, Kemsdorf D, Paterlini P, Thiers V (1991) Hepatitis B virus DNA in HBsAg-negative patients. J Hepatol 13: S49–S35
5. Chemin I, Baginski MA, Petit MA, Zoulim F, Pichoud C, Caped F, Hantz O, Trépo C (1991) Correlation between HBV DNA detection by polymerase chain reaction and pre-S1 antigenemia in symptomatic and asymptomatic hepatitis B virus infections. J Med Virol 33: 51–57

6. Coursaget P, Yvonnet B, Chotard J, Sarr M, Vincelot P, N'Doye R, Diop-Mar I, Chiron JP (1986) Seven years study of hepatitis B vaccine efficacy in infants from an endemic area (Senegal). Lancet ii: 1143–1145

7. Coursaget P, Le Cann P, Leboulleux D, Diop-Mar I, Bao O, Coll AM (1991) Detection of hepatitis B virus DNA by polymerase chain reaction in HBsAg negative Senegalese patients suffering from cirrhosis or primary liver cancer. FEMS Microbiol Lett 83: 35–38

8. Coursaget P, Leboulleux D, Soumare M, Le Cann P, Yvonnet B, Chiron JP, Coll-Secka M, Diop-Mar I (1993) Twelve-year follow-up study of hepatitis B vaccination of Senegalese infants. J Hepatol (in press)

9. Diamantis ID, Mc Gandy C, Pult I, Bühler H, Schmid M, Gudat F, Bianchi L (1992) Polymerase chain reaction detects hepatitis B virus DNA in paraffin-embedded liver tissue from patients sero and histo-negative for active hepatitis B. Virchows Archiv A Pathol Anat Histopathol 420: 11–15

10. Fortuin M, Karthigesu V, Allison L, Howard C, Hoare S, Mendy M, Whittle H (1993) Break-through infections and identification of a viral mutant in Gambian children immunised with hepatitis B vaccine (submitted)

11. Hadler SC, Francis DP, Maynard JE, Thompson SE, Juobon FN, Echenberg DF, Ostrow DG, O'Malley PM, Penley KA, Altman NL (1986) Long term immunogenicity and efficacy of hepatitis B vaccine in homosexual men. N Engl J Med 315: 209–214

12. Katychaki JN, Siem TH, Brouver R (1978) Serological evidence of presence of HBsAg undetectable by conventional radioimmunoassay in anti-HBc positive blood donnors. J Clin Pathol 31: 837–839

13. Kuhns M, McNamara A, Mason A, Campbell C, Perrillo R (1992) Serum and liver hepatitis B virus DNA in chronic hepatitis B after sustained loss of surface antigen. Gastroenterology 103: 1649–1656

14. Lai ME, Farci P, Figus A, Balestrieri A, Arnone M, Vyas GN (1989) Hepatitis B virus DNA in the serum of Sardinian blood donors negative for the hepatitis B surface antigen. Blood 73: 17–18/30

15. Lee HS, Ulrich PP, Vyas GN (1991) Mutations in S gene affecting the immunologic determinants of the envelope protein of hepatitis B virus. J Hepatol 13: S97–S101

16. Luo KX, Zhou R, He C, Liang ZS, Jiang S (1991) Hepatitis B virus DNA in sera of virus carriers positive exclusively for antibodies to the hepatitis B core antigen. J Med Virol 35: 55–59

17. Maupas P, Coursaget P, Chiron JP, Goudeau A, Barin F, Perrin J, Denis F, Diop-Mar I (1981a) Active immunization against hepatitis B in an aera of high endemicity. Part I: field design. Prog Med Virol 27: 168–184

18. Maupas P, Chiron JP, Barin F, Coursaget P, Goudeau A, Perrin J, Denis F, Diop-Mar I (1981b) Efficacity of hepatitis B vaccine in prevention of early HBsAg carrier state in children controlled trial in an endemic area (Senegal). Lancet i: 289–292

19. Paterlini P, Gerken G, Nakajima E, Terre S, D'Errico A, Grigioni W, Franco D, Nalpas B, Wands J, Kew M, Pisi E, Tiollais P, Bréchot C (1990) Polymerase chain reaction to detect hepatitis B virus DNA and RNA sequences in primary liver cancers from patients negative for hepatitis B surface antigen. N Engl J Med 323: 80–85

20. Shafritz DA, Lieberman HM, Isselbacher KJ, Wands JR (1982) Monoclonal radioimmunoassays for hepatitis B surface antigen: demonstration of hepatitis B virus

DNA or related sequences in serum and viral epitopes in immune complexes. Proc Natl Acad Sci USA 79: 5675–5679

21. Shih LN, Sheu JC, Wang JT, Huang GI, Yang PM, Lee HS, Sung JL, Wang TH, Chen DS (1990) Serum hepatitis B virus DNA in healthy HBsAg negative Chinese adults evaluated by polymerase chain reaction. J Med Virol 32: 257–260

22. Thiers V, Nakajima E, Kremsdorf K, Mack K, Schellekens H, Driss F, Goudeau A, Wands J, Sninsky J, Tiollais P, Bréchot C (1988) Transmission of hepatitis B from hepatitis B seronegative subjects. Lancet ii: 1273–1276

23. Wands JR, Bruns RR, Carlson RI, Ware A, Menitove JE, Isselbacher KJ (1982) Monoclonal IgM radioimmunoassay for hepatitis B surface antigen: high binding activity in serum that is unreactive with conventional antibodies. Proc Natl Acad Sci USA 79: 1277–1281

24. Wang JT, Wang TH, Sheu JC, Shih LN, Lin JT, Chen DS (1990) Detection of Hepatitis B virus DNA by polymerase chain reaction in plasma volunteer blood donors negative for hepatitis B surface antigen. J Infect Dis 163: 397–399

Authors' address: Dr. P. Coursaget, Institut de Virologie de Tours, 2 bis boulevard Tonnelle', F-37042 Tours Cedex, France.

Arch Virol (1993) [Suppl] 8: 133–139

Duck hepatitis B virus (DHBV) as a model for understanding hepadnavirus neutralization

Sylvie Chassot[1], **Véronique Lambert**[1], **A. Kay**[2], **C. Trépo**[1], and **Lucyna Cova**[1]

[1] INSERM U 271, Lyon, and [2] UPR 41-CNRS, Centre Hayem,
Hôpital Saint-Louis, Paris, France

Summary. The role of the immune response to the human hepatitis B virus (HBV) envelope proteins in neutralization of viral infectivity has been well documented. The similarity between HBV, prototype member of the hepadnavirus family, and the closely related duck hepatitis B virus (DHBV) has allowed, use of the latter as a convenient model for the study of molecular mechanisms of HBV replication and neutralization. In this brief review, we will examine the HBV and DHBV envelope proteins and their role as targets for virus neutralization.

HBV envelope proteins: structure and role in viral neutralization

The HBV envelope consists of three related proteins termed preS1, preS2 and S, all of which are translated within the preS/S open reading frame from three distinct AUG codons [8] (Fig. 1). These envelope proteins, which share a common carboxy-terminal sequence, are involved in the development of protective immunity by naturally infected or vaccinated persons. The S protein elicits protective immunity against HBV [21]. However, antibodies to the preS region, which are induced prior to the emergence of anti-S antibodies, are considered to play a major role in neutralization and clearance of HBV [2, 10, 13, 27]. Indeed, the preS1 protein is known to bear a binding site (aa 21–47) to a hepatocyte receptor and antibodies against this attachment site are virus-neutralizing [19, 20]. In addition, the pre-S2 region was shown to elicit antibodies that offer effective protection against HBV in chimpanzee [7, 11]. However, the inability of HBV to infect cultured cells and the narrow host range of this virus have limited experimental studies to primates. Thus, the exact role of both of these proteins in HBV neutralization has not yet been clearly elucidated.

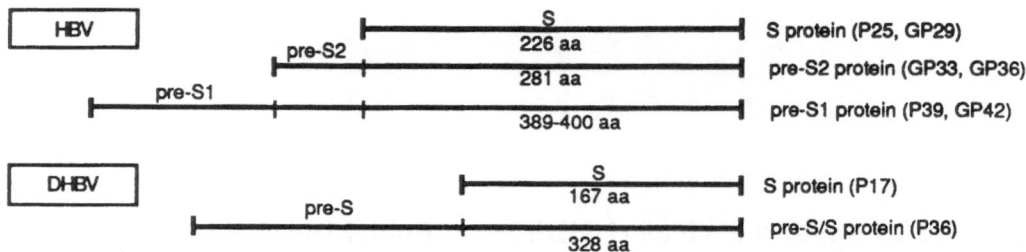

Fig. 1. Schematic representation of HBV and DHBV envelope proteins. *HBV* env proteins: the S protein (226 aa) exists in a nonglycosylated (P25) and glycosylated (GP29) form. The pre-S2 protein occurs in two distinct glycosylated forms, GP33 and GP36. The pre-S1 protein (389–400 aa) exists in a nonglycosylated (P39) and glycosylated (GP42) form. *DHBV* env proteins: the 17 kDa S protein (167 aa) and the 36 kDa pre-S/S protein (328 aa) are not glycosylated, in contrast to HBV env proteins

Envelope proteins of DHBV

A useful animal model that appears suitable for such studies is duck hepatitis B virus (DHBV) [9, 25]. DHBV is closely related to HBV with regard to genomic organization, hepatotropism and mode of replication [8, 25]. In addition, the natural host of DHBV is the domestic Pekin duck, an animal easy to handle and to keep in colony. Finally, DHBV infected ducks permit the study of neutralization mechanisms both in vitro in primary duck hepatocytes [5, 14, 23] and in vivo in ducklings [4, 15].

Little is known, however, about the structural proteins of DHBV. In contrast to HBV, the DHBV envelope contains only two major unglycosylated cocarboxyterminal proteins, the S protein (17 kDa), consisting of 167 amino acids and the preS/S or L protein (36 kDa), consisting of 328 amino acids, which are encoded by the preS/S open reading frame (Fig. 1). The S protein is synthesized from an internal AUG codon at position 1284 and represents the major surface antigen, like the S protein of HBV [18]. The L protein is myristilated at its N-terminus [17] and this myristilation signal is conserved in the L proteins of all hepadnaviruses [22]. According to several studies on preS/S transcripts [3], molecular sizes of the L protein [14, 16, 23, 24], and ATG mutants [26], the synthesis of the L protein may be initiated from the start codon at position 801 of the preS/S open reading frame which contains five or six start codons (depending on the subtypes). In addition to this major L protein species of 36 kDa, several other minor preS proteins ranging from 28 to 37 kDa have been detected in some sera and livers from infected ducks [4, 5, 14, 23, 24]. Moreover, in liver extracts, the 28 kDa protein sometimes appeared as the major preS species [4, 14,

29]. As suggested by mutational analysis, this 28 kDa protein may be generated by proteolysis from the 36 kDa protein and not initiated from an internal start codon of the DHBV preS/S open reading frame [26].

Biological roles of DHBV L envelope protein

As for HBV, the envelope proteins seem to be involved in infectivity of DHBV and carry neutralization epitopes. A study by Vickery et al. [28] indicated that adult ducks repeatedly inoculated with DHBV remained non-viremic but developed neutralizing antibodies to envelope proteins. More recently, Cheung et al. [6] demonstrated that there may be a more frequent and extensive response to the L than to S protein during convalescence of infected ducks. Similarly, Schlicht et al. [24] found that after immunization of rabbits with undenatured S particles, consisting of S and preS antigen, the major immune response was directed against the preS determinants. These findings are consistent with the computer prediction which indicates that the DHBV preS region is rather hydrophilic whereas the S region contains two major hydrophobic peaks [14]. Thus, the DHBV preS region appears to play an important role in neutralization and clearance of this virus. Moreover, we demonstrated that a polyclonal antiserum raised against the bacterially expressed first 131 amino acids of the preS region, i.e. without the 30 C-terminal amino acids completely abolished the infectivity of DHBV in vivo [15]. These data suggest that antibodies specific to the preS region lacking the S and the C-terminal portion of the preS, induce 100% protection in ducklings. Similarly, it has been shown for HBV that antibodies directed to preS1 or preS2 peptides protect chimpanzees against infection [7, 11, 20]. Thus, DHBV offers an attractive model for studying the role of preS epitopes in hepadnavirus neutralization.

Pre-S epitopes involved in DHBV neutralization

The preS epitopes involved in DHBV neutralization were investigated by P. Marion's group [18] and ours [4] using murine monoclonal anti-preS antibodies (Mab). Work of Cheung et al. [5], based on in vitro competitive bindings assays, identified three nonoverlapping preS epitopes (termed types IV, II and V) recognized by neutralizing Mabs, but no precise localization was attempted. In a recent study with preS/S fusion proteins, the same group [30] reported peptide mapping of the three epitopes on the DHBV preS sequence. These epitopes IV, II and V, which induced neutralizing Mabs, were respectively localized to amino acids 58–66, 91–99, and 139–145 (using our numbering system). However, although the epitope V was recognized by a neutralizing Mab, it did not seem to be directly involved in viral neutralization since we demonstrated that antibodies against a preS peptide lacking this epitope were able to completely neutralize DHBV infectivity [15] and mutants carrying deletions (aa 138–141 and 143–147) within this epitope were still infectious [16].

In addition to the data of Marion's group, we have identified two other preS epitopes recognized by the Mabs 900 and SD20 which reduced DHBV infectivity in vivo by 90% and 75%, respectively [4, 14]. These epitopes were initially roughly mapped between amino acids 77 and 100 on DHBV preS protein and fine mapping was then performed using overlapping peptides (Pepscan) spanning the preS sequence from amino acids 64 to 115. Interestingly, it appeared from the Pepscan results that the polyclonal antiserum raised against the N-terminal portion of the DHBV pre-S/S protein and inducing 100% protection against DHBV infection in vivo [15], recognized a region exclusively limited to the residues E^{82}-K^{95}, suggesting the immunodominance of this region within the sequence scanned. Furthermore, the epitope recognized by the Mab 900 was mapped in Pepscan within the same area whereas the epitope recognized by the Mab SD20 was localized downstream from this region. In addition, the best binding to Mabs 900 and SD20 was obtained with the peptides ^{83}IPQPQWTP^{90} and ^{100}FRRYQEER^{107}, respectively, whereas the shortest sequence common to all peptides recognized by these Mabs was found to be ^{88}WTP^{90} and ^{100}FRRYQE^{105}, respectively. Incidentally, the neutralization epitope II (aa 91–99) described by Yuasa et al. [30] was located between our two neutralization epitopes. Furthermore, the preS domain containing the three neutralization epitopes has been shown to be highly conserved among all cloned DHBV isolates [14] and to be immunodominant in infected ducks [6]. In addition, it is located within the main antigenic and hydrophilic site (aa 75–110) of DHBV, as predicted by computer algorithms [14]. Thus, all these data suggest that the preS domain encompassing these three neutralization epitopes is of functional importance in neutralization of DHBV. Moreover, the epitope recognized by the Mab 900 appeared of particular interest since this Mab almost completely protected ducklings against DHBV infection. Using single amino acid replacements, we also demonstrated that W^{88} is a key residue for binding to this Mab 900 since it could not be replaced by any other naturally occurring amino acids in Pepscan analysis [4]. This result is in good agreement with other studies describing the importance of aromatic residues in the antigenic determinants of peptides [1].

Although the studies of Marion's group and ours have identified preS epitopes important in DHBV neutralization, they do not demonstrate that these epitopes are involved in the viral attachment to hepatocyte receptors. However, Schaller's group [12] recently demonstrated that the preS sequence aa 81–120 was important for the in vitro binding of DHBV to hepatocyte membranes. This would suggest that some of the previously described neutralization epitopes [4, 5, 14, 30] could be a part of the cell receptor binding site on DHBV since the preS sequence 81–120 encompassed these epitopes. The ongoing mutagenesis of the sequence coding for these neutralization epitopes will be informative in this respect.

References

1. Appel JR, Pinilla C, Niman H, Houghten R (1990) Elucidation of discontinuous linear determinants in peptides. J Immunol 144: 976–983
2. Budkowska A, Dubreuil P, Capel F, Pillot J (1986) Hepatitis B virus pre-S gene-encoded antigenic specificity and anti-pre-S antibody: relationship between anti-pre-S response and recovery. Hepatology 6: 360–368
3. Büscher M, Reiser W, Will H, Schaller H (1985) Transcripts and the putative RNA pregenome of duck hepatitis B virus: implications for reverse transcription. Cell 40: 717–724
4. Chassot S, Lambert V, Kay A, Godinot C, Roux B, Trépo C, Cova L (1993) Fine mapping of neutralization epitopes on duck hepatitis B virus (DHBV) pre-S protein using monoclonal antibodies and overlapping octapeptides. Virology 192: 217–223
5. Cheung RC, Robinson WS, Marion PL, Greenberg HB (1989) Epitope mapping of neutralizing monoclonal antibodies against duck hepatitis B virus. J Virol 63:2445–2451
6. Cheung RC, Trujillo DE, Robinson WS, Greenberg HB, Marion PL (1990) Epitope-specific antibody response to the surface antigen of duck hepatitis B virus in infected ducks. Virology 176: 546–552
7. Emini EA, Larson V, Eichberg J, Conard P, Garsky VM, Lee DR, Ellis RW, Miller WJ, Anderson CA, Gerety RJ (1989) Protective effect of a synthetic peptide comprising the complete preS2 region of the hepatitis B virus surface protein. J Med Virol 28: 7–12
8. Ganem D, Varmus HE (1987) The molecular biology of the hepatitis B viruses. Annu Rev Biochem 56: 651–693
9. Gust ID, Burrell CJ, Coulepis AG, Robinson WS, Zuckerman AJ (1986) Taxonomic classification of human hepatitis B virus. Intervirology 25: 14–29
10. Heermann KH, Kruse F, Seifer M, Gerlich WH (1987) Immunogenicity of the gene S and pre-S domains in hepatitis B virions and HBsAg filaments. Intervirology 28: 14–25
11. Itoh Y, Takai E, Ohnuma H, Kitajima K, Tsuda F, Machida A, Mishiro S, Nakamura T, Miyakawa Y, Mayumi M (1986) A synthetic peptide vaccine in-

volving the product of the preS(2) region of hepatitis B virus DNA: protective efficacity in chimpanzees. Proc Natl Acad Sci USA 83: 9174–9178

12. Klingmüller U, Schaller H (1992) A highly conserved region in the pre-S domain of the duck hepatitis B virus is essential for binding and infection: implications from in vitro binding data that the DHBV receptor is a complex molecular biology of hepatitis B viruses. In: Chiseri FV, Gowans EJ (eds) Molecular biology of hepatitis B viruses, abstracts. University of California Press, San Diego, p 53

13. Klinkert MQ, Theilmann L, Pfaff E, Schaller H (1986) Pre-S1 antigens and antibodies early in the course of acute hepatitis B virus infection. J Virol 58: 522–525

14. Lambert V, Fernholz D, Sprengel R, Fourel I, Deléage G, Wildner G, Peyret C, Trépo C, Cova L, Will H (1990) Virus-neutralizing monoclonal antibody to a conserved epitope on the duck hepatitis B virus pre-S protein. J Virol 64: 1290–1297

15. Lambert V, Chassot S, Kay A, Trépo C, Cova L (1991) In vivo neutralization of duck hepatitis B virus by antibodies specific to the N-terminal portion of pre-S protein. Virology 185: 446–450

16. Li JS, Cova L, Buckland R, Lambert V, Deléage G, Trépo C (1989) Duck hepatitis B virus can tolerate insertion, deletion and partial frameshift mutation in the distal pre-S region. J Virol 63: 4965–4968

17. Macrae DR, Bruss V, Ganem D (1991) Myristylation of a duck hepatitis B virus envelope protein is essential for infectivity but not for virus assembly. Virology 181: 359–363

18. Marion PL, Knight SS, Feitelson MA, Oshiro LS, Robinson WS (1983) Major polypeptide of duck hepatitis B surface antigen particles. J Virol 48: 534–541

19. Neurath AR, Kent SBH, Strick N, Parker K (1986) Identification and chemical synthesis of a host cell receptor binding site on hepatitis B virus. Cell 46: 429–436

20. Neurath AR, Seto B, Strick N (1989) Antibodies to synthetic peptides from the preS1 region of the hepatitis B virus (HBV) envelope (env) protein are virus-neutralizing and protective. Vaccine 7: 234–236

21. Neurath AR, Thanavala Y (1990) Hepadnaviruses. In: Regenmortel MHV, Neurath AR (eds) Immunochemistry of viruses, vol 2: The basis of serodiagnosis and vaccines. Elsevier, Amsterdam, pp 403–458

22. Persing DH, Varmus HE, Ganem D (1987) The preS1 protein of hepatitis B virus is acetylated at its amino terminus with myristic acid. J Virol 61: 1672–1677

23. Pugh JC, Sninsky JJ, Summers JW, Schaeffer E (1987) Characterization of a pre-S polypeptide on the surfaces of infectious avian hepadnavirus particles. J Virol 61: 1384–1390

24. Schlicht HJ, Kuhn C, Guhr B, Mattaliano RJ, Schaller H (1987) Biochemical and immunological characterization of the duck hepatitis B virus envelope proteins. J Virol 61: 2280–2285

25. Schödel F, Sprengel R, Weimer T, Fernholz D, Schneider R, Will H (1989) Animal hepatitis B viruses. Adv Viral Oncol 8: 73–102

26. Schödel F, Weimer T, Fernholz D, Schneider R, Sprengel R, Wildner G, Will H (1991) The biology of avian hepatitis B viruses. In: McLachlan A (ed) Molecular biology of the hepatitis B virus. CRC Press, Boca Raton, pp 53–80

27. Theilmann L, Burkhard HD, Galle PR, Gmelin K, Kommerell B, Pfaff E (1988) Detection of antibodies against pre-S1 proteins in sera of patients with hepatitis B virus infection by ELISA using a pre-S fusion protein expressed in E. coli. Drug Res 38: 1856–1858

28. Vickery K, Freiman JS, Dixon RJ, Kearney R, Murray S, Cossart YE (1989) Immunity in Pekin ducks experimentally and naturally infected with duck hepatitis B virus. J Med Virol 28: 231–236

29. Yokosuka O, Omata M, Ito Y (1988) Expression of pre-S1, pre-S2, and C proteins in duck hepatitis B virus infection. Virology 167: 82–86
30. Yuasa S, Cheung RC, Pham Q, Robinson WS, Marion PL (1991) Peptide mapping of neutralizing and nonneutralizing epitopes of duck hepatitis B virus pre-S polypeptide. Virology 181: 14–21

Authors' address: Dr. Lucyna Cova, INSERM U 271, 151 cours Albert Thomas, Lyon F-69003, France.

IV. Variability of hepatitis B virus

Arch Virol (1993) [Suppl] 8: 143–154

Implications of genetic variation on the pathogenesis of hepatitis B virus infection

W.F. Carman[1] and **H.C. Thomas**[2]

[1] Institute of Virology, University of Glasgow, Glasgow, and [2] Department of Medicine, St Mary's Hospital Medical School, Imperial College of Science, Technology and Medicine, London, U.K.

It is now well established that sequence variation of hepatitis B virus (HBV) occurs worldwide. Although variants have been classified into subtypes by means of antigenic specificities of the surface proteins, it may be that HBV can be better classified into genotypes by means of percentage nucleotide difference. HBV can be classified into at least four genotypes based on a divergence of >7% at the nucleotide level [1] and it is clear that, based on this definition, other genotypes will be described in the future. Whether genotyping in this fashion is a satisfactory method of classifying HBV, and whether this has any relevance to the pathogenesis of the liver disease, infectivity or spread remains to be seen. Subtypes, defined by antigens associated with specific amino acid changes, are distributed geographically and have no apparent ink with disease. Whether one subtype can infect a person previously infected with a different subtype is unclear; at least, it would seem to be quite uncommon.

If the significance of genotypes is presently unclear, then the same has to be said for genetic variation that occurs within each individual over time. Since the first descriptions of pre-core (preC) mutations, a large number of different mutations, rearrangements, deletions and insertions have been described in chronic hepatitis patients. Some of these studies are cross-sectional analyses of patients with particular clinical problems and suffer from the lack of sequential samples. The major unresolved issue is whether these changes occur before, and therefore cause severe disease, or occur as a consequence of severe disease perhaps reflecting selection under the host immune response.

Variants of the preC region are the best characterised. The pre-core region consists of 87 nucleotides upstream of the core start codon (Fig. 1). The first 19 amino acids form a signal peptide which allows

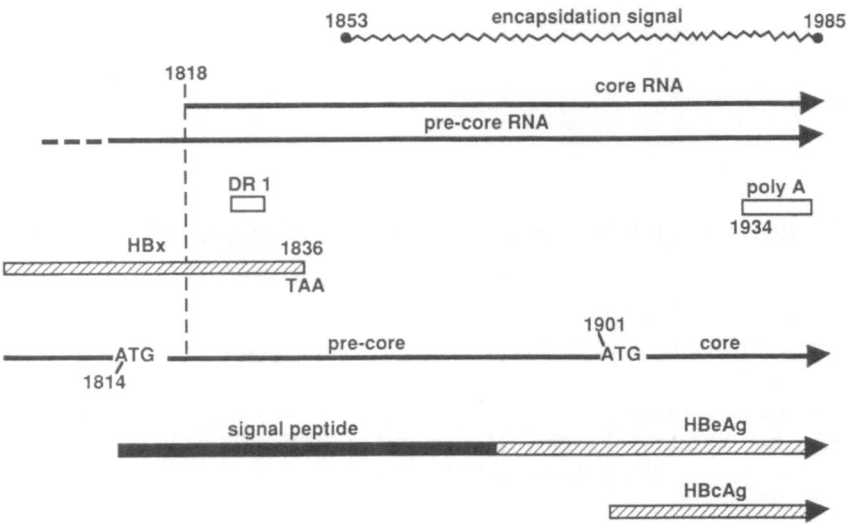

Fig. 1. Cis-acting elements and translational products within and around the precore region. Translation from the first ATG leads to a signal peptide; HBeAg is translated by reading through the second ATG. Translation from the second ATG leads to core antigen. The X protein overlaps HBeAg in a different frame. The 3.5 kb mRNA which encodes HBeAg and the 3.1 kb RNA which encodes HBcAg, polymerase and acts as the pregenome, all include areas of pre-core. The unique poly A tailing site is found within the 5′ end of core. DR1 is crucial for replication and the cis-acting packaging signal both reside within pre-core

the translated preC/core protein to be secreted into the endoplasmic reticulum; hepatitis B e antigen (HBeAg) is derived from this precursor by cleavage of amino- and carboxy-terminal fragments. Core (HBcAg) protein is the major component of the nucleocapsid and is therefore essential to virus production and infectivity. HBeAg, although secreted in large amounts in the early and chronic stages of infection, does not seem to be necessary for virus assembly or infectivity. However, the genetic material which encodes HBeAg does play an important role as it contains the 5′ end of the major mRNA, the cis acting RNA packaging signal and DR1, a crucial element in replication (Fig. 1). Thus, although it could be said that the preC region is not necessary for virion production, this would only be true at a translational level. The presence of HBeAg correlates with viral titres in serum. This is probably an indirect association but cannot be explained by a common RNA, as the mRNA that codes HBeAg is not involved in genome replication. During the early phases of infection, HBeAg parallels HBV DNA levels (as measured by dot blot hybridisation) in serum; as acute infection resolves, both decline in parallel. With the disappearance of HBeAg, seroconversion to antiHBe occurs with most patients going on to recover from their acute infection. Some patients become chronic carriers, whereas others will clear the virus entirely. Those who become chronic carriers are HBeAg positive for varying periods of time and during the latter stages of infection develop chronic hepatitis which becomes more severe as seroconversion to anti-HBe approaches.

It is believed that the immune response against HBV infected hepatocytes is responsible for the hepatitis, and that an enhanced response occurs some years after the onset of the chronic infection. This leads to high levels of serum transaminases associated with worsening of hepatitis and seroconversion to antiHBe. Whether this immune response is humoral or cellular, or both, is unclear; it is also unclear whether the response is against, inter alia, HBeAg or HBcAg. Because HBeAg and HBcAg share a large number of amino acids (from the methionine of core to the carboxy terminus of HBeAg), they also share B and T cell (3, 4) epitopes. There are at least two HBeAg B cell epitopes (termed e1 and e2); e1 overlaps the core epitope [3], the latter being conformational. Additional unique core epitopes are also present [5]. The major core epitope is coded by the middle third of the core gene (aa 76–89) [3, 5] and further studies should indicate whether this is recognised by all patients and whether other epitopes exist, linear, conformational or discontinuous. Cytotoxic T cell (CTL) epitopes (MHC class I restricted) have now been found throughout the core protein. The best characterised is an MHC class I restricted epitope between amino acids 18 and 27 [6]. Preliminary data in A2 subjects suggest that cells expressing preC/core, from which HBeAg is derived, may be lysed more efficiently than cells expressing HBcAg (unpubl. obs. by J. Waters and H.C. Thomas). Since both proteins contain the epitope 18–27, the different efficiency of CTL killing may be related to different processing of the two proteins. The influence of the signal sequence on aggregational properties, quaternary structure and antigenicity of HBeAg has been documented and similar mechanisms may be operating to produce unique peptides recognised by CTL [7].

Not all patients recover after seroconversion to antiHBe in the chronic phase of HBV infection. Some patients in Mediterranean countries [8, 9] and the Far East [10] go on to have continuing viraemia and severe disease that may progress rapidly. These patients have mutations in the preC region that stop translation of the pre-core/core product [11–15], but allow translation of core to continue unabated. PreC sequence is present, allowing replication and packaging, but the protein, HBeAg, cannot be synthesised. As the core ATG is found after these various preC mutations, it is apparent that HBc protein can be translated and nucleocapsid formation can progress. Some patients have mutations of the preC start codon, such that there is failure to initiate translation.

However, the most common mutations occur in the four G residues between 1896 and 1899. G to A mutations at 1896 and 1897 give rise to translational stop codons, whereas those at 1898 [16] and 1899 [11, 12] lead to seemingly unimportant amino acid substitutions. Those at 1898 and 1899 are only associated with, respectively, a serine at aa15 of pre-core and the preC stop codon, and further studies may reveal their importance. That A substitutions have been seen at all four G bases at the end of preC indicates that this is a region for polymerase error and probably explains the finding that the mutation at 1896 is seen in the majority of antiHBe positive patients. Mutations are selected at around the time of antiHBe seroconversion in patients with chronic hepatitis [13]. Except in the rare fulminant cases there is no evidence that these variants are transmitted between patients so that the contact of an antiHBe positive patient also becomes antiHBe positive without an HBeAg phase. Experimental evidence is lacking, but it seems reasonable to presume that one (or more) of three scenarios explains this: 1) the preC mutant is infectious, but the HBeAg producing genotype is back selected upon new infection in an immune naïve person; 2) there is a small amount of the HBeAg positive genotype present which is selected upon infection of a new host; 3) the preC mutant genotype is not very infectious.

The most likely pressure selecting these mutants, which all have in common the inability to encode the production of an important antigenic protein (HBeAg), is the immune response. One interesting question is whether antibody or CTL act as the selective pressure. Antibody selection is a well documented event in vitro and in vivo, but selection by CTL is less so. Although the latter has been seen in a transgenic mouse model of lymphochoriomeningitis virus infection [17] and in HIV positive patients [18, 19], the fact that mutations occur in CTL epitopes over time, and that such new CTL epitopes are unrecognised, does not mean necessarily that CTL are providing the selective pressure. It is also possible that this is a chance event in a random mutation process. Nevertheless, CTL escape is an interesting phenomenon which has attracted a large amount of interest and should not be discounted in studies of HBV infection. It may be that the selection of HBeAg-negative variants is a form of CTL escape by the virus.

Initially, the association between pre-core mutants and severe disease was thought to be causal rather than casual. However, it was soon found that patients who seroconvert to antiHBe without progressing to severe disease are also sometimes infected by such a mutant. It now seems that this selection process is common to most people who seroconvert to antiHBe during both acute and chronic infection. Animal studies have failed to show that the preC mutant gives rise to a more severe disease per se [20], but chimpanzee experiments using infectious cDNA whereby the infecting sequence can be rigidly controlled, need to be performed.

Fulminant hepatitis has a 50% mortality rate and occurs in less than 1% of acute hepatitis patients. These patients often become infected from antiHBe positive carriers. Similarly, infants born to antiHBe positive carrier mothers can develop a severe acute or fulminant disease. The most compelling evidence that a viral factor is responsible for this illness is that clusters of fulminant hepatitis cases have been described with the index case usually being antiHBe positive. Predictably, a number of groups have now found that antiHBe positive or HBeAg negative fulminant hepatitis patients are usually infected with the pre-core mutant [21, 24]. However, one group has shown that HBeAg positive patients are infected by the wild type [21] (and even have a worse outcome) and other groups have failed to show an association between the pre-core stop codon mutant and fulminant hepatitis in British, Hong Kong and American patients [25]. Because samples are not available from very early in the infection (before disease) it is not clear whether preC mutants are selected within each individual patient, or whether it is the preC mutant, transmitted as a single species, which causes severe disease.

AntiHBe positive patients with severe liver disease have a variable response to α-interferon. This may be due to an inability of the immune response in certain patients infected with pre-core mutants to clear the virus, even with the immune enhancing effects of α-interferon. One group has, however, reported similar results (approximately 40% responding) in antiHBe and HBeAg positive groups [26]. The presence of preC mutants in the antiHBe positive group does not seem to predict non-responsiveness although a trend towards poor response is seen [27, 28]. One group [28] has postulated that the higher the percentage of non-mutated virus in an antiHBe positive patient, the more likely a response to interferon.

The influence of pre-core variation on response of HBeAg positive patients has also been investigated. If it is the immune response which is clearing HBV, then it would be expected that there would be mixtures of preC mutant and wild type during the HBeAg clearance process, and that the presence of such a mixture would predict a good response to interferon. Although a small initial trial showed that the presence of preC mutant as a mixed population with wild type during the HBeAg positive phase predicted a good response to interferon [29], two other groups [27, 30] have not shown this association in larger numbers of patients. The overall picture is therefore unclear.

Some pre-core variants have unknown significance. For example, two groups have described a mutation at codon 15 of preC that changes the sole proline of the region to a serine. This mutation was found in an HIV positive HBV carrier with a poor antiHBc response [31]. It was hypothesised that the variant may have had some bearing on the poor response. A second group found such a mutant in 50% of Chinese antiHBe positive chronic hepatitis patients [16], the others having the preC stop codon. These two mutations were mutually exclusive, leading to the conclusion that they must have similar functions as the disease in the two patient groups was similar. Perhaps there is an effect on pre-core structure. However, the codon 15 mutation was present during both the HBeAg and antiHBe positive phases of the illness, quite unlike the situation with the preC stop codon. Thus, here is a mutation which allows secretion of HBeAg in a form detectable by the current assays, is also present during the antiHBe positive phase of continuing viraemia, with severe disease (thus an active immune response), yet the stop codon is never selected. Why this codon 15 mutation protects against selection of the preC codon 28 mutation is unclear but may constitute an important clue to the pathogenesis of this phase of the disease.

Another interesting situation is an insertion of 36 base pairs at the end of preC that gives rise to the preC stop codon, a new core ATG at the correct position and loss of the original core ATG [32, 33]. That such an insertion is found in several patients indicates that there is a common mechanism. However, the inserted sequence is not related to HBV and it is unclear from where the DNA has originated.

The prevailing theory about the function of HBeAg is that it is in some way an immune modulator. It is known that HBeAg can induce tolerance in transgenic mice [34], but many proteins could have this effect. It is a highly conserved protein, is found in all hepadnaviruses, is not part of the virion and it has now been shown that a virus which cannot encode the cell to secrete HBeAg is selected at the time of increased immune activity against hepatocytes. This suggests that the protein is a target for immune attack. Babies that are infected around the time of birth by HBeAg positive mothers become HBeAg positive themselves and initially exhibit very little immune activity against HBV. Why these subjects later develop activity and 50% die of cirrhosis or HCC is unknown. One hypothesis is that immune tolerance is lost over time, as shown in the transgenic mouse model. Although there is no evidence for this, it is proposed that HBeAg crosses the placenta and tolerises the infant's immune system to HBeAg. In this way HBV is not recognised in hepatocytes for a number of years. A similar mechanism could equally apply to adults with chronic hepatitis B. As the tolerising effect of HBeAg is lost, so immune activity increases, peaks of transaminases occur, hepatitis worsens and, in an attempt to escape the immune response, a virus which cannot express this antigenic protein is selected. CTL responses have been found against epitopes present in HBe and HBc antigens only during acute infection indicating that it is the CTL response which is clearing these hepatocytes. In the majority of patients, this leads to cure. However, some go on to become HBeAg positive carriers and it is several years before they clear the infection. Why some patients should have a delayed response to HBV is not known but may be related to the genetic makeup of the individual and their ability to overcome the tolerising effect of HBeAg.

In chronic carriers who have minimal liver disease after seroconversion to antiHBe, pre-core mutants are also selected and thus attention has become focused on the core gene to explain why some develop severe disease. The first two studies involved HBeAg positive Japanese patients [35, 36]. When comparing patients with raised ALT to those without hepatitis, a significant number of amino acid substitutions were found in the former group. This ranged between one and four changes, but these were clustered between amino acids 84–101, previously predicted to be a region bearing B cell epitopes [3]. The other Japanese group found deletions in patients with severe disease, all of which were in multiples of three bases. This latter finding would have predicted shorter, but possibly still functional, proteins. Many different deletions were found in multiple clones from a single patient, implying that these clones were under a similar selection pressure. Sequences adjacent to the deletions were not similar in all the clones. Deletions have been confirmed, often multiples of three bases but sometimes not, in British patients [37].

Core sequencing studies have now been performed in Mediterranean antiHBe positive patients, with normal or raised ALT [38]. The number of substitutions away from the consensus sequence, of all the patients in the study, was least in the antiHBe positive group with no liver disease. HBeAg positive patients, even those with raised ALT, had a similar number to the first patient group, but the antiHBe positive patients with raised ALT had five times as many amino acid substitutions as the HBeAg positive group. Two fulminant hepatitis patients had ten times as many amino acid changes. It thus seems not only to be the presence or absence of hepatitis which predicts amino acid variability in the core gene, but also the presence or absence of HBeAg. Because of the known association of antiHBe positivity with the preC stop codon, pre-core sequence mutations were correlated with core sequence. It became apparent that only patients with a preC mutation had large numbers of substitutions in core; antiHBe positivity alone was not sufficient. In particular, a mutation at amino acid 12 of core (a threonine to serine) was only found in those who already had a pre-core stop codon and was significantly associated with severe disease [39].

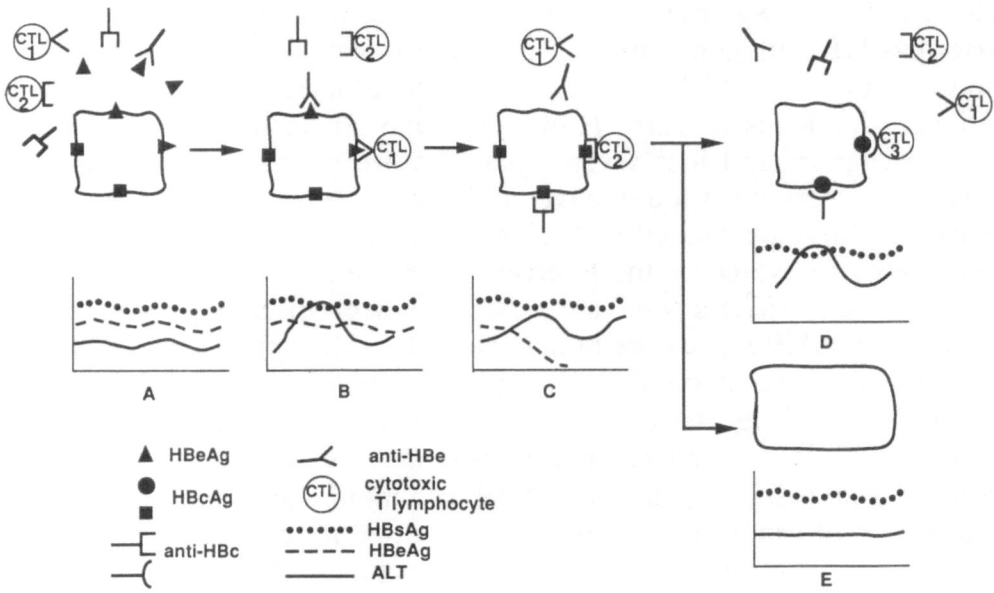

Fig. 2. Hypothetical representation of the serological correlation with HBV pre-core/ core sequence in hepatocytes. Initially (**A**) HBeAg tolerises the immune response to itself and HBcAg. With loss of tolerisation (**B**), immune attack ensues, resulting in selection of pre- core mutants (**C**) and attack on core epitopes. Either ongoing attack occurs (**D**) with new core epitopes resulting from escape from humoral and/CTL pressure of (**E**) the infection is cleared

It thus would seem that patients who have selected the HBeAg ne- gative virus go on to select mutations in core. The mutations were spread evenly in the antiHBe positive group but tended to cluster, as with previous studies, in the middle of the core gene in the HBeAg positive group. It is proposed (Fig. 2) that when the host selects this HBeAg negative virus, viraemia continues in the absence of the immune mod- ulatory effect of HBe protein, resulting in an enhanced immune response against core. Virus containing multiple mutations is then selected in the antiHBe positive group, leading to divergence from the consensus sequence. These studies do not resolve the basic problem of cause or effect. Sequential studies found that over a period of between one and five years in those with severe disease, very few amino acid changes occur in core [39]. This either indicates that the amino acid changes are present before the severe disease and are not selected by the immune response, or that the selection occurs over a short period of time early in the disease and that further change is much slower. Studies in patients with a defined onset will be needed to address these issues. Another issue is whether these changes relate to immune epitopes, and if they affect recognition. If so, are these humoral or CTL epitopes?

Mutation is also widespread in other genes of HBV. Deletions in the X protein [40], mis-sense mutations in the polymerase gene (and a stop codon) [41], and, most importantly, changes in the surface gene have been described [42]. Evolution of preS1 and preS2 mutations has been shown in sequential samples [32] and it seems to be not uncommon for the preS2 ATG to be deleted [43]. The preS1 peptide, 21–47, thought to be involved in hepatocyte binding, has been conserved in all of the pre-S1 variants so far seen. Mutations of the S protein have implications for design of diagnostic assays and vaccines but do not seem to be directly related to pathogenesis. Too little information is available on variation other than preC to enable reasonable conclusions to be generated.

In summary, variants of HBV are continuously selected in patients with chronic infection: there are large numbers of substitutions, insertions, deletions and rearrangements. Very little is known about the genetic changes in acute disease, except that pre-core mutants can be selected. A picture is building up of a virus which can escape various pressures, most probably the immune response, on a number of different proteins, yet still remain functional and produce high levels of viraemia with ongoing disease. The theory that HBeAg is an immune modulator is now supported with indirect evidence and the finding that core contains important epitopes, both humoral and cellular leads us to believe that the changes seen in core and the association with severe disease are linked. Whether such genotypes are as infectious as the "wild-type" or whether these are end-stage viruses that are slowly being eliminated by the immune response is unclear. It remains to be seen whether use can be made of these changes in devising new therapies for hepatitis B virus infection, and whether it will be possible to find particular mutations which will predict a poor prognosis.

References

1. Okamoto H, Tsuda F, Sakugawa H, Sastrosoewignjo RI, Imai M, Miyakawa Y, Mayumi M (1988) Typing hepatitis B virus by homology in nucleotide sequence: comparison of surface antigen subtypes. J Gen Virol 69: 2575–2583
2. Junker-Niepmann M, Bartenschlager R, Schaller H (1990) A short cis-acting sequence is required for hepatitis B virus pregenome encapsidation and sufficient for packaging of foreign RNA. EMBO J 9: 3389–3396
3. Salfeld J, Pfaff E, Noah M, Schaller H (1989) Antigenic determinants and functional domains in core antigen and e antigen from hepatitis B virus. J Virol 63: 798–808
4. Milich DR, McLachlan A, Moriarty A, Thornton GB (1987) Immune response to hepatitis B virus core antigen (HBcAg): localisation of T cell recognition sites with HBcAg/HBeAg. J Immunol 139: 1223–1231

5. Waters JA, Jowett TP, Thomas HC (1986) Identification of a dominant epitope of the nucleocapsid (HBc) of the hepatitis B virus. J Med Virol 19: 79–86

6. Bertoletti A, Ferrari C, Fiaccadori F, Penna A, Margolskee R, Schlicht HJ, Fowler P, Guilhot S, Chisari FV (1991) HLA class I-restricted human cytotoxic T cells recognize endogenously synthesized HBV core antigens. Proc Natl Acad Sci USA 88: 10445–10449

7. Ferrari C, Bertoletti A, Penna A, Cavalli A, Valli A, Missale G, Pilli M, Fowler P, Giuberti T, Chisari FV, Fiaccadori F (1991) Identification of immunodominant T cell epitopes of the hepatitis B virus nucleocapsid antigen. J Clin Invest 88: 214–222

8. Bonino F, Rosina F, Rizzetto M, Rizzi R, Chiaberge E, Tardanico R, Callea F, Verme G (1986) chronic hepatitis in HBsAg carriers with serum HBV-DNA and anti-HBe. Gastroenterology 90: 1268–1273

9. Hadziyannis SJ, Lieberman HM, Karvountzis MG, Shafritz D (1983) Analysis of liver disease, nuclear HBcAg, viral replication and hepatitis B virus in liver and serum of HBeAg vs anti-HBe positive chronic hepatitis B virus infection. Hepatology 3: 652–662

10. Chu C-M, Karayiannis P, Fowler MJF, Monjardino J, Liaw Y-F, Thomas HC (1985) Natural history of chronic hepatitis B virus infection in Taiwan: studies of hepatitis B virus DNA in serum. Hepatology 5: 431–434

11. Carman WF, Jacyna MR, Hadziyannis S, Karayiannis P, McGarvey M, Makris A, Thomas HC (1989) Mutation preventing formation of e antigen in patients with chronic HBV infection. Lancet ii: 588–591

12. Brunetto MR, Stemmler M, Schodel F, Will H, Ottobrelli A, Rizzetto M, Bonino F (1989) Identification of HBV variants which cannot produce precore-derived HBeAg and may be responsible for severe hepatitis. Ital J Gastroenterol 21: 151–154

13. Okamoto H, Yotsumoto S, Akahane Y, Yamanaka T, Miyazaki Y, Sugai Y, Tsuda F, Tanaka T, Miyakawa Y, Mayumi M (1990) Hepatitis B viruses with pre-core region defects prevail in persistently infected hosts along with seroconversion to the antibody against e antigen. J Virol 64: 1298–1303

14. Fiordalisi G, Cariani E, Mantero G, Zanetti A, Tanzi E, Chiaramonte M, Primi D (1990) High genomic variability in the pre-C region of hepatitis B virus in anti-HBe, HBV-DNA positive chronic hepatitis. J Med Virol 31: 297–300

15. Tong S, Li J, Vitvitski L, Trepo C (1990) Active hepatitis B virus replication in the presence of anti-HBe is associated with viral variants containing an inactive pre-C region. Virology 176: 596–603

16. Carman WF, Ferrao M, Lok ASF, Ma OCK, Lai CL, Thomas HC (1992) Pre-core sequence variation in Chinese isolates of hepatitis B virus. J Infect Dis 165: 127–133

17. Pircher H, Moskophidis D, Rohrer U, Bürki K, Hengartner H, Zinkernagel RM (1990) Viral escape by selection of cytotoxic T cell-resistant virus variants in vivo. Nature 346: 629–633

18. Phillips RE, Rowland-Jones S, Nixon DF, Gotch FM, Edwards JP, Ogunlesi AO, Elvin JG, Rothbard JA, Bangham CRM, Rizza C, McMichael AJ (1991) Human immunodeficiency virus genetic variation that can escape cytotoxic T cell recognition. Nature 354: 453–459

19. Johnson RP, Trocha A, Buchanan TM, Walker BD (1992) Identification of over-lapping HLA class I-restricted cytotoxic T cell epitopes in a conserved region of the human immunodeficiency virus type 1 envelope glycoprotein: definition of

minimum epitopes and analysis of the effects of sequence variation. J Exp Med 175: 961–971

20. Chen HS, Kew MC, Hornbuckle WE, Tennant BC, Cote PJ, Gerin JL, Purcell RH, Miller RH (1992) The precore gene of the woodchuck hepatitis virus genome is not essential for viral replication in the natural host. J Virol 66: 5682–5684

21. Carman WF, Hadziyannis S, Karayiannis P, Fagan EA, Tassopoulos NC, Williams R, Thomas HC (1991) Association of the precore variant of HBV with acute and fulminant hepatitis B infection. In: Hollinger FB, Lemon SM, Margolis H (eds) Viral hepatitis and liver diseases. Williams and Wilkins, Baltimore, pp 216–219

22. Kosaka Y, Takase K, Kojima M, Shimizu M, Inoue K, Yoshiba M, Tanaka S, Akahane Y, Okamoto H, Tsuda F, Miyakawa Y, Mayumi M (1991) Fulminant hepatitis B: induction by hepatitis B virus mutants defective in the precore region and incapable of encoding e antigen. Gastroenterology 324: 1087–1094

23. Liang TJ, Hasegawa K, Rimon N, Wands JR, Ben-Porath E (1991) A hepatitis B virus mutant associated with an epidemic of fulminant hepatitis. N Engl J Med 324: 1705–1709

24. Omata M, Ehata T, Yokosuka O, Hosoda K, Ohto M (1991) Mutations in the precore region of hepatitis B virus DNA in patients with fulminant and severe hepatitis. N Engl J Med 324: 1699–1704

25. Lo ES-F, Lo Y-MD, Tse CH, Fleming KA (1992) Detection of hepatitis B pre-core mutant by allele specific polymerase chain reaction. J Clin Pathol 45: 689–692

26. Fattovich G, Farci P, Rugge M (1992) Chronic hepatitis B positive for antiHBe and with hepatitis B viral DNA in serum can be successfully treated with alpha interferon. Hepatology 5: 584–589

27. Carman WF, Fattovich G, McIntyre G, Alberti A, Thomas HC (1991) Pre-core HBV mutants: response to interferon therapy. J Hepatol 13: S16

28. Brunetto MR, Saracco G, Oliveri F, Demartini A, Giarin MM, Manzini P, Calvo PL, Verme G, Bonino F (1991) HBV heterogeneity and response to interferon. J Hepatol 13: S15

29. Takeda K, Akahane Y, Suzuki H, Okamoto H, Tsuda F, Miyakawa, Mayumi M (1990) Defects in the precore region of the HBV genome in patients with chronic hepatitis B after sustained seroconversion from HBeAg to anti-HBe induced spontaneously or with interferon therapy. Hepatology 12: 1284–1289

30. Xu J, Brown D, Harrison T, Lin Y, Dusheiko G (1992) Absence of hepatitis B virus precore mutants in patients with chronic hepatitis B responding to interferon-alpha. Hepatology 15: 1002–1006

31. Liang TJ, Blum HE, Wands JR (1990) Characterisation and biological properties of a hepatitis B virus isolated from a patient without hepatitis B virus serologic markers. Hepatology 12: 204–212

32. Tran A, Kremsdorf D, Capel F, Housset C, Dauguet C, Petit M-A, Brechot C (1991) Emergence of and takeover by hepatitis B virus (HBV) with rearrangements in the pre-S/S and pre-C/C genes during chronic HBV infection. J Virol 65: 3566–3574

33. Bhat RA, Ulrich PP, Vyas GN (1990) Molecular characterisation of a new variant of hepatitis B virus in a persistently infected homosexual man. Hepatology 11: 271–276

34. Milich DR, Jones JE, Hughes JL, Price J, Raney AK, McLachlan A (1990) Is a function of the secreted hepatitis B e antigen to induce immunologic tolerance in utero? Proc Natl Acad Sci USA 87: 6599–6603

35. Ehata T, Omata M, Yokosuka O, Hosodo K, Ohto M (1992) Variations in codons 84–101 in the core nucleotide sequence correlate with hepatocellular injury in chronic hepatitis B virus infection. J Clin Invest 89: 332–338
36. Wakita T, Kakumu S, Shibata M, Yoskioka K, Ito Y, Shinagawa T, Ishikawa T, Takayanagi M, Morishima T (1991) Detection of pre-C and core region mutants of hepatitis B virus in chronic hepatitis B virus carriers. J Clin Invest 88: 1793–1801
37. Ackrill AM, Naomov NV, Eddleston ALWF, Williams R (1993) Specific deletions in the hepatitis B virus core open reading frame in patients with chronic active hepatitis B. J Med Virol (in press)
38. Carman W, Thomas H, Domingo E (1993) Viral genetic variation: hepatitis B virus as a clinical example. Lancet 341: 349–354
39. Carman WF, McIntyre G, Hadziyannis S, Fattovich G, Alberti A, Thomas HC (1993) The development of multiple HBV core gene mutations after loss of HBeAg is associated with severe chronic hepatitis. Lancet (in press)
40. Feitelson M, McIntyre G, McCruden EAB, Blumberg B, Thomas HC, Carman WF (1992) Mutated HBV DNA in HBsAg negative dialysis patients. J Hepatol 16: S11
41. Blum HE, Galun E, Liang J, von Weizsäcker F, Wands JR (1991) Naturally occurring missense mutation in the polymerase gene terminating hepatitis B virus replication. J Virol 65: 1836–1842
42. Carman WF, Zanetti AR, Karayiannis P, Waters J, Manzillo G, Tanzi E, Zuckerman AJ, Thomas HC (1990) Vaccine-induced escape mutant of hepatitis B virus. Lancet 336: 325–329
43. Santantonio T, Jung MC, Schneider R, Pastore G, Pape GR, Will H (1992) Hepatitis- B virus genomes that cannot synthesise pre-S2 proteins occur frequently and as dominant virus populations in chronic carriers in Italy. Virology 188: 948–952

Authors' address: Dr. W.F. Carman, Institute of Virology, University of Glasgow, Church Street, Glasgow G11 5JR, Scotland, U.K.

Arch Virol (1993) [Suppl] 8: 155–169

Hepatitis B virus C-gene variants

S. Miska and **H. Will**

Heinrich-Pette-Institut für Experimentelle Virologie und Immunologie an der
Universität Hamburg, Hamburg, Federal Republic of Germany

Summary. The heterogeneity of hepatitis B virus (HBV) is increasingly
believed to play a role in viral persistence, pathogenesis, and the type of
response to antiviral therapy. One of the best studied parts of the HBV
genome is the C-gene which codes for the nucleocapsid protein (HBc)
and the e-antigen (HBeAg). Here we attempt to review the recent data on
the sequence heterogeneity of this region and its possible implications.

Introduction

Hepatitis B virus (HBV) infection can result in acute, fulminant or
chronic hepatitis. Chronic infection is often associated with the development of liver cirrhosis and hepatocellular carcinoma. Despite the availability of a vaccine, there are still about 200 to 300 million chronic carriers
world wide who are in need of antiviral treatment. Resolution of chronic
hepatitis and elimination of the virus can be achieved by treatment with
interferon alpha (IFN) in only a minority of chronic carriers, and no
other similar or more successful treatment is known. Mechanisms that
lead to resolution of hepatitis and HBV elimination and reasons for the
failure of most patients to respond to IFN therapy are not known. It
is generally believed that the immune system, stimulated by IFN, or
spontaneously activated during the natural course of chronic infection,
can lead to virus elimination. Changes of the antigenic make-up of the
viral proteins are likely to play a role in immune escape. In this article,
we attempt primarily to review the current knowledge of the heterogeneity of C-gene sequences, as they encode antigens that are likely to
be important immune targets, the virus encoded e-antigen (HBe) and
the nucleocapsid protein (HBcAg). This speculation is supported by the
vigorous T cell response to HBc/HBe in acute, self-limiting hepatitis,

in contrast to the significantly milder T-cell response in patients with chronic infection [3, 15, 20, 60].

The basis of HBV sequence heterogeneity

One step in replication of HBV involves reverse transcription of a so-called RNA pregenome (for review see [51]). The reverse transcriptase encoded in the P-gene of HBV propably lacks proof-reading activity, similar to that of retroviruses, and therefore the continuous emergence of HBV sequence variants during chronic infection is expected. Whether or not such variants will ever represent a major viral population in the serum of a patient will depend on their biological and immunological properties. The small size of the HBV genome with tightly packed signals for transcription, translation and replication, the lack of an intergenic region, the presence of overlapping genes, and essential functions by some viral proteins restrict the number of viable HBV variants that can become a major virus population. This restriction is particularly obvious for the C-gene.

Coding capacity of the C-gene

The C-gene is divided by the first two in-frame AUG codons into two regions, the preC- and the C-region. Translation initiation at the second AUG codon leads to synthesis of the only known nucleocapsid protein (HBc, P21) (Fig. 1). This protein has the property of being able to self assemble into viral cores, and it probably also interacts with the RNA pregenome, with the viral polymerase/RTase, and with envelope proteins during virus maturation [66]. Any mutation that interferes with such functions is likely to block or hamper virus production. During infection, virtually all patients develop antibodies to HBc (antiHBc). The HBc-specific T-cell response is believed to be most important for virus elimination [29], although there is only little direct evidence for this belief [3, 38]. Translation initiation at the first AUG results in synthesis of a precursor protein (precore protein) which is posttranslationally processed, resulting in several proteins with HBe-antigenicity (Fig. 1). The preC sequence functions as a signal sequence which directs the precore protein to the secretory pathway. Sequence motifs within the 10 distal amino acids of the preC sequence (a cysteine and a hydrophobic tripeptide trp-leu-trp) prevent assembly of the precore protein and processing products thereof into nucleocapsid particles ([62, 63], see also Schlicht et al., this volume). In some cells and in yeast, however, unprocessed

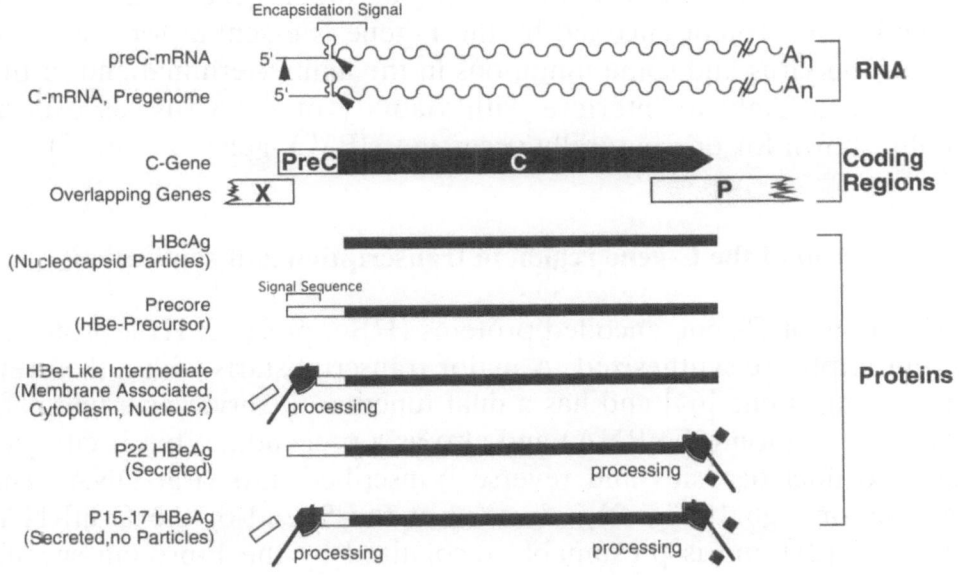

Fig. 1. Transcripts, open reading frames, and proteins expressed from the C-gene

precore protein is able to form nucleocapsid-like particles [28, 37]. From most of the precore protein molecules part of the signal sequence (19 amino acids) is cleaved off [65] and some of the resulting intermediates appear to end up transiently in the cell membrane [49]. Some of the molecules may remain in the cytoplasm or even enter the nucleus [36]. Most or all of the membrane-associated intermediates are further processed by proteolytic cleavage at the carboxy-terminus (at slightly different positions). The cleavage probably occurs in the Golgi or at the cell membrane. The amino- and carboxy-terminally truncated form of the precore protein is the major fraction of the so-called e-antigen (HBe or HBeAg). However, as discovered recently [54], there is at least one further protein with HBe-antigenicity in the serum of patients: a precore protein which is only truncated at the carboxy- but not at the aminoterminus (Fig. 1). HBe expression was, until recently, regarded as an indicator of active viral replication, whereas seroconversion to antibodies to HBe (antiHBe) was thought to be a hallmark of the beginning of virus elimination. Unlike HBc, HBe has no essential function in the viral life cycle, as demonstrated in the duck and woodchuck hepatitis B virus model systems [11, 12, 48, 50]. Circumstantial evidence suggests a function as a tolerogen which may be important for the establishment of chronic infection [26]. The C-gene overlaps at its 5'-end with the 3'-end of the X-gene (Fig. 1). The 3'-end of X-gene, however, is dispensable for its putative function as transcriptional activator. Furthermore, the 3'-end of the C-gene overlaps with the 5'-end of the P-gene (Fig. 1). The genome-linked protein encoded by the P-gene is essential for the replication of the virus and some mutations in the amino-terminal end of the P-protein are likely to interfere with viability of the virus, as experimentally shown for one naturally occurring HBV P-gene mutant [5].

Function of the C-gene region in transcription and replication

For synthesis of C-gene encoded proteins (HBc, precore, HBe proteins) two transcripts are synthesized. A major transcript starts within the preC open reading frame [64] and has a dual function: it serves as a template for HBc translation (C mRNA) and also as a pregenome that is encapsidated into nucleocapsids and reverse transcribed into viral DNA. The encapsidation signal (Fig. 2) is located at the 5'-end of the C mRNA/pregenome [21] and is presumably recognized by the P-protein specifically, thus triggering its encapsidation. From the C mRNA, not only HBc is synthesized, but also the P-protein. For synthesis of precore and HBe proteins a set of mRNAs (preC mRNAs) with heterogeneous 5'-ends located upstream of the first AUG of the C-gene is used. Although these mRNAs are only slightly longer than the C mRNA/pregenome and contain the entire encapsidation signal, they are not encapsidated into viral cores. The reason for this difference is that ribosomes initiating translation at the first AUG and translating through the encapsidation signal prevent its recognition [31]. Thus, preC mRNAs are selectively excluded from encapsidation. Mutations which interfere with translation of preC mRNAs are therefore expected to lead to encapsidation of preC mRNAs (see below). One of the direct repeats which is essential for initiation of DNA minus-strand synthesis is located within the preC region and mutations of the corresponding functional nucleotides in this sequence can interfere with viral replication [53].

Mutations in the preC sequence

Viremic chronic carriers can be divided into two large groups: those whose sera are HBeAg-positive and those who are antiHBe-positive. HBeAg-positive chronic carriers are usually infected with HBV genomes containing a continuous preC open reading frame whereas antiHBe positive carriers are predominantly infected with HBV variants with mutated preC sequences [2, 4, 6, 7, 9, 10, 16, 18, 23, 24, 30, 33–35, 40–42, 44, 45, 55, 57, 59]. Most of the preC mutations identified interfere with translation of the preC region and thus prevent precore and

HBe protein synthesis (initiation codon mutations, stop codons and frame shifts due to insertions and deletions). Other mutations are either silent or change amino acids. The most predominant preC nucleotide change found is a TGG to TAG nonsense mutation at the penultimate codon (codon 28) of the preC region. An interesting preC mutant was recently found in hepatocellular carcinoma tissue. In this mutant the only cysteine residue in the preC region was changed to a tryptophan residue [25]. This mutation was experimentally shown to lead to expression of an HBe-like protein which dimerizes and acquires HBc antigenicity [63]. PreC variants which cannot express precore and HBe proteins were shown to emerge spontaneously or after interferon treatment, in sera of chronic carriers, often at the time of seroconversion from HBe to antiHBe [18, 30, 33, 42, 45]. These observations raise the following questions: why do preC variant viruses emerge, are they as viable as preC wildtype viruses, what are the driving forces of their selection, do they play a role in viral pathogenesis and persistence, and are they more resistant to interferon treatment?

Known and presumed biological properties of preC variants

For three types of DHBV preC mutants (a frame shift-, a stop codon-, and a translation initiation codon mutant [11, 48, 50], infectivity and full replication competence were demonstrated before the first HBV preC variant had been identified as such in human sera. These observations, together with the identification of preC variants as dominant virus populations in antiHBe positive carriers with virus titers as high as preC wildtype viruses, argue against a major defect caused by most naturally occurring mutations in the preC region. Two groups have directly analyzed the viability of HBV preC variants in cell culture ([56] and Miska and Will, unpubl.). They found that the major preC mutant viruses with a stop codon at the penultimate codon of the preC region with or without associated point mutations at the ultimate codon are fully replication competent. Taking into account our current knowledge of viral replication this is now well understandable. Neither of these types of mutations would obviously interfere with the stability of the stem-loop structure which is an essential part of the encapsidation signal (Fig. 2) and should therefore not hamper viral replication. In contrast, most other mutations in the preC region can be predicted to destroy or functionally weaken the stem-loop structure of the encapsidation signal. For some of these mutations, a negative or detrimental effect on pregenomic encapsidation and thus on replication has in fact been found ([56, 58] and Miska and Will, unpubl.). These data can explain in

160   S. Miska and H. Will

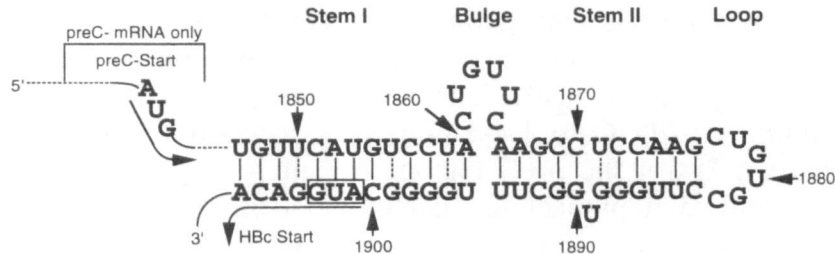

Fig. 2. The stem-loop structure of part of the encapsidation signal. Translation initiation codons for HBc and the precore protein are indicated

part why some preC mutations are more frequently found than others and why some mutations do not occur at all in vivo. It cannot yet be excluded, however, that one or more of the mutations in the preC sequence close to the second AUG codon positively or negatively affects synthesis of HBc on the level of translation. Our own preliminary evidence suggests that the most prevalent mutations (TAG stop codon and GAC mutation) do not significantly alter HBc expression (Miska and Will, unpubl. data).

There is a basic difference between the most prevalent preC variants with a stop codon at the penultimate codon and those with a mutated preC translation initiation codon. The first type of variant should still be able to produce a preC peptide, whereas the second cannot, and this may have several consequences. Provided the preC peptide carries T-cell epitopes it may still be expressed by the stop codon mutant and represent a target for immune recognition (note: a functional cytotoxic T-cell epitope in a signal sequence was mapped for lymphocytic choriomeningitis virus [8]). In contrast, the translation initiation codon mutation should abolish expression of the preC sequences and this would therefore be the safest way to escape immune recognition. What may be an even more important difference is the fact that mutation of the translation initiation codon will lead to encapsidation of the preC mRNAs and also to reverse transcription of this RNA into DNA minus strands, as shown by in vitro experiments [31]. The DNA minus strands that were reverse transcribed from preC mRNAs, however, can probably not be copied into plus strands and the corresponding virions are defective. These virions may interfere with efficient replication of wildtype virus and lower viremia or may have new biological, immunological and pathogenic properties.

Another type of preC variant is represented by frame-shift mutations in the proximal preC region which partially overlaps with the X-gene. The carboxy-terminal end of the X-gene is not essential for the transactivating function of the corresponding protein and therefore, frameshifts which hit this gene should not effect viral viability. In fact, a particular preC frameshift mutant genome was recently reported to produce functional particles [22]. In this genome, a mutant X-mRNA was shown to encode the proximal amino acid sequence deduced from the C-gene (aa 4–46, beginning with the preC ATG) carboxy-terminally fused to the X reading frame. Other frameshift mutations predict expression of preC peptides fused to normally non-coding sequences (hidden frames). If such preC fusion proteins are immunogenic, it may be possible to establish novel diagnostic immuno assays which could allow immunological detection of some frameshift mutants.

Driving forces for preC variant selection

Although HBeAg itself is not essential for virus formation and replication in animal models such as DHBV and woodchuck hepatitis virus (WHV), naturally occurring preC deficient variants have not been identified for any animal hepadnavirus. This observation suggest that HBe expression is an advantage for hepadnaviruses. Preliminary evidence argues that HBe may play a role in establishing the chronic carrier status in woodchucks [12]. This may be similar in humans since children born to HBeAg positive mothers almost invariably enter the chronic course of disease, while those born to antiHBe positive women do so less frequently [52]. The interpretation of elegant experiments with mice [26] offers, as a possible explanation, that HBeAg acts as a tolerogen in utero. If true, this can nicely explain the prevalence of preC wildtype sequences in short-term chronic carriers. Assuming that HBe can represent a critical target for immune-mediated virus elimination, it also becomes conceivable why HBe is no longer needed or may even be detrimental in long-term chronic carriers: once the chronic-carrier status is established, HBe would no longer be needed. Formally, it has not yet been shown directly whether an HBV preC mutant virus which cannot express HBe antigen is infectious. So far, only serum containing a heterogeneous virus population with a predominant preC mutant subpopulation was shown to cause viremia in a chimpanzee [35]. The presence of antiHBe in sera of patients does not prove the prior expression of HBe protein because degraded HBc also has HBe antigenicity. The induction of antiHBe without prior expression of HBe protein has been directly demonstrated in woodchucks infected with a cloned WHV preC mutant virus [12].

HBe-antibody-dependent cellular cytotoxicity, complement-mediated cytotoxicity, or HBe-specific cytotoxic T-cells could be responsible for elimination of HBe expressing preC wildtype viruses. Two major arguments support the concept of immune-mediated mechanisms being responsible for positive selection of preC mutant viruses. First, such variants emerge during seroconversion from HBe to antiHBe, both during the natural course of infection and, even more frequently, during or shortly after interferon treatment. Second, both preC wildtype and at least the most prevalent preC stop codon mutant appear similarly affected by cellular antiviral genes turned on by interferon both in vitro (Miska and Will, unpubl. data) and in vivo as visualized by a similar immediate fall in virus titers [18]. These observations strongly argue for the immune-stimulatory effect of interferon being the major driving force for preC mutant selection. If cytotoxic T-cells are the major driving force one can make at least two predictions: the first 10 amino acids of the 15–17 kDa HBe protein should be at least part of one or several CTL-epitopes, and these should not be exposed and recognized when expressed as a short peptide from the most prevalent preC stop codon mutant; alternatively, one must postulate the existence of HBe-specific CTL-epitopes which cannot be created from HBc, despite of the fact that the sequences of both proteins are largely overlapping (Fig. 1). Since HBc and HBe differ in conformation and enter in part different cellular compartments (secretory and non-secretory), it is conceivable that both proteins meet different proteases, are cleaved differently, and are therefore not completely cross reactive on the T-cell level. Experiments are necessary to investigate this possibility.

The possible role of HBc sequence variation

There is increasing evidence that loss of HBe expression is not the only trick the virus can use to escape immune-mediated elimination. HBcAg is thought to be one decisive factor for immune-mediated clearance of infected hepatocytes and HBc-specific cytotoxic T-cells have recently been identified [3, 19, 38]. In contrast to HBe, HBc is essential for virus viability and has to fullfill several functions: it needs to assemble into nucleocapsids, it must interact with the viral polymerase, the pregenome, and with envelope proteins. Loss of HBc expression can therefore not be tolerated and the number of HBc mutations that can be introduced without loss of functions is severely limited. In spite of all of these essential functions, there is increasing evidence for the occurrence of mutations in the HBc coding region [1, 13, 27, 32, 61] which may all be a consequence of the virus' strategy to escape immune-mediated elimination. A systematic study is needed to investigate this possibility. The number of HBc sequence variants that are viable can be tested in cell culture systems; some may turn out to be cytopathogenic and carcinogenic (as previously speculated in [39, 43]), others may lower production of coinfecting wildtype virus. The selective elimination of preC wildtype and survival of preC mutant viruses in many interferon-treated patients already indicates that HBc (which is expressed by both viruses) cannot be the major target for virus elimination in these patients. Future studies will have to show whether preC mutations often accumulate together with changes of the HBc coding sequence. The emergence of variants with mutations both in preC and preS were recently described [14, 17, 40, 47, 59].

The role of C-gene variants in viral persistence and pathogenesis

PreC variants are usually not present as a major species in the HBe positive phase of chronic hepatitis but represent a very small, hardly detectable subpopulation. As a dominant virus population they frequently emerge during or some time after seroconversion to antiHBe while concomitantly the preC wildtype virus is eliminated. This takeover of the preC mutant is most frequently, although not always, seen during or shortly after treatment of patients with interferon. These observations indicate that the presence of preC mutant viruses is one factor of HBV persistence and represents one obstacle to spontaneous or interferon-mediated HBV elimination. It becomes increasingly clear, that preC mutant virus can enter a nonproductive latent stage in the liver with hardly detectable viremia in the serum. Particularly, interferon-treated patients who seroconvert to antiHBe have frequently no or almost no detectable virus in the serum shortly after treatment. Months up to several years thereafter the virus can reactivate and high viremia and liver disease can again develop [46].

Whether preC mutant viruses are more pathogenic per se than preC wildtype virus has been speculated. This speculation derives from the fact that viremic carriers often have more severe liver disease or progress more rapidly to chronic active hepatitis. There is also indirect evidence for an association of preC mutant virus infection with fulminant hepatitis [9, 23, 24, 34]. However, in none of these studies was it excluded that another mutation in the HBV genome was present in addition to the preC mutation, and contributed to or caused the fulminant hepatitis. Furthermore, infections with other microorganisms were not ruled out. Experiments with preC mutants of animal viruses argue against a causal role of preC mutants in fulminant hepatitis. A WHV and several DHBV preC mutants produced in cell culture did infect the corresponding animals but did not lead to severe liver disease or fulminant hepatitis. Whether these data are relevant for the human system is however unclear. If relevant, we would also have to conclude that preC viruses are not more pathogenic than preC wildtype viruses. This conclusion is supported by the fact that healthy chronic carriers infected predominantly with preC mutant viruses exist. In fact, if we assume that HBe protein is a target for immune cells, preC mutants which do not express HBe should be less pathogenic. As long as the preC mutant virus replicates there will however be a chance to revert to preC wildtype virus. In this case, HBe should be reexpressed at least on the surface of the hepatocytes and these cells could be immediately attacked by previously activated HBe-specific immune cells or antibodies. This scenario would predict that it is the potential of the preC mutant to revert to an HBe-expressing preC wildtype virus which leads to the fluctuating hepatitis in viremic antiHBe positive carriers. The waxing and waning of HBe presentation on the hepatocyte surface would be the major driving force for liver injury. A direct cytopathogenic effect of preC mutant virus on hepatocytes in vivo is not likely since it has not been observed in cell culture experiments. The less-well studied HBc sequence heterogeneity is more likely to play a direct role in pathogenicity. Nonfunctional HBc proteins created by mutations may accumulate within cells and could thus become directly cytopathogenic as shown in vitro [43] or increase the pool of peptides presented by HLA class I molecules (enhanced cytotoxic T-lymphocyte liver damage) as speculated previously [1]. A final answer concerning the possible role of mutant viruses in hepatopathogenesis requires a detailed knowledge of interaction of the virus with the immune system and the study of biological characteristics of mutants in cell culture.

Perspectives

Much work remains to be done in order to understand the role of HBV sequence variation in immune escape, viral persistence and pathogenicity. The current knowledge of variants and their biological and clinical characteristics is only the beginning of a new avenue in hepadnavirus research. We consider it likely that all viral proteins are potential targets for immune elimination and if a critical number of B and/or T-cell epitopes escape recognition by sequence variation, the hepatitis B virus will have an increased chance of survival. Both, the number and type of viral epitopes exposed and the B- and T-cell repertoire of the patient will determine whether or not the virus will survive or be eliminated.

Acknowledgements

This work was supported by grants from the Deutsche Forschungsgemeinschaft (SFB217 and Wi 664/3-4) and the Wilhelm Sander Stiftung. We greatly appreciate the critical reading of the manuscript by Carol Stocking.

References

1. Ackrill AM, Naoumov NV, Eddleston ALWF, Williams R (1993) Specific deletions in the hepatitis B virus core open reading frame in patients with chronic active hepatitis B. J Med Virol (in press)
2. Akahane Y, Yamanaka T, Suzuki H, Sugai Y, Tsuda F, Yostumoto S, Omi S, Okamoto H, Miyakawa Y, Mayumi M (1990) Chronic active hepatitis with hepatitis B virus DNA and antibody against e-antigen in the serum. Disturbed synthesis and secretion of e antigen from hepatocytes due to a point mutation in the precore region. Gastroenterology 99: 1113–1119
3. Bertoletti A, Ferrari C, Fiaccadori F, Penna A, Margolskee R, Schlicht HJ, Fowler P, Guilhot S, Chisari FV (1991) HLA class I restricted human cytotoxic T cells recognize endogenously synthesized hepatitis B virus nucleocapsid antigen. Proc Natl Acad Sci USA 88: 10445–10449
4. Bhat R, Ulrich P, Vyas G (1990) Molecular characterization of a new variant of hepatitis B virus in a persistently infected homosexual man. Hepatology 11: 271–276
5. Blum HE, Galun E, Liang TJ, von Weizsäcker F, Wands J (1991) Naturally occurring missense mutation in the polymerase gene terminating hepatitis B virus replication. J Virol 65: 1836–1842
6. Brunetto MR, Stemler M, Bonino F, Schödel F, Oliveri F, Rizzetto M, Verme G, Will H (1990) Identification of HBV variants which cannot produce precore derived e-antigen and may be responsible for severe hepatitis. J Hepatol 10: 258–261
7. Brunetto MR, Stemler M, Schödel F, Will H, Ottbrelli A, Rizzetto M, Verme G, Bonino F (1989) A new hepatitis B virus strain in a subset of patients with severe chronic hepatitis. Ital J Gastroenterol 21: 151–154
8. Buchmeier MJ, Zinkernagel RM (1992) Immunodominant T cell epitope from signal sequence. Science 257: 1142
9. Carman WF, Fagan E, Hadziyannis S, Karayiannis P, Tassopoulos NC, Williams R, Thomas HC (1990) Association of a precore genomic variant of hepatitis B virus with fulminant hepatitis. Hepatology 14: 219–222
10. Carman WF, Jacyna MR, Hadziyannis S, Karayannis P, McGarvey MJ, Makris M, Thomas HC (1989) Mutations preventing formation of hepatitis B e antigen in patients with chronic hepatitis B infection. Lancet 2: 588–591
11. Chang C, Enders G, Sprengel R, Peters N, Varmus H, Ganem D (1987) Expression of the precore region of an avian hepatitis B virus is not required for viral replication. J Virol 61: 3322–3325
12. Chen HS, Kew MC, Hornbuckle WE, Tennant BC, Core PJ, Gerin JL, Purcell RH, Miller RH (1992) The precore gene of the woodchuck hepatitis virus genome is not essential for viral replication in the natural host. J Virol 66: 5682–5684

13. Ehata T, Omata M, Yokusuka O, Hosoda K, Ohto M (1992) Variations in codons 84–101 in the core nucleotide sequence correlate with hepatocellular injury in chronic hepatitis B virus infection. J Clin Invest 89: 332–338

14. Fernholz D, Galle PR, Stemler M, Brunetto M, Bonino F, Will H (1993) Infectious hepatitis B virus variant defective in pre-S2 protein expression in a chronic carrier. Virology 194: 134–148

15. Ferrari C, Penna A, Bertoletti A, Valli A, Degli-Antoni T, Giuberti A, Cavalli A, Petit MA, Fiaccadori F (1990) Cellular immune response to hepatitis B virus-encoded antigens in acute and chronic hepatitis B virus infection. J Immunol 145: 3442–3449

16. Fiordalisi G, Cariani E, Mantero G, Zanetti A, Tanzi E, Chiaramonte M, Primi D (1990) High genomic variability in the pre C region of hepatitis B virus in anti-HBe positive chronic hepatitis. J Med Virol 31: 297–300

17. Gerken G, Kremsdorf D, Capel F, Petit MA, Dauguet C, Manns MP, Meyer zum Büschenfelde K-H, Brechot C (1991) Hepatitis B defective virus with rearrangements in the pre-S gene during chronic HBV infection. Virology 183: 555–565

18. Günther S, Meisel H, Reip A, Miska S, Krüger D, Will H (1992) Frequent and rapid emergence of mutated preC sequences in HBV from e-antigen positive carriers who seroconvert to anti-HBe during interferon treatment. Virology 187: 271–279

19. Guilhot S, Fowler P, Portillo G, Margolskee RF, Ferrari C, Bertoletti A, Chisari FV (1992) Hepatitis B virus (HBV)-specific cytotoxic T-cell response in humans: production of target cells by stable expression of HBV-encoded proteins in immortalized human B-cell lines. J Virol 66: 2670–2678

20. Jung MC, Spengler U, Schraut W, Hoffman R, Zachoval R, Eisenburg J, Eichenlaub D, Riethmüller G, Paumgartner G, Ziegler-Heitbrock HWL, Will H, Pape GR (1991) Hepatitis B virus antigen specific T-cell activation in patients with acute and chronic hepatitis B. J Hepatol 13: 310–317

21. Junker-Niepmann M, Bartenschlager R, Schaller H (1990) A short cis acting sequence is required for hepatitis B virus pregenome encapsidation and sufficient for packaging of foreign RNA. EMBO J 9: 3389–3396

22. Kim SH, Hong SP, Kim SK, Lee WS, Rho HM (1992) Replication of a mutant hepatitis B virus with a fused X-C reading frame in hepatoma cells. J Gen Virol 73: 2421–2424

23. Kojima M, Shimizu M, Tsuchimochi T, Koyasu M, Tanaka S, Iizuka H, Tanaka T, Okamoto H, Tsuda F, Miyakawa Y, Mayumi M (1991) Posttransfusion fulminant hepatitis B associated with precore-defective HBV mutants. Vox Sang 60: 34–39

24. Liang TJ, Hasegawa, K, Rimon N, Wands JR, Ben Porath E (1991) A hepatitis B virus associated with an epidemic of fulminant hepatitis. N Engl J Med 324: 1705–1709

25. Manzin A, Menzo S, Bagnarelli P, Varaldo PE, Bearzi I, Carloni G, Galibert F, Clementi M (1992) Sequence analysis of the hepatitis B virus pre-C region in hepatocellular carcinoma (HCC) and nontumoral liver tissues from HCC patients. Virology 188: 890–895

26. Milich DR, Jones JE, Hughes JL, Price J, Raney AK, McLachlan A (1990) Is a function of the secreted e antigen to induce T-cell tolerance in utero? Proc Natl Acad Sci USA 87: 6599–6603

27. Miska S, Günther S, Vassilev M, Meisel H, Pape GR, Will H (1993) Heterogeneity of hepatitis B virus C-gene sequences: implications for amplification and sequencing. J Hepatol 18: 53–61

28. Miyanohara A, Imamura T, Araki M, Sugawara K, Ohtomo N, Matsubara K (1986) Expression of hepatitis B virus core antigen gene in Saccharomyces cerevisiae: synthesis of two polypeptides translated from different initiation codons. J Virol 59: 176–180

29. Mondelli MU, Chisari FV, Ferrari C (1990) The cellular immune response to nucleocapsid antigens in hepatitis B virus infection. Springer Semin Immunopathol 12: 25–31

30. Naoumov N, Schneider R, Grötzinger T, Jung MC, Miska S, Pape GR, Will H (1992) Precore mutant hepatitis B virus infection and liver disease. Gastroenterology 102: 538–543

31. Nassal M, Junker-Niepmann M, Schaller H (1990) Translational inactivation of RNA function: discrimination against a subset of genomic transcripts during HBV nucleocapsid assembly. Cell 63: 1357–1363

32. Okamoto H, Tsuda F, Mayumi M (1987) Defective mutants of hepatitis B virus in the circulation of symptom-free carriers. Jpn J Exp Med 57: 217–221

33. Okamoto H, Yotsumoto S, Akahane Y, Yamanaka T, Miyazaki Y, Sugai Y, Tsuda F, Tanaka T, Miyakawa Y, Mayumi M (1990) Hepatitis B viruses with precore defects prevail in persistently infected hosts along with seroconversion to the antibody against e antigen. J Virol 64: 1298–1303

34. Omata M, Ehata T, Yokosuka O, Hosoda K, Ohto M (1991) Mutations in the precore region of hepatitis B virus DNA in patients with fulminant and severe hepatitis. N Engl J Med 324: 1699–1704

35. Omi S, Okamoto H, Tsuda F, Mayumi M (1990) Defects in the precore region of hepatitis B virus DNA in a plasma pool from carriers seropositive for antibodies against e antigen and with infectivity in chimpanzees. J Gastoenterol Hepatol 5: 646–652

36. Ou J-H, Yeh C-T, Yen TSB (1989) Transport of hepatitis B virus precore protein into the nucleus after the cleavage of its signal sequence. J Virol 63: 5238–5243

37. Ou J-H, Bell KD (1990) Comparative studies of hepatitis B virus precore and core particles. Virology 174: 185–191

38. Penna A, Chisari FV, Beroletti A, Missale G, Fowler P, Giuberti T, Fiaccadori F, Ferrari C (1991) Cytotoxic T lymphocytes recognize an HLA-A2 restricted epitope within the hepatitis B virus nucleocapsid antigen. J Exp Med 174:1565–1570

39. Raimondo G, Burk RD, Lieberman HM, Muschel J, Hadzijannis S, Will H, Kew MC, Dusheiko GM, Shafritz DA (1988) Interrupted replication of hepatitis B virus in liver tissue of HBsAg carriers with hepatocellular carcinoma. Virology 166: 103–112

40. Raimondo G, Campo S, Smedile V, Rodino G, Sardo MA, Brancatelli S, Villari D, Pernice M, Longo G, Squadrito G (1991) Hepatitis B virus variant with a deletion in the preS2 and two translational stop codons in the precore regions in a patient with hepatocellular carcinoma. J Hepatol 13 [Suppl 4]: 74–77

41. Raimondo G, Schneider R, Stemler M, Smedile V, Rodino G, Will H (1990) A new hepatitis B virus variant in a chronic carrier with multiple episodes of viral reactivation and acute hepatitis. Virology 179: 64–68

42. Raimondo G, Stemler M, Schneider R, Wildner G, Squadrito G, Will H (1990) Latency and reactivation of a precore mutant hepatitis B virus in a chronically infected patient. J Hepatol 11: 374–380

43. Roingeard P, Romet-Lemonne JL, Leturcq D, Goudeau A, Essex M (1990) Hepatitis B virus core antigen (HBc) accumulation in an HBV nonproducer clone

of HepG2-transfected cells is associated with cytopathic effect. Virology 179: 113–120

44. Santantonio T, Jung MC, Miska S, Pastore G, Pape G, Will H (1991) Prevalence and type of pre-C mutants in anti-HBe positive carriers with chronic liver disease in a highly endemic area. Virology 183: 840–844

45. Santantonio T, Jung MC, Schneider R, Pastore G, Pape G, Will H (1991) Selection for a stop codon mutation in a hepatitis B virus variant with a pre-C initiation codon mutation during interferon treatment. J Hepatol 13: 368–371

46. Santantonio T, Jung, MC, Monno L, Milella M, Iaccovazzi T, Pastore G, Will H (1993) Long term response to interferon therapy in chronic hepatitis B: importance of hepatitis B virus heterogeneity. Arch Virol [Suppl] 8: 171–178

47. Santantonio T, Jung MC, Schneider R, Fernholz D, Milella M, Nonno L, Pastore G, Pape GR, Will H (1991) Hepatitis B virus genomes that cannot synthesize pre-S2 proteins occur frequently as dominant virus populations in chronic carriers in Italy. Virology 188: 948–952

48. Schlicht HJ, Salfeld J, Schaller H (1987) The duck hepatitis B virus pre-C region encodes a signal sequence which is essential for synthesis and secretion of processed core proteins, but not for virus formation. J Virol 61: 3701–3709

49. Schlicht HJ, von Brunn A, Theilmann L (1991) Antibodies in anti HBe positive patient sera bind to an HBe protein expressed on the cell surface of human hepatoma cells: implications for virus clearance. Hepatology 13: 57–61

50. Schneider R, Fernholz D, Wildner G, Will H (1991) Mechanism, kinetics and role of duck hepatitis B virus e antigen expression in vivo. Virology 182: 503–512

51. Seeger C, Summers J, Mason W (1991) Viral DNA Synthesis. Curr Top Microbiol Immunol 1168: 41–60

52. Shiraki K, Tanimoto K, Yamada K, Kasaogi T, Yoshihara N (1987) Is fulminant hepatitis more common among infants born to e antigen negative mothers? Hepatology 7: 974–976

53. Strapans S, Loeb DD, Ganem D (1991) Mutations affecting hepadnavirus plus strand synthesis dissociate the primer cleavage from translocation and reveal the origin of linear viral DNA. J Virol 65: 1255–1262

54. Takahashi K, Kishimoto S, Ohori K, Yoshizawa H, Machida A, Ohnuma H, Tsuda F, Munekata E, Miyakawa Y, Mayumi M (1991) Molecular heterogeneity of e antigen polypeptides in sera from carriers of hepatitis B virus. J Immunol 147: 3156–3160

55. Takeda K, Akahane Y, Suzuki H, Okamoto H, Tsuda F, Miyakawa Y, Mayumi M (1990) Defects in the precore region of the HBV genome in patients with chronic hepatitis B after sustained seroconversion from HBeAg to anti-HBe induced spontaneously or with interferon therapy. Hepatology 12: 1284–1289

56. Tong S, Diot C, Gripon P, Li J, Vitvitski L, Trépo C, Guguen-Guillouzo G (1991) In vitro replication competence of a cloned hepatitis B virus variant with a nonsense mutation in the distal pre-C region. Virology 181: 733–737

57. Tong SP, Li J, Vitvitski L, Trépo C (1990) Active hepatitis B virus replication in the presence of anti-HBe is associated with viral variants containing an inactive pre-C region. Virology 176: 596–603

58. Tong SP, Li JS, Vitvitski L, Trépo C (1992) Replication capacities of natural and artificial precore stop codon muants of hepatitis B virus: relevance of pregenome encapsidation signal. Virology 191: 237–245

59. Tran A, Kremsdorf T, Capel F, Housset C, Dauguet C, Petit MA, Brechot C (1991) Emergence and takeover by hepatitis B virus (HBV) rearrangements in the pre-S/S and pre-C/C genes during chronic HBV infection. J Virol 65: 3566–3574

60. Tsai SL, Chen PJ, Lai MY, Yang PM, Sung JL, Huang JH, Hwang TH, Chang TH, Chen DS (1992) Acute exacerbations of chronic type B hepatitis are accompanied by increased T cell responses to hepatitis B core and e antigens. J Clin Invest 89: 87–96

61. Wakita T, Kakumu S, Shibata M, Yoshioka K, Ito Y, Shinagawa T, Takayanagi M, Morishima T (1991) Detection of pre-C and core region mutants of hepatits B virus in chronic hepatitis B virus carriers. J Clin Invest 88: 1793–1801

62. Wasenauer G, Köck J, Schlicht HJ (1992) A cysteine and a hydrophobic sequence in the noncleaved portion of the preC leader peptide determine the biophysical properties of the secretory core protein (HBe protein) of human hepatitis B virus. J Virol 66: 5338–5346

63. Wasenauer G, Köck J, Schlicht HJ (1993) Relevance of cysteine residues for biosynthesis and antigenicity of human hepatitis B virus e protein. J Virol 67: 1315–1321

64. Will H, Reiser W, Weimer T, Pfaff E, Büscher M, Sprengel R, Cattaneo R, Schaller H (1987) Replication strategy of human hepatitis B virus. J Virol 61: 904–911

65. Yang SQY, Walter M, Standring DN (1992) Hepatitis B virus p25 precore protein accumulates in Xenopus oocytes as an untranslocated phosphoprotein with an uncleaved signal peptide. J Virol 66: 37–45

66. Yu M, Summers J (1991) A domain of the hepadnavirus capsid protein is specifically required for DNA maturation and virus assembly. J Virol 65: 2511–2517

Authors' address: Dr. H. Will, Heinrich-Pette-Institut für experimentelle Virologie und Immunologie, Martinistrasse 52, D-20251 Hamburg, Federal Republic of Germany.

Arch Virol (1993) [Suppl] 8: 171–178

Long-term response to interferon therapy in chronic hepatitis B: importance of hepatitis B virus heterogeneity

Teresa Santantonio[1,2], Maria-Christina Jung[1,3], Laura Monno[2], M. Milella[2], Tiziana Iacovazzi[2], G.R. Pape[3], G. Pastore[2], and H. Will[1]

[1] Heinrich-Pette-Institut für Experimentelle Virologie und Immunologie an der Universität Hamburg, Hamburg, Federal Republic of Germany
[2] Clinica della Malattie Infettive, Universita' di Bari, Bari, Italy
[3] Institut fur Immunologie, Universität München, München, Federal Republic of Germany

Summary. The long-term therapeutic efficacy of αIFN and the influence of preC variants on the type of response were evaluated in 25 patients with chronic hepatitis B, 14 HBeAg and 11 antiHBe positive patients, treated with αIFN and monitored for at least four years after discontinuing therapy. In both groups of patients, serum HBV-DNA became frequently undetectable by DNA dot blot during treatment, suggesting that αIFN has an antiviral effect both on HBeAg and antiHBe positive chronic carriers. However, long term follow up showed that the loss of viral DNA in antiHBe carriers was only transient, because all responder patients relapsed from 1 to 48 months after IFN withdrawal. In the HBeAg positive carriers, selection for preC mutants was observed at the end of follow up in 2 patients who seroconverted to antiHBe and remained viremic. Both the frequent occurrence of reactivations in antiHBe compared to HBeAg carriers, and the association of IFN therapy with preC mutant virus selection during long term post-treatment follow up observed in this study, indicate that preC variants are more resistent to IFN therapy than preC wild type HBV. Our data suggest therefore, that IFN therapy may be less frequently able to induce a permanent remission in patients infected with preC mutants.

Introduction

Alpha interferon (αIFN) can inhibit viral replication and induce a bio-chemical remission in both HBeAg and antiHBe positive carriers and, at present, is the most effective therapy for chronic hepatitis B [2, 6, 11, 12, 14, 15, 19]. Strong evidence has recently indicated that in HBeAg positive carriers the infection is generally due to a virus with a continuous preC open reading frame (designated as wildtype) which expresses HBeAg. In antiHBe chronic carriers, on the other hand, HBV preC mutants were found containing mutations in the preC region that prevent precore and HBeAg expression [3, 4, 21]. Although αIFN is partially effective both in HBeAg and antiHBe positive carriers, the response very often does not persist because of viral reactivation after therapy is discontinued, more frequently observed in antiHBe positive patients [2, 11, 15].

The objectives of this study were to evaluate the long-term therapeutic efficacy of αIFN in chronic hepatitis B and the influence of preC variants on the type of response to IFN.

Materials and methods

A total of 25 patients, 14 HBeAg and 11 antiHBe positive, with histological evidence of chronic liver disease were studied. The first group received recombinant αIFN at a dose of 4.5 MU, three times weekly for 16 weeks and to the second group, lymphoblastoid IFN was administered at a dose of 10 MU, three times weekly for six months. After discontinuing therapy, all patients were monitored for at least four years with periodic examinations, biochemical liver tests in serum and hepatitis B virus markers. Responders are defined as those patients who lost HBV-DNA (as detected by DNA dot blot) during IFN treatment.

Serological HBV markers were determined by commercial immunoenzymatic assays (Sorin Saluggia, Vc., Italy). Serum HBV-DNA was detected by spot hybridization using a biotin-labeled 3.2-Kb HBV genome [20]. Viral DNA from sera of the 25 patients was extracted and then the HBV-DNA preC sequence was amplified by polymerase chain reaction using synthetic oligonucleotide primers located upstream from the preC initiation codon (5'-GTCAACGACCGACCTTGAGGC-3', nucleotide map position 2964–2984) [16] and downstream from the C-gene stop codon (5'-CCCACCTTATGAGTCCAAGG-3', map position 575–556). Symmetric and asymmetric amplification was performed as previously described [21] for 35 cycles in a programmable DNA thermal cycler (Perkin-Elmer/Cetus, Norwalk, CT, U.S.A.). The amplified DNA was directly sequenced by the dideoxynucleotide chain termination method using a 5' 32p-labeled primer and a dideoxy sequencing kit (USB, Denver, CO, U.S.A.).

Results

Before IFN therapy, 13 out of 14 HBeAg positive patients were found to be infected with a preC wildtype virus and only one was infected with a mixture of wildtype and preC mutant viruses. At the end of therapy, seven HBeAg positive patients (50%) became HBV-DNA negative with normalized alanine aminotransferase (ALT) levels. During follow up, four of the responder patients continued to have no detectable HBV-DNA in the serum and two lost HBsAg and seroconverted to anti-HBs. The remaining three responder patients relapsed with reappearance of HBV-DNA in the serum, loss of antiHBe antibodies and an increase in ALT levels. In two cases, relapse occurred within one year of follow up (early relapse) and the sera again became HBeAg positive (data shown for one patient in Fig. 1A). In both patients a preC wildtype virus population was detected both before therapy and during reactivation. The third patient relapsed after four years of follow up (late relapse), with reappearance in serum of HBV-DNA and antiHBe, but not HBeAg (Fig. 1B). Analysis of the preC region demonstrated a preC wildtype virus population before therapy and a mixture of wildtype and preC mutant viruses during reactivation.

Of the seven HBeAg positive non-responder patients who remained HBV-DNA positive during IFN treatment, three became HBV-DNA negative during follow up, one of whom underwent a late relapse with seroconversion to antiHBe (Fig. 1C). In this patient, there was only a minor subpopulation of preC mutants before IFN therapy; however, after reactivation, a preC mutant became obviously dominant by selection.

With regard to the antiHBe-positive patients, 10 of 11 were HBV-DNA negative and had normal ALT levels at the end of therapy. During the four year post treatment follow up, all responder patients relapsed from 1 to 48 months after discontinuing IFN, with persistent or recurrent reappearance of serum HBV-DNA. Only in one patient did the HBV-DNA become undetectable after reactivation and ALT was normalized for the rest of the follow-up period. After 60 months of follow up, a loss of HBsAg was demonstrated (Fig. 1D).

Before IFN therapy, eight antiHBe positive patients were infected predominantly or exclusively with preC mutants and three patients with a mixture of wildtype and preC mutants. In 2 patients with a preC stop codon mutation before treatment, a mixture of wildtype and preC mutated sequences was found before or during reactivation. Moreover, when several consecutive samples were analyzed, as in one patient with a fluctuating course of ALT levels, a continuous fluctuation in the proportion of wildtype and preC mutants was observed; wildtype virus increased during the periods of remission, while preC mutant prevailed during reactivation.

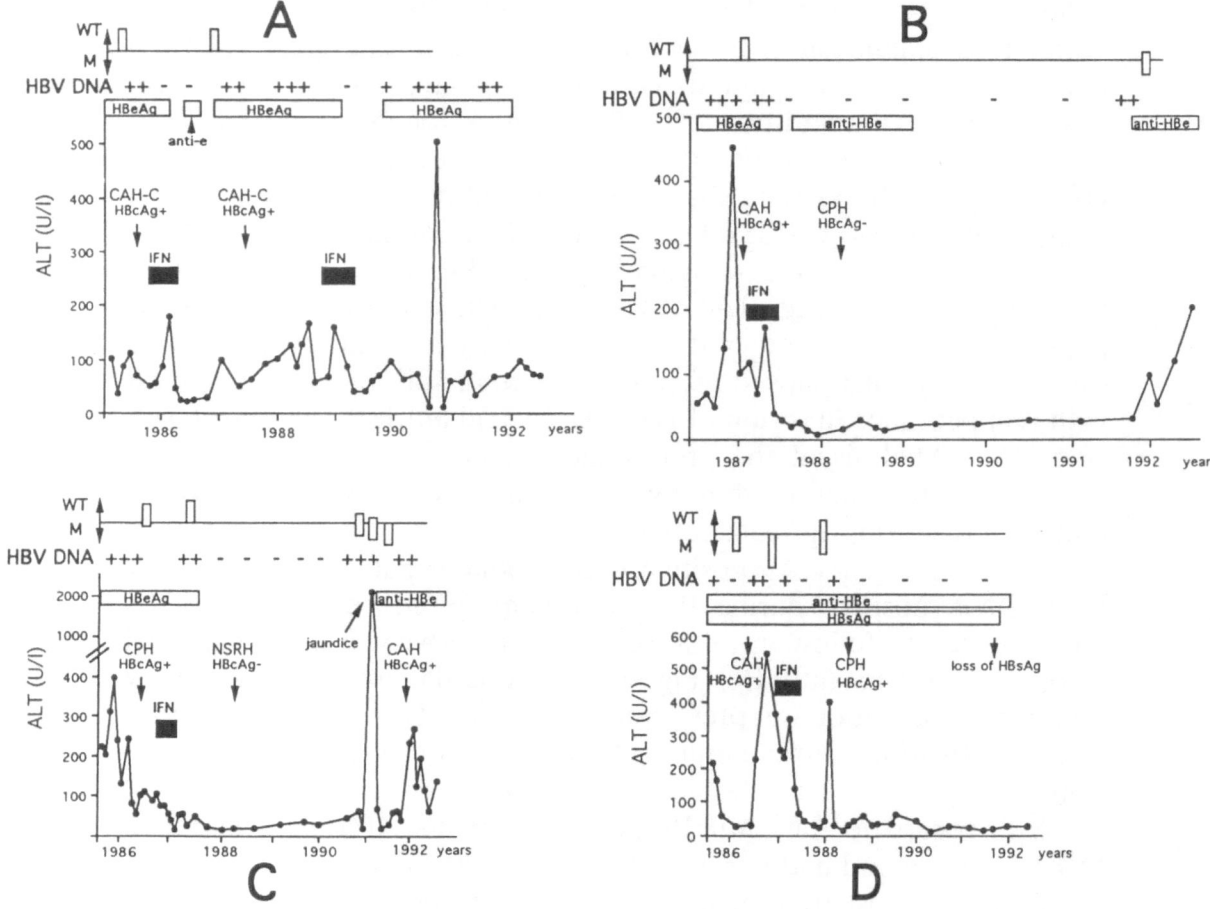

Fig. 1. Serological markers of four chronic HBV carriers treated with interferon. Top: the approximate ratio between wildtype (*WT*) and preC mutant (*M*) viruses as determined by direct sequencing is indicated. *NSRH* Non-specific reactive hepatitis, *CPH* chronic persistent hepatitis, *CAH* chronic active hepatitis, *CAH-C* chronic active hepatitis with cirrhosis. Arrows indicate time points when liver biopsies were taken and examined. **A** Reactivation of WT virus and reappearance of HBeAg in the serum in one HBeAg+ patient with early relapse. **B** Appearance of a preC mutant virus during reactivation in a HBeAg+ patient with late relapse. **C** Gradual selection for a preC HBV mutant and seroconversion to antiHBe in a HBeAg+ patient with late relapse. **D** HBV reactivation in an antiHBe+ patient who subsequently lost HBV DNA and HBsAg. The preC mutant was the dominant virus population before IFN therapy, while a mixture of WT and preC mutant viruses was found before the ALT peak

Discussion

The results presented in this study demonstrate that in HBeAg and antiHBe positive patients, HBV-DNA levels become frequently un-detectable by DNA dot blot during or shortly after IFN treatment. The

ALT levels are similarly decreased in both groups of chronic carriers. At the end of therapy, loss of viral DNA in serum is seen more often in antiHBe than in HBeAg positive carriers; whether this is statistically significant remains to be clarified. The lower levels of viraemia usually seen in antiHBe positive individuals could explain the more frequent loss of detectable amounts of HBV-DNA in this group. In fact, in these patients, IFN eventually reduces the expression of viral antigens to such low levels that immune lysis of infected hepatocytes could no longer take place. However, long term follow up demonstrated that the loss of viral DNA in antiHBe positive carriers was only transient, because reactivation occurred in all patients from 1 to 48 months after IFN withdrawal. The occurrence of late relapses both in HBeAg and antiHBe positive patients indicates that a correct evaluation of response to IFN requires a long-term follow up. However, a stable virological and biochemical remission can still occur after reactivation, as observed in one antiHBe positive patient who showed negativization of serum HBV-DNA and the clearance of HBsAg even more markedly after a single episode of reactivation. One possible explanation is that IFN administration results not only in immediate but also in late antiviral effects.

Both the more frequent occurrence of reactivations in antiHBe compared to HBeAg positive carriers, and the association of IFN treatment with preC mutant virus selection in HBeAg positive patients, suggest that preC variants are more resistent to IFN-mediated immune elimination than preC wildtype HBV. It is likely that hepatocytes infected with preC defective HBV, in virtue of their inability to synthesize and secrete HBeAg, escape immune elimination in the late phase of chronic hepatitis more easily. Thus, preC mutants appear to have a greater tendency to survive and hide in a non-productive state in the liver as previously reported [17]. It is likely that preC defective viruses occasionally revert to a preC wildtype sequence. In this case, HBeAg may transiently be expressed on liver cells which would then be immediately attacked by the previously activated HBe-specific lymphocytes and/or by antibodies, resulting in exacerbation of hepatitis. The elimination of these cells before HBeAg is secreted very efficiently, could explain why HBeAg never reappears in the serum. However, this does not account for the occasional presence of preC wildtype viruses predominantly or exclusively found in antiHBe carriers. Three explanations are possible; first, mutations elsewhere in the HBe-coding region (C-gene) interfere with HBeAg synthesis and secretion or alter antigenicity; second, HBeAg/antiHBe complex formation prevents HBeAg detection; and third, the titer of preC wildtype is too low for synthesis of amounts of HBeAg needed for detection. Evidence has been published for all three possibilities. First, major mutations within the C gene have been discovered [1, 13, 25] which could interfere with proper folding and secretion of HBeAg. Second, in all antiHBe patients, HBeAg/antiHBe complexes appear to be present [5]. Third, some preC mutants replicate less well than wildtype ([24], see also Miska and Will, this volume). In addition, mutations often coexist in the preS-region [7, 9, 21], the X-gene [8, 18] and probably also in the P-gene, and can potentially interfere with efficient replication and gene expression. These latter mutations in association with preC mutations are also likely to facilitate the escape of the virus from immune-elimination when important B and/or T-cell epitopes of the corresponding viral protein are changed.

In conclusion, the results presented in this study demonstrate that presence or selection for HBV preC mutants in patients with chronic hepatitis B is associated with a transient response to IFN therapy, and therefore with a persistent liver disease. This becomes more evident after long-term follow up.

These data are not in agreement with previous studies in which preC mutants appear to have no prognostic value for virus elimination in HBeAg positive patients [10] or are considered to be necessary for response to treatment [23]. A larger number of patients will be required to clarify the clinical relevance of preC mutants and their influence on long term response to IFN.

Acknowledgements

This work was supported by grants from the Deutsche Forschungsgemeinschaft (SFB217 and Wi664/3-4) and the Wilhelm-Sander Stiftung and from Ministero della Ricerca Scientifica e Tecnologica, Roma, Italy. We thank P. Maselli Campagna for her careful review of the manuscript.

References

1. Ackrill AM, Naoumov NV, Eddleston ALWF, Williams R (1992) Comparison of pre-core/core hepatitis B virus region in liver tissue and serum from patients with chronic hepatitis B infection. J Hepatol 16: 224–227
2. Brunetto MR, Oliveri F, Demartini A, Calvo P, Manzini P, Torrani Cerenzia M, Bonino F (1991) Treatment with interferon of chronic hepatitis B associated with antibody to hepatitis B e antigen. J Hepatol 13: S8–S11
3. Brunetto MR, Stemler M, Schödel F, Will H, Ottobrelli A, Rizzetto M, Verme G, Bonino F (1989) Identification of HBV variants which cannot produce precore derived HBeAg and may be responsible for severe hepatitis. Ital J Gastroenterol 21: 151–154
4. Carman WF, Jacyna MR, Hadziyannis S, Karayiannis P, McGarvey MJ, Makris A, Thomas HC (1989) Mutation preventing formation of hepatitis B e antigen in patients with chronic hepatitis B infection. Lancet ii: 558–591

5. Castillo I, Bartolomé J, Quiroga JA, Porres JC, Carreno V (1990) Detection of HBeAg/anti-HBe immune complexes in the reactivation of B virus replication among antiHBe chronic carriers. Liver 10: 79–84

6. Fattovich G, Farci P, Rugge M, Brollo, L, Mandas A, Pontisso P, Giustina G, Lai ME, Belussi F, Busatto G, Balestrieri A, Ruol A, Alberti A (1992) A randomized controlled trial of lymphoblastoid interferon-α in patients with chronic hepatitis B lacking HBeAg. Hepatology 15: 584–589

7. Fernholz D, Galle PR, Stemler M, Brunetto M, Bonino F, Will H (1993) Infectious hepatitis B variant defective in Pre-S2 protein expression in a chronic carrier. Virology 194: 137–148

8. Feitelson M, Duan LX, Horiike N, Clayton M (1991) Hepatitis B X open reading frame deletion mutants isolated from atypical hepatitis B virus infections. J Hepatol 13 [Suppl 4]: S58–S60

9. Gerken G, Kremsdorf D, Chapel F, Petit M, Dauget C, Manns M, Meyer zum Büschenfelde K, Brechot C (1991) Hepatitis B defective virus with rearrangements in the pre-S gene during chronic HBV infection. Virology 183: 555–565

10. Günther S, Meisel M, Reip A, Miska S, Kruger DH, Will H (1992) Frequent and rapid emergence of mutated pre-C sequences in HBV from e-antigen positive carriers who seroconvert to anti-HBe during interferon treatment. Virology 187: 271–279

11. Hadziyannis S, Bramou T, Makris A, Moussoulis G, Zignego L, Papaioannou C (1990) Interferon alfa-2b treatment of HBeAg negative/serum HBV-DNA positive chronic active hepatitis type B. J Hepatol 11: S133–S136

12. Hoofnagle JH, Peters M, Mullen KD, Jones DB, Rustgi V, Di Bisceglie A, Hallahan C, Park Y, Meschievitz C, Jones EA (1988) Randomized, controlled trial of recombinant human α-interferon in patients with chronic hepatitis B. Gastroenterology 95: 1318–1325

13. Miska S, Günther S, Vassilev M, Meisel H, Pape GR, Will H (1993) Heterogeneity of hepatitis B virus C-gene sequences: implications for amplification and sequencing. J Hepatol 18: 53–61

14. Pastore G, Santantonio T, Milella M, Monno L, Sforza E, Moschetta R, Maladorno D, Criscuolo D (1990) Changes of HBV markers in serum and liver tissue in patients with chronic hepatitis B treated with recombinant alpha-interferon (rIFN-α): results of a controlled study. Antivir Chem Chemother 1: 329–331

15. Pastore G, Santantonio T, Milella M, Monno L, Mariano N, Moschetta R, Pollice L (1992) Anti-HBe-positive chronic hepatitis B with HBV-DNA in the serum: response to a 6-month course of lymphoblastoid interferon. J Hepatol 14: 221–225

16. Pasek M, Goto T, Gilbert W, Zink B, Schaller H, MacKay P, Leadbetter G, Murray K (1979) Hepatitis B virus genes and their expression in E. coli. Nature 282: 575–579

17. Raimondo G, Stemler M, Schneider R, Wildner G, Squadrito G, Will H (1990) Latency and reactivation of a precore mutant hepatitis B virus in a chronically infected patient. J Hepatol 11: 374–380

18. Repp R, Keller C, Borkhardt A, Csecke A, Schaefer S, Gerlich WH, Lampert F (1992) Detection of a hepatitis B virus variant with a truncated X gene and enhancer II. Arch Virol 125: 299–304

19. Saracco G, Mazzella G, Rosina F, Cancellieri C, Lattore W, Raise E, Rocca G, Giorda L, Verme G, Gasbarrini G, Barbara L, Bonino F, Rizzetto M, Roda E (1989) A controlled trial of human lymphoblastoid interferon in chronic hepatitis B in Italy. Hepatology 10: 336–341

20. Santantonio T, Pontisso P, Chemello L, Milella M, Luchena N, Pastore G (1990) Detection of hepatitis B virus DNA in serum by spot hybridization technique: sensitivity and specificity of radiolabeled and biotinylated probes. Res Clin Lab 20: 29–35
21. Santantonio T, Jung MC, Miska S, Pastore G, Pape GR, Will H (1991) Prevalence and type of pre-C HBV mutants in anti-HBe positive carriers with chronic liver disease in a highly endemic area. Virology 183: 840–844
22. Santantonio T, Jung MC, Schneider R, Fernholz D, Milella M, Monno L, Pastore G, Pape GR, Will H (1992) Hepatitis B virus genomes which cannot synthesize pre-S2 proteins occur frequently and as dominant virus populations in chronic carriers in Italy. Virology 188: 948–952
23. Takeda K, Akahane Y, Suzuki H, Okamoto H, Tsuda F, Miyakawa Y, Mayumi M (1990) Defects in the precore region of the HBV genome in patients with chronic hepatitis B after sustained seroconversion from HBeAg induced spontaneously or with interferon therapy. Hepatology 12: 1284–1289
24. Tong S, Diot C, Gripon P, Li J, Vivitski L, Trepo C, Guguen-Guillouzo C (1991) In vitro replication competence of a cloned hepatitis B virus variant with a nonsense mutation in the distal pre-C region. Virology 181: 733–737
25. Wakita T, Kakumu S, Shibata M, Yoshioka K, Ito Y, Shinagawa T, Takayanagi M, Morishima T (1991) Detection of pre-C and core region mutants of hepatitis B virus in chronic hepatitis B virus carriers. J Clin Invest 88: 1793–1801

Authors' address: Dr. Teresa Santantonio, Clinica Malattie Infettive – Policlinico, Piazza G. Cesare 11, I-70124 Bari, Italy.

Arch Virol (1993) [Suppl] 8: 179–187

_Archives___
Virology
© Springer-Verlag 1993

Significance and relevance of serum preS1 antigen detection in wild-type and variant hepatitis B virus (HBV) infections

Marie-Anne Petit[1], F. Capel[1], G. Gerken[2], S. Dubanchet[1], C. Bréchot[3], and C. Trépo[4]

[1] INSERM U 131, Clamart, France
[2] Medizinische Klinik und Poliklinik der Johannes Gutenberg-Universität,
Mainz, Federal Republic of Germany
[3] INSERM U 75, CHU Necker, Paris Cedex, France
[4] INSERM U 271, Lyon, France

Summary. These studies assessed whether the serum expression of preS1 antigen could be a useful HBV marker for monitoring the progress of antiviral therapy in the treatment of chronic active hepatitis B (CAH-B) virus infections. Our findings indicate that: 1) the rearrangements we observed in the preS region of mutated HBV DNA molecules during chronic infection did not effect the preS1 sequence (21-47) critical for HBV infectivity; 2) the persistence or even the rebound of preS1 antigen expression during follow-up in responders to antiviral therapy may indicate virus persistence, suggesting the possibility of relapse through wild-type HBV or the emergence of HBV variants following the immunoelimination phase.

Introduction

The envelope of hepatitis B virus (HBV) contains polypeptide sequences coded by the preS regions of the viral genome expressing two additional preS2 and preS1 antigenic specificities (preS2Ag and preS1Ag) distinct from HBsAg. The preS2 and preS1 domains have been shown to contain viral attachment sites to human hepatocytes. HBV can bind indirectly via polymerized human serum albumin (pHSA) as an intermediate molecule through the preS2 domain, or directly to the host-cell membrane through the preS1 domain 21-47. Previously, we developed polyclonal-monoclonal radioimmunoassays (PAb-MAb RIAs) to accurately assess the preS2- and preS1-epitopes expressed on the HBV/-HBsAg particles in serum samples from HBV-infected individuals.

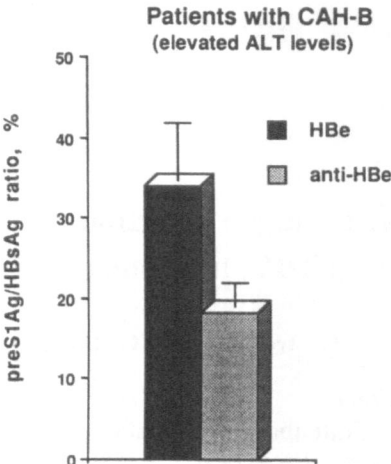

Fig. 1. PreS1 serologic status in patients with chronic active hepatitis B (CAH-B)

PreS2-specific monoclonal antibody (MAb) F124 has been shown to recognize the pHSA-binding site on the middle HBs (MHBs) proteins and on defective 22-nm HBsAg particles. The preS1-specific MAb F35.25 recognizes the hepatocyte-receptor binding site on the large HBs (LHBs) proteins and on complete virions. Antigen recognition was shown to be independent of d/y changes in HBsAg [1–3]. Using this method, the serum expression of preS1Ag correlated well with the level of HBV replication (serum HBV DNA detected by Polymerase Chain Reaction – PCR – and/or liver hepatitis B core antigen – HBcAg – detected by immunofluorescence) in patients with chronic active hepatitis B (CAH-B), especially among antiHBe carriers [4, 5] (Fig. 1).

Conservation of preS sequences critical for HBV infectivity in HBV mutants

The in vivo occurrence of genetic HBV variants has recently been reported [6]. In particular, mutations in the precore (preC) coding sequence were described in antiHBe-positive patients with persistent HBV replication and, often, severe chronic liver disease (CLD). These mutations result in the lack of hepatitis Be antigen (HBeAg) secretion. It has been hypothesized that emergence of the HBeAg-minus mutant (nucleotide 1896) might be involved in the establishment of the chronic HBV carrier state after acute infection and/or in the development of severe forms of chronic hepatitis.

In a first study, a defective form of HBV was identified in an HBsAg- and antiHBe-positive patient whose chronic HBV infection rapidly evolved to hepatocellular carcinoma and death within 5 years [7]. PCR

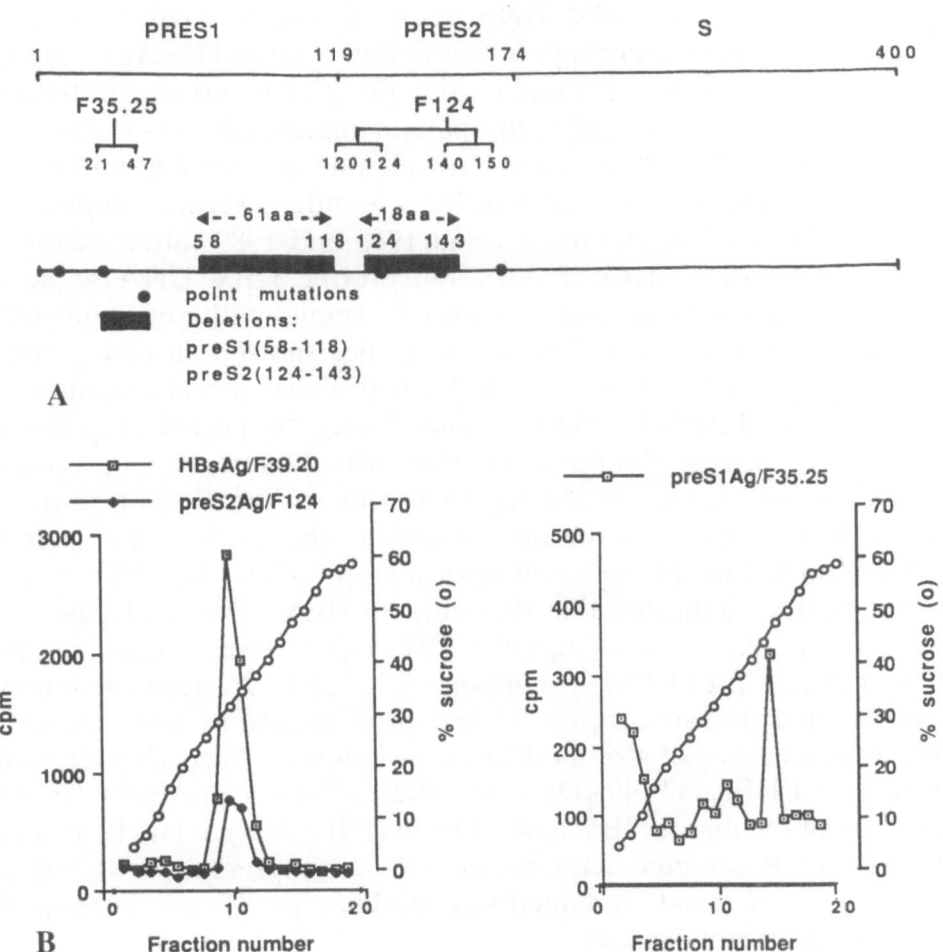

Fig. 2. Immunological analysis of a defective form of HBV in a HBsAg- and anti-HBe-positive patient with liver cancer. **A** Rearrangements in the preS1/S2 coding sequence of defective HBV after cloning and sequencing. **B** Expression of preS2(F124)- and preS1(F35.25)-epitopes on serum HBV-associated particles. The preS-deleted HBV DNA fragments were identified in fraction 9/10 by PCR analysis

analysis of HBV DNA sequences showed the presence of marked rearrangements in the preS1/S2 coding sequence detected after cloning and sequencing (Fig. 2A). However, immunological analysis of serum HBV envelope antigens by PAb-MAb RIAs revealed that the epitopes recognized by F124 in the central part of preS2 (aa 140–150) and by F35.25 in the 5′ preS1 (aa 21–47) could be detected. After an isopycnic sucrose gradient centrifugation, the preS-deleted HBV DNA fragments were identified by PCR analysis in fractions 9/10 (35–40% sucrose) where circulating HBV-associated envelope antigens activities, HBsAg-F39.20, preS2Ag-F124, and preS1Ag-F35.25, were recovered (Fig. 2B).

In a second study, HBV DNA sequences were analyzed in serial serum samples obtained during a 6-year follow-up of an HBsAg-, HBeAg-, and HBV DNA-positive chronic carrier [8]. Follow-up was initiated in 1983. Between 1985 and 1987, the patient underwent two unsuccessful therapeutic trials (a 7-week course of adenine arabinoside and a combination of acyclovir and interferon for 3 months). Serum samples from 1983 and 1985 (before treatment) and 1988 and 1989 (after treatment) were available for analysis. PreS/S and preC/C HBV DNA sequences were amplified by PCR and sequenced. Semiquantitative analysis of wild-type and mutated HBV DNA molecules showed an emergence of and takeover by HBV DNA molecules with marked rearrangements of both preC/C and preS/S coding regions during the period of follow-up. In the preC/C region, the point mutation which induces a stop codon in the preC region was identified together with an insertion of 36 nucleic acids in the core gene. In the preS/S region, nucleotide deletions in the preS1 region led to the appearance of a stop codon (Fig. 3A). Sucrose gradient analysis indicated that the mutated HBV DNA molecules were present in circulating viral particles (VPs) that were positive for both preS2-F124 and preS1-F35.25 epitopes (Fig. 3B). Western immunoblot analysis revealed that the preC/C and preS mutations were associated with the appearance of size-modified translation products (higher molecular weight LHBs, 42–49 kDa and HBe/c, 25 kDa proteins). In vitro expression of the major HBV mutant form (HBV mut) in HuH7 liver cells showed that VPs secreted in the culture medium expressed preS1(F35.25)-epitopes (Fig. 3C) and contained size-modified preS1- and core-specific proteins (results not shown).

Altogether these results indicate that the preS mutations identified in mutated HBV strains did not eliminate envelope-specific epitopes that were expressed on the preS2 and preS1 sequences and that were critical for HBV infectivity. Thus, the modified virus might potentially be infectious. Therefore, our method for detecting preS1Ag in serum can be used to follow emergence of HBV mutants in chronic HBV carriers undergoing antiviral therapy.

---→

Fig. 3. Protein analysis of the mutated HBV strain identified in an HBsAg-, HBeAg- and HBV DNA-positive chronic carrier. **A** Rearrangements in the preS/S and preC/C regions. **B** HBV envelope protein analysis of circulating VPs. The mutated HBV DNA molecules were identified in fraction 12 by PCR analysis. **C** In vitro expression of the original strain (HBV wt) and the major HBV mutant form (HBV mut): RIA analysis of HBV envelope and core proteins secreted in the culture medium

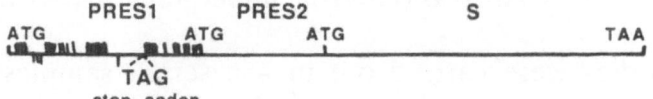

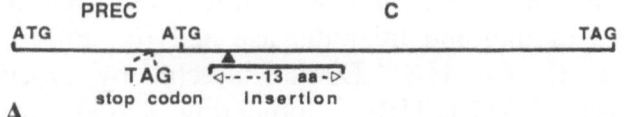

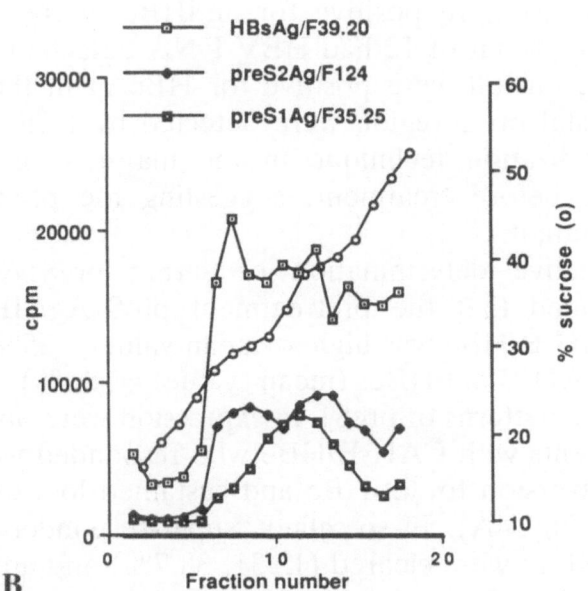

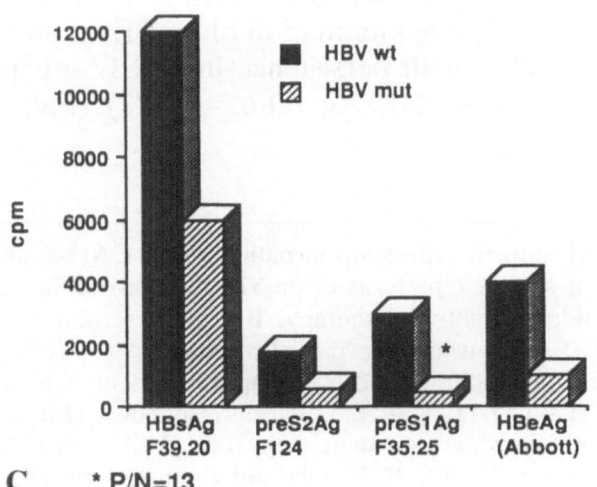

Serum preS1 antigen expression during the follow-up of patients with chronic active hepatitis B (CAH-B) undergoing antiviral therapy

Follow-up studies were carried out in 456 serum samples from 70 patients with CAH-B diagnosed on the basis of classical clinical, biochemical and histological criteria. All chronic hepatitis B patients had elevated serum levels of alanine (ALT) and aspartate (AST) aminotransferases, showing evidence of ongoing liver disease activity. Fifty eight patients were positive for HBeAg, HBV DNA (detected by hybridization) and DNA polymerase (CAH-B/HBe), indicating active viral replication before antiviral therapy. Twelve patients (with histologically confirmed cirrhosis in 9 of 12) were positive for antiHBe in the serum (CAH-B/anti-HBe). Only 4 out of 12 had HBV DNA detectable in the serum by hybridization, but all were positive for HBcAg in the liver. Mutation(s) in the distal preC region were detected by PCR and a specific oiigoprobe hybridization technique in the majority of patients with CAH-B/antiHBe before treatment, suggesting the preexistence of a HBeAg-minus mutant.

The retrospective determination of serum preS1Ag throughout follow-up confirmed that the pretreatment preS1Ag/HBsAg ratio of patients with CAH-B/HBe was higher (mean value = 25%) than that of patients with CAH-B/antiHBe (mean value = 14%). Interestingly, different serologic patterns of preS1Ag expression were observed at final follow-up in patients with CAH-B/HBe who responded well to antiviral therapy (seroconversion to antiHBe and sustained loss of HBV DNA) (31/58, 53.45%, Fig. 4A). In so-called "super-responders" (R+), both HBsAg and preS1Ag were cleared (12/31, 38.7%) and antiHBs production was observed in 50% of these cases (6/12). The remaining good responders (19/31, 61.3%) were referred to as partial responders (PR) because they exhibited persistence of HBsAg associated in rare cases (3/19, 15.8%) with complete elimination of preS1Ag, and in the majority of cases (16/19, 84.2%) with persistence in preS1Ag expression at final follow-up (mean preS1Ag/HBsAg ratio = 16%) (Fig. 4A). Transient

---

Fig. 4. Serum PreS1 antigen expression in patients with CAH-B undergoing antiviral therapy. **A** Different serologic patterns of preS1Ag expression in patients with CAH-B/HBe who responded to antiviral therapy. **B** Typical serologic preS1Ag profile in patients with CAH-B/HBe who were partial-responders. **C** PreS1 status (preS1Ag/-HBsAg ratio) before and following IFN therapy in patients with CAH-B/anti-HBe. *Super-responder (R+) (1/12, 8.3%). **Partial-responders (PR) with a rebound of preS1Ag expression following treatment with IFN (7/12, 58%). The bulk of these patients (11/12, 91.7%) did not eliminate preS1Ag

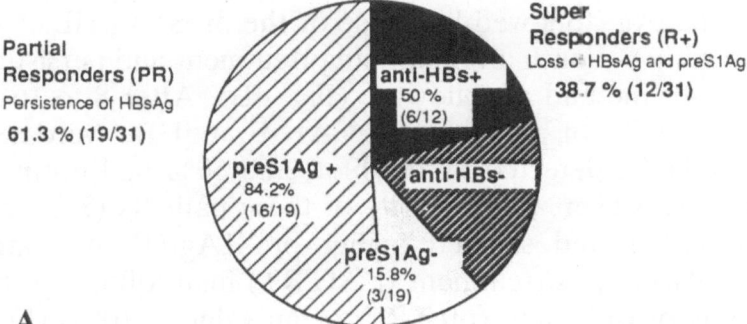

RESPONDERS (R)
Loss of HBeAg and HBV DNA
53.45 % (31/58)

Partial
Responders (PR)
Persistence of HBsAg
61.3 % (19/31)

Super
Responders (R+)
Loss of HBsAg and preS1Ag
38.7 % (12/31)

anti-HBs+
50 %
(6/12)

anti-HBs-

preS1Ag +
84.2%
(16/19)

preS1Ag-
15.8%
(3/19)

A

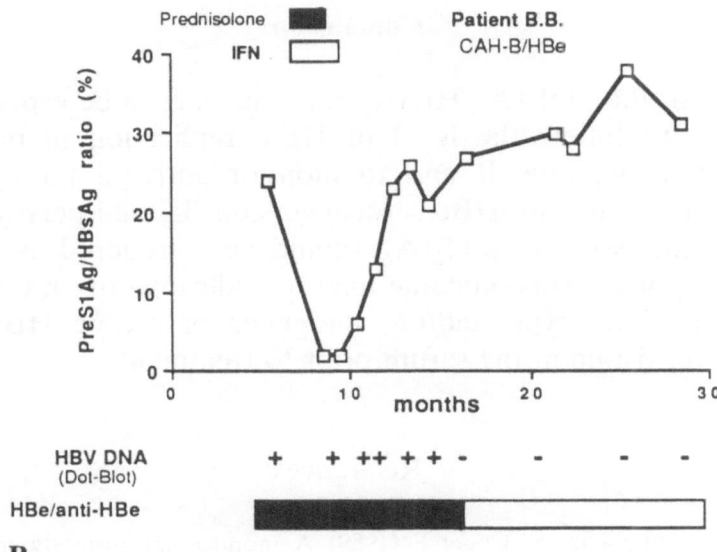

Prednisolone
IFN

Patient B.B.
CAH-B/HBe

PreS1Ag/HBsAg ratio (%)

months

HBV DNA
(Dot-Blot)   +   + ++ + -   -   -   -

HBe/anti-HBe

B

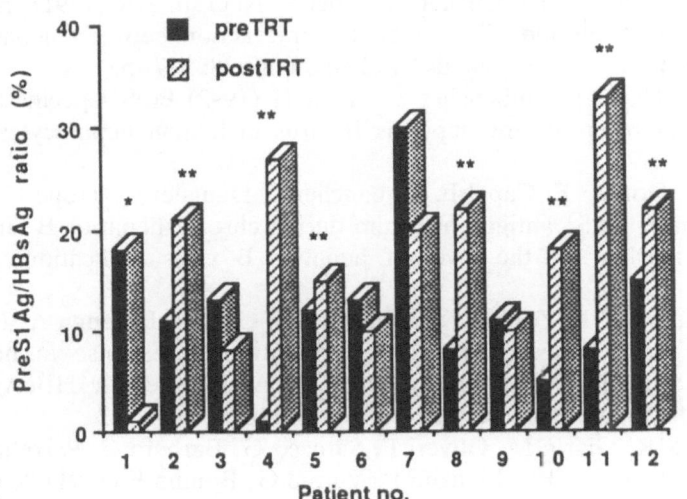

preTRT
postTRT

PreS1Ag/HBsAg ratio (%)

Patient no.

C

clearance of preS1Ag (preS1Ag/HBsAg ratio < 5%) could be observed at the end of antiviral therapy preceding the disappearance of HBV DNA and HBeAg, followed by a rise of the preS1Ag/HBsAg ratio (up to 30%) starting 1-month after stopping treatment and persisting or even increasing until the end of follow-up (Fig. 4B). After 8 to 16 months of IFN therapy, 100% of patients with CAH-B/antiHBe were negative for serum HBV DNA detected by dot-blot, and 50% had normal levels of serum ALT. However, the majority of these patients (7/12, 58%) were partial-responders and showed higher preS1Ag/HBsAg ratios (mean value = 22.6%) at posttreatment (postTRT) final follow-up than before the beginning of treatment (preTRT, mean value < 10%) (Fig. 4C).

Conclusion

Monitoring of the preS1Ag/HBsAg ratio appears to be especially useful as a means to follow the level of HBV replication in patients with chronic active hepatitis B and to monitor antiviral therapy in good responders following antiHBe seroconversion. Indeed persistent expression or re-expression of preS1Ag should be considered as a partial or transient response. This outcome may be indicative of viral reactivation through HBV wild-type and/or emergence of escape HBV mutants, probably preexisting in the serum prior to treatment.

References

1. Petit MA, Dubanchet S, Capel F (1989) A monoclonal antibody specific for the hepatocyte receptor binding site on hepatitis B virus. Mol Immunol 26: 531–537
2. Petit MA, Strick N, Dubanchet S, Capel F, Neurath AR (1991) Inhibitory activity of monoclonal antibody F35.25 on the interaction between hepatocytes (HepG2 cells) and preS1-specific ligands. Mol Immunol 28: 517–521
3. Petit MA, Capel F, Dubanchet S, Mabit H (1992) PreS1-specific binding proteins as potential receptors for hepatitis B virus in human hepatocytes. Virology 187: 211–222
4. Petit MA, Zoulim F, Capel F, Dubanchet S, Dauguet C, Trepo C (1990) Variable expression of preS1 antigen in serum during chronic hepatitis B virus infection: an accurate marker for the level of hepatitis B virus replication. Hepatology 11: 809–814
5. Petit MA, Capel F, Zoulim F, Dubanchet S, Chemin I, Penna A, Ferrari C, Trépo C (1992) PreS antigen expression and anti-preS response in hepatitis B virus infections: relationship to serum HBV DNA, intrahepatic HBcAg, liver damage and specific T-cell response. Arch Virol [Suppl] 4: 105–112
6. Brunetto MR, Giarin M, Oliveri F, Saracco G, Barbera C, Parrella T, Abate ML, Chiaberge L, Calvo PL, Manzini P, Verme G, Bonino F (1991) "e" Antigen defective hepatitis B virus and course of chronic infection. J Hepatol [Suppl] 13: S82–S86

7. Gerken G, Kremsdorf D, Capel F, Petit MA, Dauguet C, Manns MP, Meyer Zum Büschenfelde KH, Bréchot C (1991) Hepatitis B defective virus with rearrangements in the preS gene during chronic HBV infection. Virology 183: 555–565
8. Tran A, Kremsdorf D, Capel F, Housset C, Dauguet C, Petit MA, Bréchot C (1991) Emergence of and takeover by hepatitis B virus (HBV) with rearrangements in the preS/S and preC/C genes during chronic HBV infection. J Virol 65: 3566–3574

Authors' address: Dr. Marie-Anne Petit, INSERM Unité 131, Immunopathology and Viral Immunology, 32 rue des Carnets, F-92140 Clamart, France.

Arch Virol (1993) [Suppl] 8: 189–199

Complete nucleotide sequences of six hepatitis B viral genomes encoding the surface antigen subtypes *ayw4*, *adw4q⁻*, and *adrq⁻* and their phylogenetic classification

Heléne Norder[1], **Anne-Marie Couroucé**[2], and **L.O. Magnius**[1]

[1] Department of Virology, The National Bacteriological Laboratory,
Stockholm, Sweden
[2] Institute National Transfusion Sanguine, Paris, France

Summary. The complete nucleotide sequences of six hepatitis B viral (HBV) genomes were determined by dideoxy chain termination sequencing of ten overlapping nucleotide fragments obtained by the polymerase chain reaction. Four of the genomes belonged to the two genomic groups E and F of HBV which have been previously identified by us on the basis of sequence divergences within the S gene. Genomic group E encodes the HBsAg subtype *ayw4*, group F adw4q⁻. The other two genomes were of Pacific origin within group C and encoded *adrq⁻*. The relationship of these complete human HBV genomes to 21 that have been previously published, together with one chimpanzee virus and four rodent hepadnaviral genomes, was investigated by constructing a phylogenetic tree utilizing a combination of distance matrix and approximate parsimonious methods. Thereby, the previously demonstrated segregation of human HBV strains into six genomic groups was confirmed. Both of the representatives of the groups E and F were found to differ by 8.1–13.6% and by 12.8–15.5% from the genomes of the other genomic groups and by 1.5 and 3.7% from each other. Since they differed by more than 8% from the genomes in the other groups, the limit originally used to define HBV genomic groups, their status as new genomic groups was confirmed. The two Pacific group C strains were found to differ by 2.7% from each other and by 4.1 to 5.4% from other group C genomes, suggesting that they diverged early from the other group C genomes. According to both the overall similarity and the phylogenetic dendrogram the F strains formed the most divergent cluster of HBV genomes favoring the concept that they represented the original HBV strains of the New World. The next split in the dendrogram segregated the A, D, E and the chimpanzee strains from the Asian B and C strains. Information on the nucleotide sequences and their encoded products of HBV strains of different genomic groups will provide a basis to understand biological variations of the HBV infection in different parts of the world.

Introduction

The human hepatitis B virus (HBV) is the first representative to be described, of a group of viruses that have only recently received the status of a new family, the *hepadnaviridae*. This family also includes three rodent viruses, woodchuck hepatitis B virus [35], groundsquirrel hepatitis virus [15] and tree squirrel hepatitis virus [7], as well as two avian members, duck hepatitis B virus [16] and heron hepatitis B virus [33]. Roughly speaking, the genetic relatedness of these viruses parallells that of the hosts. Thus, in a phylogenetic tree the mammalian viruses were separated from the avian ones and among the mammalian viruses those that infect rodents formed a separate cluster [25]. Whether two non-human primate HBV strains, one chimpanzee strain [39] and one gibbon strain [17], represent original non-human members of this family indigenous to these species, or rather are more recent acquisitions from man, has not been definitely settled.

Human HBV genomes have been shown to be classifiable into at least six genomic groups, A to F [19, 23]. The geographical distribution of these groups has been shown in a study on 122 HBV genomes derived worldwide [20]. Group A genomes mainly originated from Northern Europe and from Sub-Saharan Africa. The group B and C genomes were confined to original populations of South-East Asia and the Far East. The group D genomes were found worldwide, but were the predominating strains in the Mediterranean Area, the Near and Middle East, and in South Asia. Group E genomes were indigenous to the Western part of Sub-Saharan Africa as far South as Angola. There were indications that the F group represented the genomic group of the original populations of the New World.

The occurrence of nine different subtypes of HBsAg reflecting genetic variations of HBV has long been documented. These are *ayw1*, *ayw2*, *ayw3*, *ayw4*, *ayr*, *adw2*, *adw4*, *adrq*$^+$ and *adrq*$^-$ [4, 5]. The relation of these subtypes to the genetic classification was recently agreed upon [19]. Thus, genomes specifying *adw2* were found in groups A, B and C. African HBV strains specifying *ayw1* were allocated to group A, while their Vietnamese counterparts were placed in group B [19]. The strains specifying *ayw2* and *ayw3* were both allocated to group

D. It was also shown that the strains specifying *ayw4* and *adw4* differed to an extent from each other and from other groups suggesting their classification into two new genomic groups, E and F [19]. Strains expressing r have so far only been found in group C [19, 23, 30].

In the present study we have sequenced the complete genomes of representatives of the two new genomic groups, E and F. These new groups encode the HBsAg subtypes *ayw4* and *adw4q⁻*, from which sequence data on the complete genomes have previously not been available. We have also sequenced two complete genomes within group C of Pacific origin encoding the subtype *adrq⁻*.

Materials and methods

Materials

Serum samples from six HBV carriers of known geographic origin were used as sources of HBV DNA for sequencing. Two serum samples, Bas and Kou, contained HBV genomes representing the group E specifying subtype *ayw4*. These derived from a French carrier with West-African contact and a Senegalese carrier, respectively. Two other serum samples, Fou and 9203/85, derived from a French and a Colombian carrier, respectively, contained HBV genomes representing group F, both specifying subtype *adw4*. The remaining two genomes within group C specifying *adrq−*, HMA and Cha, derived from New Caledonia and French Polynesia, respectively. The S gene sequences of five of these six HBV genomes have previously been published [19].

Methods

Amplification by the polymerase chain reaction (PCR) and sequencing was performed as described previously [18–19]. The HBV genomes were amplified in ten overlapping PCR fragments with the following nucleotide positions: 3198–486, 323–748, 636–986, 636–1285, 1175–1797, 1430–1890, 1778–2284, 1778–2485, 2379–3098 and 2816–205 according to the genome published by Valenzuela et al. [38] with the enumeration of Okamoto et al. [21]. The oligonucleotides and excess of dNTPs were removed from the amplified product by using Magic PCR Preps (Promega Corporation, Madison, WI, U.S.A.). The products from at least one run of amplification with each of these primer pairs were used as template in the sequencing reaction. This was performed with the dideoxychain termination method. Each PCR product was sequenced in both directions using the primers used in the PCR as sequencing primers.

Construction of the phylogenetic tree

Parsimony tree reconstruction and pairwise alignment of the sequences were performed according to Hein [11] with the computer program Tree Align. This program for multiple sequence alignment of DNA or protein sequences uses a combination of distance matrix and approximate parsimony methods. A special computer program was written for calculating the inter- and intragroup differences of the 28 HBV genomes.

Results and discussion

The complete nucleotide sequences of the six HBV genomes that we investigated are shown in Fig. 1, where they are aligned with the HBV sequence of Valenzuela et al. [38]. The length of the group C and F genomes was 3,215 nucleotides, which is the same length as the HBV genomes belonging to genomic groups B and C, with the exception of pHBr330, which has a 27 nucleotide-long deletion in the X gene [24]. The length of the group E genomes was 3,212 nucleotides due to a deletion in the pre-S1 region.

The overall genetic relatedness of the six genomes that were sequenced here and 22 previously sequenced HBV genomes, obtained by pairwise comparisons, are shown in Table 1. Each of the two representatives of the genomic groups designated E and F, were found to differ by 8.1 to 13.6% and by 12.8 to 15.5% from the genomes of the other groups, and by 1.5 and 3.7% from each other, respectively. Since they differed by more than 8% from the genomes of the other groups, the figure originally used to define the genomic groups [23], their status as new genomic groups were confirmed. The two Pacific strains within group C were found to differ by 2.7% from each other and by 4.1 to 5.4% from other group C genomes.

The results of the phylogenetic analysis are shown in Fig. 2. The dendrogram obtained from results of the six HBV genomes sequenced here and the 22 previously sequenced HBV clones confirmed the previously observed segregation of human hepadnaviral genomes into six genetic groups designated with A to F. The two strains within genomic group F formed the most divergent group of HBV genomes supporting the concept that this group might represent the original HBV strains of the New World. The second split in the tree segregated the genomes into one branch with the A, D, E and the chimpanzee strains, and another branch with the Asian B and C strains. The F strains not only represented the first split from the human hepadnaviral ancestor but also the branch that showed the lowest number of substitutions compared to the common ancestor of the primate hepadnaviral genomes.

If the evolution of HBV would parallel that of man, and considering man's rather recent migration to the Americas, at least from the standpoint of human evolution, it is not understood why the first split in the phylogenetic dendrogram segregates the New World HBV strains from those of the Old World, rather than the African from the non-African strains. The latter alternative would better agree with the evolution of the host as implicated by studies on nuclear as well as on mitochondrial DNA [2]. The New/Old World split of HBV genomes might be explained by either more divergent strains in the Old World being replaced by more advanced strains with a higher ability to spread, or

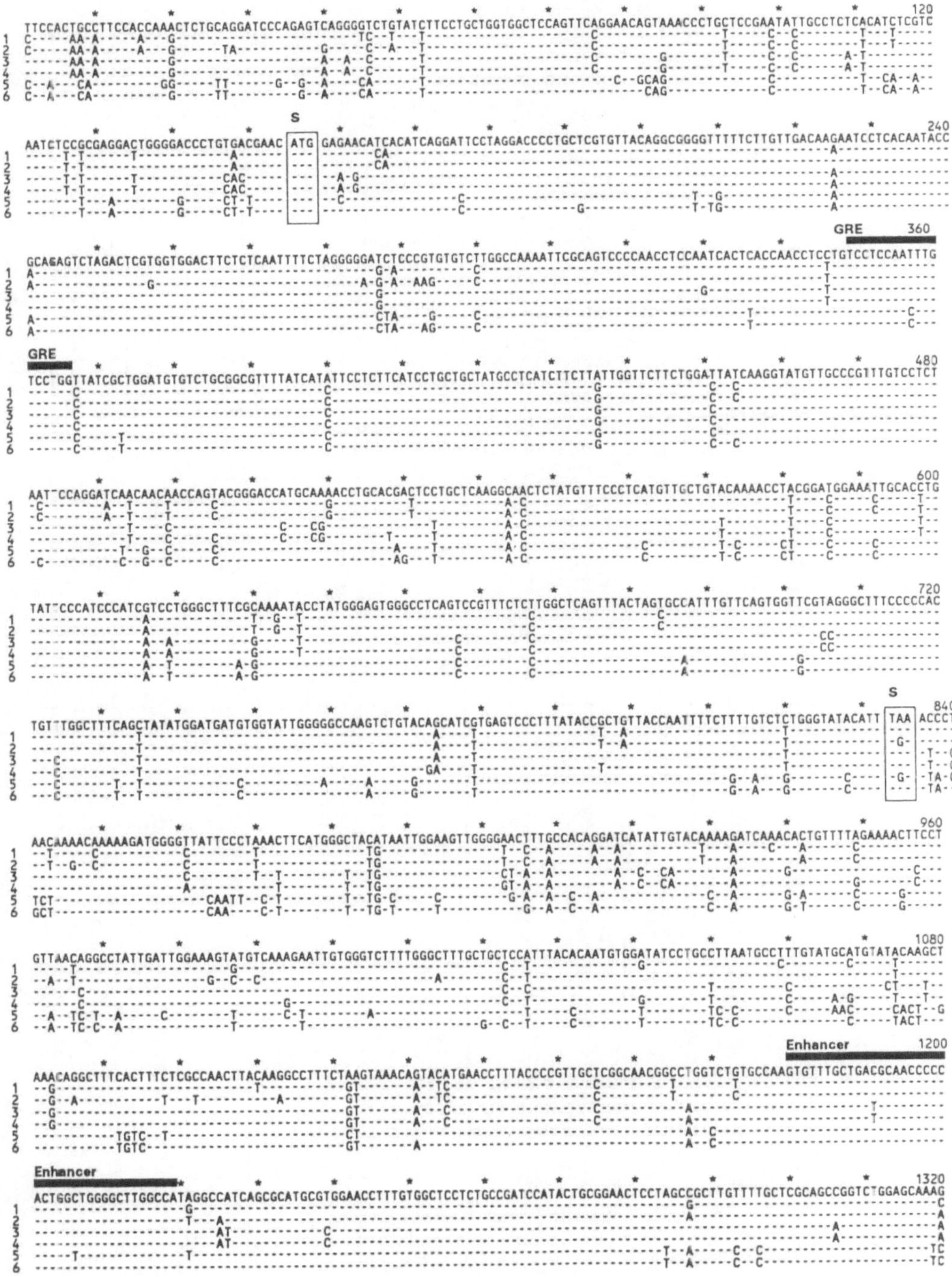

Fig. 1 (continued)

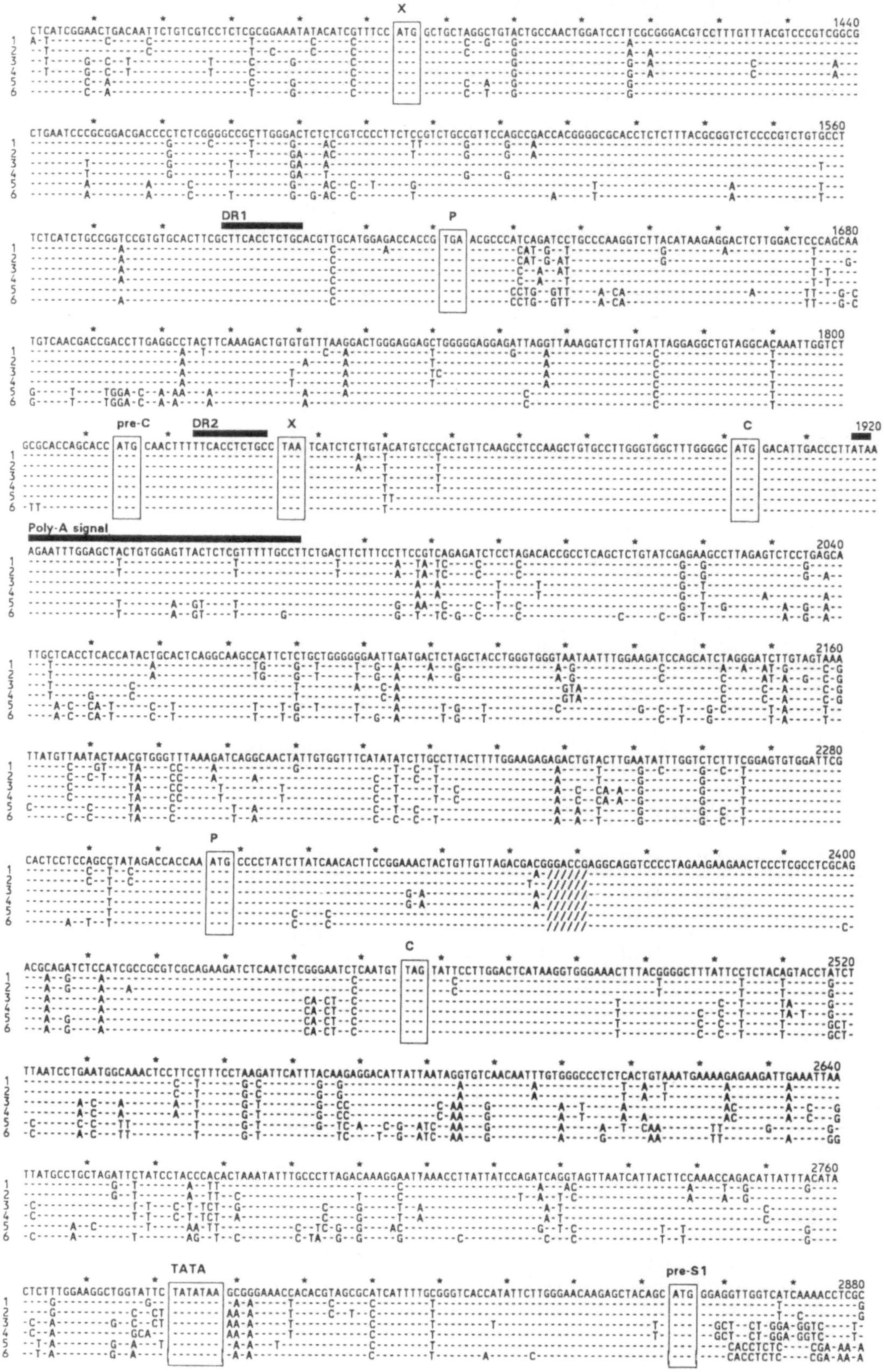
X
1440
1560
DR1
P
1680
1800
pre-C
DR2
X
C
1920
Poly-A signal
2040
2160
2280
P
2400
C
2520
2640
2760
TATA
pre-S1
2880

```
                                   *         *         *         *         *         *         *         *         *         *       3000
     AAACGCATGGGGACGAATCTTTCTGTTCCCAATCCTCTGGGATTCTTTCCCGATCATCAGTTGGACCCTGCATTCGGAGCCAACTCAAACAATCCAGATTGGGACTTCAACCCCGTCAAG
   1 ------------------------G--CACA---------------------------C------------G-------------------------C-------------------AA----
   2 -----------------------------------------------------------C-----------------------------------------------------------AA----
   3 G--///TG-----A----A---CACCA------------------T--------C--C--------T--A------A---A-A-C-GA---------------CA---T---AA----A
   4 G--///TG-----A----A---CACCA------------------T--------C--C--------T--A------A---A-A-C-GA---------------CA---T---AA----A
   5 -GG---------ACA----C----G-------------------A--C------AC--C---T--CTT------A---A-T--C-G--G---C-----------------AAAAA----
   6 -GG---------ACA---------G-------A------C----G-A--C-------C--T--GCT------A---A-T--C-G--G--C--C----------------A-AAA----

                                   *         *         *         *         *         *         *         *         *         *       3120
     GACGACTGGCCAGCAGCCAACCAAGTAGGAGTGGGAGCATTCGGGCCAAGGCTCACCCCTCCACACGGCGGTATTTTGGGGTGGAGCⁱCTCAGGCTCAGGGCATATTGACCACAGTGTCA
   1 -TC-------AG--A-T--G----C----------------G-T------A-----A--C-------------------------C---------------A------C--
   2 -TC-A-------AG--A-T--G----------------------G-T------A-----------------------------A------A----C--
   3 --C----A----A------A-G-----------------TG-T---T--C------A--CC----------------------A------GC-A-AA----T--C--
   4 --C---A-G-A------A-G-------------C------GG--T---T-C------A--CC---------------------A------GC-A-AA----T--C--
   5 -ACT-------ATG--A--A-G---------------GT-A--T--G-T---A--C----------------------A------TG-TC-A--A--CT--C--
   6 -AGT-------ATG--A--A-G----------------GC-A--T--G-T---A--C-------------------T--CC-GC------A------TG-T--A--A--CT--C--

                                   *         *         *         *         *         *         *         *   pre-S2        3221
     ACAATTCCTCCTCCTGCCTCCACCAATCGGCAGTCAGGAAGGCAGCCTACTCCCATCTCTCCACCTCTAAGAGACAGTCATCCTCAGGCC│ATG│CAGTGGAA
   1 G--GC-------T--------------------------A----------G----------------------------│---│------
   2 G--GC-------T--------------------------A----------G----------------------------│---│------
   3 G--GA---G------------------------------------C-A--A----------T-G---------C------│---│------
   4 G--GAC--G------------------------------------C-A--A----------T-G---------C------│---│------
   5 G--GA------G------------------T--C--G--A----A--C-AG------------------CA----A----A│---│------
   6 G--GA---G-------------------G------T--------T--C-G---A----A--C-AAG-------------CA----│---│------
```

Fig. 1. Nucleotide sequences of six HBV genomes: *1* HMA, *2* Cha, *3* Bas, *4* Kou, *5* Fou, and *6* 9203/85, aligned with the clone pHBV-3200 [38]. Identical nucleotides are marked with a line and deletions are indicated by slashes. Important genetic regulatory sites as the start and stop codons for translation, the TATA box, the glucocorticoid receptor binding site, the enhancer, the polyadenylation site, DR1 and DR2 are marked

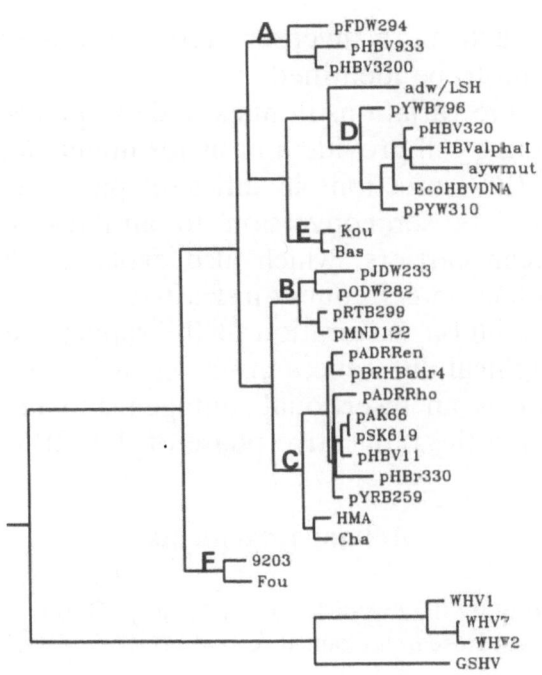

Fig. 2. Dendrogram based on the complete sequence of 28 human HBV, one ground-squirrel virus and three woodchuck hepatitis virus genomes. Sequences of the previously published clones used in the comparison are: Chimpanzee adw/LSH [39]; pYWB796, pAK66, pPYW310, pJDW233, pODW282 [23]; pHBV320 [1]; HBValphal [37]; aywmut [14]; EcoHBVDNA [9]; pHBV1-1 [12]; pFDW294 [6]; pHBV933, pHBr330 [24]; pHBV3200 [38]; pRTB299 [28]; pMND122 [29]; adrRen [26]; pBRHBadr4 [8]; adrm [27]; pSK619 [22]; pYRB259 [21]; GSHV pBA131 [31]; WHV1 [10]; WHV7 [3]; WHV2 [13]

Table 1. Mean number and percentage differences in nucleotide sequences of the complete genome of 21 HBV clones within groups A to D, six HBV strains within groups C, E and F and a chimpanzee strain, clone adw/LSH [39]

Group	Nucleotide differences							
	group							
Group	A	B	C	adrq−	D	E	F	adw/LSH
A	4.0%	9.5%	9.4%	9.1%	10.8%	10.2%	13.9%	10.6%
B	307	4.2%	8.8%	8.8%	11.2%	11.1%	13.7%	11.0%
C	303	283	2.5%	4.6%	11.1%	10.6%	13.8%	10.9%
adrq−	293	284	149	2.7%	11.0%	9.7%	13.0%	10.5%
D	347	362	358	355	4.2%	8.8%	14.5%	9.8%
E	329	358	341	312	285	1.5%	13.4%	10.2%
F	447	442	443	420	467	432	3.7%	13.3%
adw/LSH	341	354	351	337	315	330	430	−

alternatively, that such more divergent HBV strains of the Old World still exist and remain to be identified.

Information on the variations in nucleotide sequence between strains of the genomic groups will provide a basis for understanding the biological variations of HBV infections in different parts of the world. Such variations are the late seroconversion to antiHBe in East-Asian as compared to African carriers, which also explains why vertical transmission of HBV infection is frequent in East-Asia [34] but uncommon in Africa [36]. The variation in duration of the replicative state might also explain the geographical differences in oncogenicity of HBV, since HBV is presumed to act as an insertional mutagen and integration is more likely to occur during the replicative phase of the HBV infection [32].

Acknowledgements

This work was supported by the Swedish Cancer Society, Grant no. 3312-B92-01X-AA and the Swedish Medical Research Council, Grant no. B93-16X-10385-01.

References

1. Bichko V, Pushko P, Dreilina D, Pumpen P, Grens E (1985) Subtype ayw variant of hepatitis B virus DNA primary structure analysis. FEBS Lett 185: 208−212
2. Cann RL, Stoneking M, Wilson AC (1987) Mitochondrial DNA and human evolution. Nature 1: 31−36

3. Cohen JI, Miller RH, Rosenblum B, Denniston K, Gerin JL, Purcell RH (1988) Sequence comparison of woodchuck hepatitis replicative forms shows conservation of the genome. Virology 162: 12–20

4. Courouce A-M, Holland PC, Muller JY, Soulier JP (1976) HBs antigen subtypes. Bibl Hematol 42: 1

5. Courouce-Pauty A-M, Lemaire JM, Roux JF (1978) New hepatitis B surface antigen subtypes inside the ad category. Vox Sang 35: 304–308

6. Estacio RC, Chavez CC, Okamoto H, Lingao AL, Reyes MT, Domingo E, Mayumi M (1988) Nucleotide sequence of a hepatitis B virus genome of subtype adw isolated from a Philippino: comparison with the reported three genomes of the same subtype. J Gastroenterol Hepatol 3: 215–222

7. Feitelson MA, Millman I, Halbherr T, Simmons H, Blumberg S (1986) A new identified hepatitis B type virus in tree squirrels. Proc Natl Acad Sci USA 83: 2233–2237

8. Fujiyama A, Miyanohara A, Nozaki C, Yoneyama T, Ohtomo N, Matsubara K (1983) Cloning and structural analyses of hepatitis B virus DNAs subtype adr. Nucleic Acids Res 11: 4601–4610

9. Galibert F, Mandart E, Fitoussi F, Tiollais P, Charnay P (1979) Nucleotide sequence of the hepatitis B virus genome (subtype ayw) cloned in *E. coli*. Nature 281: 646–650

10. Galibert F, Chen TN, Mandart E (1982) Nucleotide sequence of a cloned woodchuck hepatitis virus genome: comparison with the hepatitis B virus sequence. J Virol 41: 51–65

11. Hein J (1990) Unified approach to alignment and phylogenies. Methods Enzymol 183: 626–645

12. Kobayashi M, Koike K (1984) complete nucleotide sequence of hepatitis B virus DNA of subtype adr and its conserved gene organization. Gene 30: 227–232

13. Kodama K, Ogasawara N, Yoshikawa H, Murakami S (1985) Nucleotide sequence of a cloned woodchuck hepatitis virus genome: evolutional relationship between hepadnaviruses. J Virol 56: 978–986

14. Lai ME, Melis A, Mazzoleni AP, Uccheddu P, Balestrieri A (1991) Sequence analysis of Hepatitis B virus genome of a new mutant of ayw subtype isolated in Sardinia. Nucleic Acids Res 19: 5078–5078

15. Marion PL, Knight SS, Salazar FH, Popper H, Robinson WS (1980) A virus in Beechey ground squirrels that is related to hepatitis B virus of humans. Proc Natl Acad Sci USA 77: 2941–2945

16. Mason WS, Seal G, Summers J (1980) Virus of Pekin ducks with structural and biological relatedness to human hepatitis B virus. J Virol 36: 829–836

17. Mimms L, Floreani M, Fields H, Decker R (1991) Differentiation of hepatitis B virus subtypes from humans and a gibbon (Hylobates lar) using monoclonal antibodies against the preS domain. In: Hollinger BF, Lemon SM, Margolis HS (eds) Viral hepatitis and liver disease. Williams and Wilkins, Baltimore, pp 202–207

18. Norder H, Hammas B, Magnius LO (1990) Typing of hepatitis B virus genomes by a simplified polymerase chain reaction. J Med Virol 31: 215–221

19. Norder H, Hammas B, Löfdahl S, Courouce A-M, Magnius LO (1992) Comparison of the aminoacid sequence of nine different serotypes of HBsAg and genomic classification of the corresponding HBV strains. J Gen Virol 73: 1201–1208

20. Norder H, Hammas B, Lee S-D, Bile K, Courouce A-M, Mushahwar I, Magnius LO (1993) Genetic relatedness of hepatitis B viral strains of diverse geographic origin and natural variations in the primary structure of the surface antigen. J Gen Virol 74: 1341–1348

21. Okamoto H, Imai M, Shimozaki M, Hoshi Y, Iizuka H, Gotanda T, Tsuda F, Miyakawa Y, Mayumi M (1986) Nucleotide sequence of a cloned hepatitis B virus genome subtype ayr: comparison with genomes of the other three subtypes. J Gen Virol 67: 2305–2314

22. Okamoto H, Imai M, Kametani M, Nakamura T, Mayumi M (1987) Genomic heterogeneity of hepatitis B virus in a 54-year old woman who contracted the infection through materno-fetal transmission. Jpn J Exp Med 57: 231–236

23. Okamoto H, Tsuda F, Sakugawa H, Sastrosoewignjo RI, Imai M, Miyakawa Y, Mayumi M (1988) Typing hepatitis B virus by homology in nucleotide sequence: comparison of surface antigen subtypes. J Gen Virol 69: 2575–2583

24. Ono Y, Ohda H, Sasada H, Igarashi K, Sugino Y, Nishioka K (1983) The complete nucleotide sequences of the cloned hepatitis B virus DNA; subtype adr and adw. Nucleic Acids Res 11: 1747–1757

25. Orito E, Mizokami M, Ina Y, Moriyama EN, Kameshima N, Yamamoto M, Gojobori T (1989) Host-independent evolution and a genetic classification of the hepadnavirus family based on nucleotide sequences. Proc Natl Acad Sci USA 86: 7059–7062

26. Renbao G, Meijin C, Lueping S, Suwen Q, Zaiping L (1987) The complete nucleotide sequence of the cloned DNA of hepatitis B virus subtype adr in pADR-1. Sci Sin Ser B 30: 507–521

27. Rho HM, Kim K, Hyun SW, Kim YS (1989) The nucleotide sequence and reading frames of a mutant hepatitis B virus subtype adr. Nucleic Acids Res 17: 2124

28. Sastrosoewignjo R, Okamoto H, Mayumi M, Warsa UC, Sujudi (1985) The complete nucleotide sequence of an HBV DNA clone of subtype as (pRTB299) from Indonesia. ICMR Ann 5: 39–50

29. Sastrosoewignjo RI, Omi S, Okamoto H, Mayumi M, Rustam M, Sujudi (1987) The complete nucleotide sequence of HBV DNA. clone of subtype adw (pMND122) from Menado in Sulawesl Island Indonesla. ICMR Ann 7: 51–60

30. Sastrosoewignjo RI, Sandjaja B, Okamoto H (1991) Molecular epidemiology of hepatitis B virus in Indoesia. J Gastroenterol Hepatol 6: 491–498

31. Seeger C, Ganem D, Varmus HE (1984) Nucleotide sequence of an infectious molecularly cloned genome of ground squirrel hepatitis virus. J Virol 51: 367–375

32. Shafritz DA, Shouval D, Sherman HI, Hadziyannis SJ, Kew MC (1981) Integration of hepatitis virus DNA into the genome of liver cell in chronic liver disease and hepatocellular carcinoma. N Engl J Med 305: 1067–1073

33. Sprengel R, Kalets EF, Will H (1988) Isolation and characterization of a hepatitis B virus endemic in herons. J Virol 62: 3832–3839

34. Stevens CE, Neurath RA, Beasly RP, Szmuness W (1979) HBeAg and anti-HBe detection by radioimmunoassay. Correlation with vertical transmission of hepatitis B virus in Taiwan. J Med Virol 3: 237–241

35. Summers J, Smolec JM, Snyder R (1978) A virus similar to human hepatitis B virus associated with hepatitis and hepatoma in woodchucks. Proc Natl Acad Sci USA 75: 4533–4537

36. Tabor A, Bayley AC, Cairns J, Peeleu L, Gerety RJ (1985) Horizontal transmission of hepatitis B virus among children and adults in five rural villages in Zambia. J Med Virol 15: 113–120

37. Tong S, Li J, Vitvitski L, Trepo C (1990) Active hepatitis B virus replication in the presence of anti-HBe is associated with viral variants containing an inactive pre-C region. Virology 176: 596–603
38. Valenzuela P, Quiroga M, Zaldivar J, Gray P, Rutter WJ (1980) The nucleotide sequence of the hepatitis B viral genome and the identification of the major viral genes. In: Fields B, Jalnisch R, Fox C (eds) ICN-UCLA symposia on animal virus genetics. Academic Press, New York, pp 57–70
39. Vaudin M, Wolstenholme AJ, Tsiquaye KN, Zuckerman AJ, Harrison TJ (1988) The complete nucleotide sequence of the genome of a hepatitis B virus isolated from a naturally infected chimpanzee. J Gen Virol 69: 1383–1389

Authors' address: Dr. L.O. Magnius, Department of Virology, The National Bacteriological Laboratory, S-10521, Stockholm, Sweden.

V. Diagnosis of chronic viral hepatitis

Arch Virol (1993) [Suppl] 8: 203–211

Role of IgM antibody to hepatitis B core antigen in the diagnosis of hepatitis B exacerbations

G. Colloredo Mels[1], **G. Bellati**[5], **G. Leandro**[2], **Maurizia Rossana Brunetto**[6],
O. Vicari[4], **P. Piantino**[6], **M. Borzio**[3], **G. Angeli**[1], **G. Ideo**[5], and **F. Bonino**[6]

Departments of Internal Medicine of the [1] Bolognini Hospital, Seriate, Bergamo,
[2] Saverio De Bellis Hospital, IRCCS, Castellana Grotte, Bari, [3] Fatebenefratelli
Hospital, Milano, [4] Transfusion Center, OOMM Hospital, Bergamo, [5] Hepatology
Center "Crespi", Niguarda Hospital, Milano, and [6] Laboratory of the Department
of Gastroenterology, Molinette Hospital, Torino, Italy

Summary. IgM anti-HBc levels were measured by the IMx Core-M Abbott assay in 939 serum samples in order to define a specific and sensitive cut-off value for diagnosis of chronic hepatitis B. The sera used were obtained from 52 chronic HBV patients and 10 HBV carriers with HCV or HDV co-infections and 155 asymptomatic subjects without evidence of liver disease. A Youden index value of 95.4% with 98% sensitivity and 97.4% specificity was obtained for an IMx Index value of 0.204 as cut-off. A one-year follow-up study with monthly tests has shown that quantitative analysis of IgM anti-HBc can serve as a non-invasive tool for monitoring HBV infection, and provides an accurate diagnosis of hepatitis B exacerbations. Significant elevations of IgM anti-HBc levels were associated with hepatitis B excerbations in 96.2% of the cases but with none of the ALT flare-ups observed in HCV or HDV infected individuals. These results suggest that quantitative analysis of IgM anti-HBc provides the highest degree of confidence in definition of spontaneous and therapy-induced exacerbations or remissions of hepatitis B.

Introduction

Hepatitis exacerbations are frequent in chronic hepatitis B surface antigen (HBsAg) carriers and they can be related either to spontaneous hepatitis B exacerbations [3, 5, 6, 9, 11, 14, 17, 20–24, 26–29, 33] or hepatitis caused by super-infections with other hepatotropic viruses or toxic/iatrogenic injuries. Their frequency, etiology and severity influence the natural history and prognosis of the liver disease [25, 30].

The main related clinical problems are detection of the flare-ups, since the majority are clinically asymptomatic [7] and the aetiological differentiation of hepatitis B exacerbations from episodes of liver cell necrosis due to other factors.

Although IgM antiHBc is considered to be a marker for HBV-induced liver disease [1, 12, 13, 16, 19, 31, 32, 36], only two studies have indicated the usefulness of IgM anti-HBc detection for diagnosis of HBV exacerbations [15, 29] in chronic HBsAg carriers.

To test the diagnostic significance of IgM antiHBc in hepatitis exacerbations, we studied the fluctuations of serum IgM antiHBc, HBV-DNA, aminotransferase levels during hepatitis exacerbations in 62 chronic HBsAg+ carriers, followed-up with monthly biochemical monitoring for a period of at least one year.

Materials and methods

Patients

There were three groups of patients. The first group (group A) consisted of 52 asymptomatic HBsAg carriers with chronic hepatitis B. 40 were men and 12 women, with a mean age of 40.2 years (range 18–70). Of these, 11 were HBeAg+ and 41 antiHBe+. Seven patients had chronic persistent hepatitis, 34 had chronic active hepatitis and 11 had liver cirrhosis. Intrahepatic HBcAg was detected in all of them by immunochemistry. Anti-HD, anti-HCV and anti-HIV was not detected in any of these patients at the beginning or at the end of the study.

The second group (group B) consisted of 10 chronic HBsAg+/anti-HBe+ carriers, 7 with HCV and 3 with HDV chronic co-infections. Seven were men, 3 were women, their mean age was 43.7 years (range 16–69). One patient had chronic persistent hepatitis, 3 had chronic active hepatitis and 3 had liver cirrhosis. Intrahepatic HBcAg and HDAg were detected in all the patients of this group by immunochemistry.

None of the patients of group A and B admitted homosexuality or alcohol abuse or had received blood transfusions or blood products or antiviral immunosuppressive or hepatotoxic drug therapy before or during the follow up period (mean 13.1 months, range 12–21 months). Only two patients of group B admitted to drug abuse.

Group C consisted of 155 asymptomatic subjects without biochemical and/or histological evidence of liver disease. Ninety six were men, 59 were women, the mean age was 37.5 years (range 18–57); 41 of them were antiHBs and antiHBc positive.

Blood samples were taken from all the HBsAg+ patients every month during the follow-up (mean 12.6 months, range 12–20 months). Only one serum sample was obtained from each group C individual. A total of 939 serum samples were obtained.

Laboratory tests

Alanine-aminotransferase (ALT) and aspartate-aminotransferase (AST), were assayed by standard procedures. Serological markers of HBV (HBsAg, antiHBs, antiHBc,

HBeAg, antiHBe, HBV-DNA, IgM antiHBc) were tested for with commercial kits (AUSRIA II, CORAB, HBeKit, HBV-DNA RIA [18], IgM antiHBc-IMx [2, 8], Abbott Laboratories, North Chicago, IL, U.S.A.). A radioimmunological method was used for anti-HDV antibodies (Sorin Biomedica, Saluggia, Italy). HBV-DNA and IgM antiHBc were measured semiquantitatively using reference standard curves.

The cut-off values were 40 IU/L for ALT and 1pg/ml for HBV-DNA. For the cut-off level of IgM antiHBc, a Receiver Operating Characteristic (ROC) curve [34] was calculated by a computerized program, using 62 values (baseline sample) of the chronic HBsAg carriers (groups A+B) and 155 serum samples of group C. The best Youden index (J = sensitivity + specificity − 1) [10, 35] was then calculated and the corresponding cut-off value of the IMx index was identified.

Histological sections were stained with hematoxylin-eosin and examined for HBcAg and HDAg using the indirect peroxidase-antiperoxidase technique (PAP) (rabbit antiHBc and human anti-HDAg antisera).

Statistical analyses

We evaluated the median values of ALT, HBV-DNA and IgM anti-HBc during the follow-up. A sudden elevation of serum ALT levels equal to or greater than twice the median value was defined as a hepatitis exacerbation. For each exacerbation the following phases were considered:

1. basal value of the parameter considered;
2. peak value;
3. the % increase from the basal value: [(peak value − basal value)/ basal value] × 100;
4. trough value
5. the % decrease after the peak: [(trough value − basal value)/basal value − % increase)] × 100
6. the total duration (months) of the flare-up: (duration of the increase + duration of the decrease).

Values are expressed as means and standard errors of means (SEM). The Student t test, the Fisher exact test and linear regression analysis were applied, using BMDP software, and when necessary data were transformed logarithmically.

Results

Validation of IgM anti-HBc cut-off for chronic hepatitis B

Figure 1 shows the ROC curve of the IgM anti-HBc assay. An IMx index of 0.169 identified chronic hepatitis B patients with 100% sensitivity (91% specificity) while an IMx index of 0.224 provided 100% specificity (94% sensitivity). The best Youden index (J) (95.4%) was identified at the IMx cut-off value of 0.204, at which the sensitivity was 98% (c.i. 88%–99%) and the specificity 97.4% (c.i. 93%–99%).

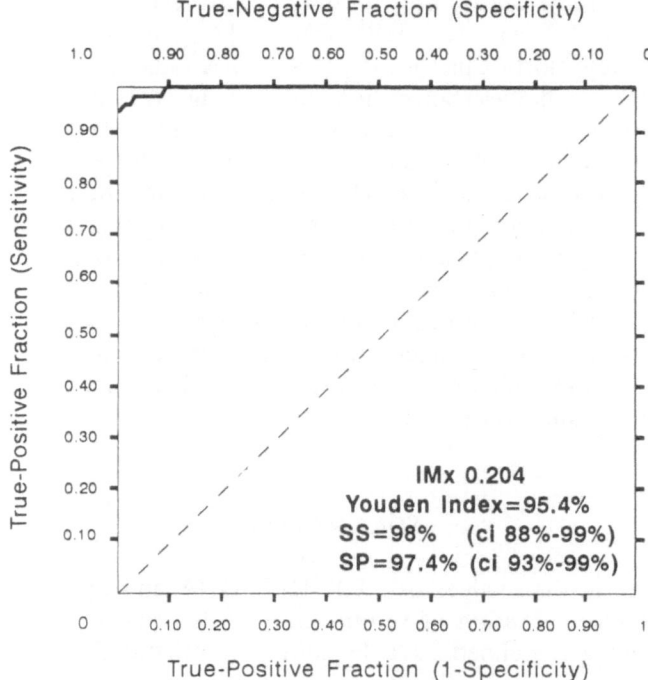

Fig. 1. ROC curve of IgM antiHBc. IMx cut-off level, with best Youden index for discrimination between healthy subjects and chronic hepatitis B patients

Monitoring study of chronic HBV carriers

Thirty-nine asymptomatic hepatitis exacerbations occurred during the follow up, in 23 out of 52 patients of group A. In 27 exacerbations, the period of follow-up (before, during and after the flare-up) was long enough to extensively analyze the kinetics of the fluctuations of viremia, aminotransferases and IgM anti-HBc.

There was a temporal relationship between the fluctuations (increase and decrease) of the IgM antiHBc and those of viremia and ALT. In 96.2% of the cases, the HBV-DNA peak occurred before the ALT peak or at the same time and the ALT peak preceded or occurred at the same time as the IgM antiHBc index peak in 96.2% of the cases. The details of this study have been published elsewhere [4].

IgM antiHBc fluctuations during the flare-ups were smaller (increase = +111.1 ± 24.5%; decrease = − 98.6 ± 20.6%) than those of viremia (increase = +1843.2 ± 439.8%; decrease = − 1909.4 ± 505.1%) and of aminotransferases (increase = +604.6 ± 109.6%; decrease = − 583.8 ± 114.4%) and in only 12/27 (44.4%) of the cases was the maximum peak value of IMx at the positive cut-off level (1.2 IMx) recommended by the

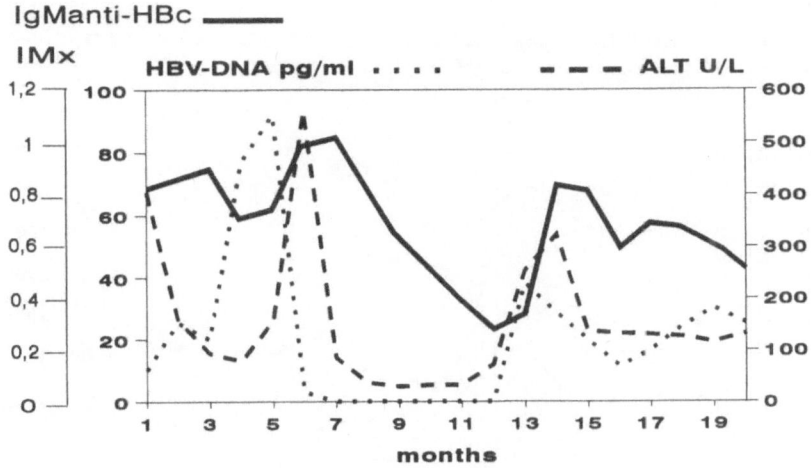

Fig. 2. Relationships between the fluctuations of HBV-DNA, ALT and IgM anti-HBc serum levels during the episodes of hepatitis B exacerbations in a chronic hepatitis B (antiHBe+) patient

manufacturer to distinguish acute from chronic hepatitis B. In 8/27 (29.6%) the level was within the acute hepatitis B "gray zone" (0.8–1.2 IMx) and in 7/27 (25.9%) it was below the "gray zone", of 0.2 to 0.8 IMx index. This result emphasizes the usefulness of measuring low levels of IgM anti-HBc in chronic hepatitis B patients by highly sensitive quantitative assays.

HBV-DNA fluctuation had the shortest duration (3.7 ± 0.2 months), ALT flares lasted for 4.8 ± 0.3 months, IgM antiHBc for 5.8 ± 0.5 months.

Examples of the temporal relationships of the fluctuations of viremia, aminotransferases and IgM antiHBc during hepatitis B exacerbations in a chronic HBsAg+ patient are shown in Fig. 2.

In group B, two clinically asymptomatic hepatitis exacerbations were observed, one in an HBsAg+ carrier with superimposed chronic HCV infection (anti-HCV positive for structural and non-structural HCV proteins -RIBA II°- and HCV-RNA positive by single step RT-PCR) (Fig. 3) and one in an HBsAg+ carrier with superimposed chronic HDV infection (anti-HD and IgM anti-HD positive with intrahepatic HDAg) (Fig. 4). In both exacerbations, we found none of the typical chronological patterns of the kinetics of HBV-DNA, ALT and IgM anti-HBc observed in hepatitis B exacerbations. Fluctuations of HBV-DNA did not precede, nor did fluctuations of IgM anti-HBc follow the ALT flare-up, indicating that both hepatitis episodes were unrelated to HBV, but probably to HCV in the former and HDV in the latter patient.

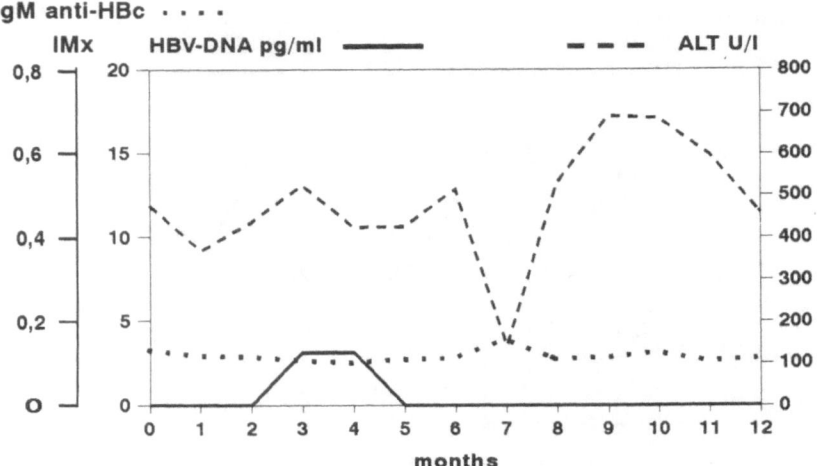

Fig. 3. Hepatitis flare-up unrelated to hepatitis B in a chronic hepatitis C patient without the typical sequence of fluctuations of HBV-DNA, ALT and IgM antiHBc seen in Fig. 2

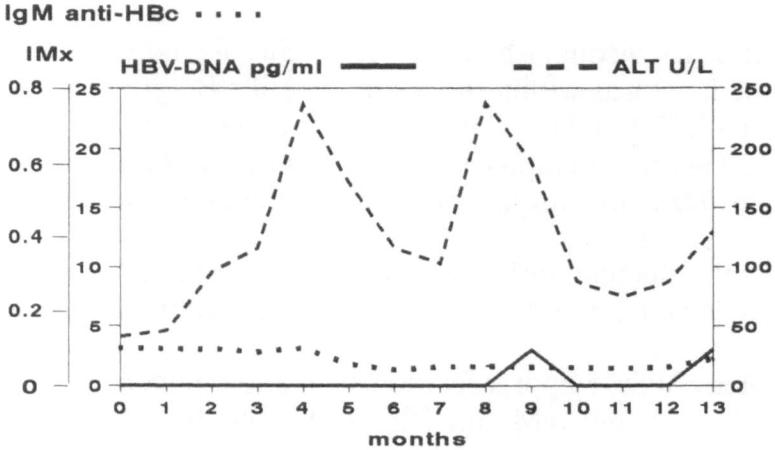

Fig. 4. Hepatitis flare-ups unrelated to hepatitis B in a chronic hepatitis D patient. Lack of fluctuations of IgM antiHBc after the hepatitis episodes. A significant increase of serum IgM antiHD levels after each ALT flare-up suggests that both exacerbations were caused by HDV reactivations

Discussion

We measure IgM anti-HBc in 939 serum samples obtained from 62 chronic HBV patients or carriers and 155 asymptomatic subjects without clinical evidence of liver disease. Statistical analysis confirmed that 0.200 (corresponding to 10 IgM antiHBc units of the Paul-Ehrlich-Institute,

PEI) can be used as a specific and sensitive cut-off value between positive and negative IMx indexes in chronic HBsAg carriers. The Youden index was 95.4% with 98% sensitivity and 97.4% specificity. The follow-up study showed that quantitative analysis of IgM antiHBc is a useful non-invasive tool for monitoring HBV infection and provides an accurate diagnosis of hepatitis B exacerbations. More than 96% of hepatitis B flare-ups were associated with IgM anti-HBc increments. In contrast, none of the ALT flare-ups observed in HCV infected individuals and none of hepatitis D exacerbations (ALT flare-ups associated with IgM anti-HD level increments) were concomitant with variations of serum IgM antiHBc levels. The duration of IgM antiHBc fluctuations was about twice as long as those of viremia, suggesting a significant difference in the kinetics of clearance of the antibody and HBV-DNA. This observation extends the diagnostic importance of IgM antiHBc, which as a long lasting marker can be used as a screening test for chronic hepatitis B. Use of the IgM antiHBc test can avoid misdiagnosis of chronic hepatitis B in asymptomatic patients who have normal serum levels of ALT and undetectable HBV-DNA.

In conclusion, quantitative analysis of serum IgM antiHBc by a highly sensitive and specific assay is the most accurate tool for diagnosis and monitoring of both acute and chronic hepatitis B. With this test, we can obtain the highest degree of confidence in the definition of spontaneous and therapy-induced exacerbations or remissions of hepatitis B. An IgM anti-HBc index below 0.200 in patients with undetectable viremia and normal serum ALT levels indicates a complete remission while persistence of values higher than 0.200 represents the hallmark of an unstable equilibrium phase, with high risk of hepatitis B reactivation.

References

1. Brunetto MR, Arrigoni A, Toti M, Almi P, Zanetti A, Ferroni P, Doris R, Aneloni V, Crovari P, Coppola RC, Bonino F (1988) The diagnostic significance of IgM antibody to B core antigen, revisited. Ital J Gastroenterol 20: 167–170
2. Brunetto MR, Torrani Cerentia M, Olivieri F, Piantino P, Randone A, Calvo PL, Bonino F (1993) Monitoring the natural course and response to therapy of chronic hepatitis B with an automated semi-quantitative assay for IgM anti-HBc. J Hepatol (in press)
3. Castillo I, Bartolomé J, Quiroga A, Porres JC, Carreno V (1990) Detection of HBeAg/antiHBe immune complexes in the reactivation of hepatitis B virus replication among antiHBe chronic carriers. Liver 10: 79–84
4. Colloredo Mels G, Bellati G, Leandro G, Brunetto MR, Vicari O, Borzio M, Piantino P, Fornaciari G, Scudeller G, Angeli G, Bonino F, Ideo G (1993) Fluctuations of viremia, aminotransferases and IgM antibody to hepatitis B core antigen in chronic hepatitis B patients with disease exacerbations. Liver (in press)

5. Davis GD, Hoofnagle JH, Waggoner JG (1984) Spontaneous reactivation of chronic hepatitis B virus infection. Gastroenterology 86: 230–235
6. Davis GL, Hoofnagle JH (1985) Reactivation of chronic type B hepatitis presenting as acute viral hepatitis. Ann Intern Med 102: 762–765
7. Davis GL, Hoofnagle JH (1987) Reactivation of chronic hepatitis B virus infection. Gastroenterology 92: 2028–2031
8. Eble K, Clemens J, Krenc C, Rynning M, Stojak J, Stuckmann J, Hutten P, Nelson L, Du Charme L, Hojvat S, Mimms L (1991) Differential diagnosis of acute viral hepatitis using rapid, fully automated immunoassays. J Med Virol 33: 139–150
9. Fattovich G, Brollo L, Alberti A, Realdi G, Pontisso P, Giustina G, Roul A (1990) Spontaneous reactivation of hepatitis B virus infection in patients with chronic type B hepatitis. Liver 10: 141–146
10. Feinstein AR (1985) Clinical epidemiology: the architecture of clinical research. W.B. Saunders, Philadelphia, pp 434–437
11. Gayno S, Marcellin P, Loriot MA, Martinot-Peignoux M, Levy P, Erlinger S, Benhamou JP (1992) Detection of serum HBV-DNA by polymerase chain reaction (PCR) in patients before reactivation of chronic hepatitis B. J Hepatol 14: 357–360
12. Gerlich WH, Lüer W, Thomssen R, The Study Group for Viral hepatitis of the Deutsche Forschungsgemeinschaft (1980) Diagnosis of acute and inapparent hepatitis B virus infection by measurement of IgM antibody to hepatitis B core antigen. J Infect Dis 142: 85–101
13. Gerlich WH, Uy A, Lambrecht F, Thomssen R (1986) Cut-off levels of IgM antibody against viral core antigen for differentiation of acute, chronic and past hepatitis B virus infections. J Clin Microbiol 24: 288–293
14. Gupta S, Govindarajan S, Fong T-L, Redeker A (1990) Spontaneous reactivation in chronic hepatitis B: patterns and natural history. J Clin Gastroenterol 12: 562–568
15. Koike K, Iino K, Kurai K, Mitamura K, Endo Y, Oka H (1987) IgM anti-HBc in antiHBe positive chronic type B hepatitis with acute exacerbations. Hepatology 7: 573–576
16. Kryger P (1985) Significance of anti-HBc IgM in the differential diagnosis of viral hepatitis. J Virol Methods 10: 283–289
17. Krogsgaard K, Aldershvile J, Kryger P, Pedersen C, Anderson P, Dalboge H, Nielsen JO, Hansson BG (1990) Reactivation of viral replication in antiHBe positive chronic HBsAg carriers. Liver 10: 54–58
18. Kuhns MC, McNamara AL, Perillo RP, Cabal CM, Campbell CR (1989) Quantitation of hepatitis B viral DNA by solution hybridisation: comparison with DNA polymerase and hepatitis B e antigen during antiviral therapy. J Med Virol 27: 274–281
19. Lavarini C, Farci P, Chiaberge E, Veglio V, Giacobbi D, Crivelli O, Rizzetto M (1983) IgM antibody against hepatitis B core antigen (IgM anti-HBc): diagnostic and prognostic significance in acute HBsAg positive hepatitis. Br J Med 287: 1254–1256
20. Levy P, Marcellin P, Martinot-Peignoux M, Degott C, Nataf J, Benhamou JP (1990) Clinical course of spontaneous reactivation of hepatitis B virus infection in patients with cronic hepatitis B. Hepatology 12: 570–574
21. Liaw Y-F, Chu C-M, Su I-J, Huang M-J, Lin D-Y, Chang-Chien C-S (1983) Clinical and histological events preceding hepatitis B e antigen seroconversion in chronic type B hepatitis. Gastroenterology 84: 216–219

22. Liaw Y-F, Yang S-S, Chen T-J, Chu C-M (1985) Acute exacerbation in hepatitis B e antigen positive chronic type B hepatitis. A clinicopathological study. J Hepatol 1: 227–233
23. Liaw Y-F, Tai D-I, Chu C-M, Pao CC, Chen T-J (1987) Acute exacerbation in chronic type B hepatitis: comparison between HBeAg and antibody-positive patients. Hepatology 7: 20–23
24. LiawY-F, Pao C-C, Chu C-M (1988) Changes of serum HBV-DNA in relation to serum transaminase level during acute exacerbation in patients with chronic type B hepatitis. Liver 8: 231–235
25. Liaw Y-F, Tai D-I, Chu C-M, Chen T-J (1988) The development of cirrhosis in patients with chronic type B hepatitis: a prospective study. Hepatology 8: 493–496
26. Lok ASF, Lai C-L, Wu P-C, Leung EKY, Lam T-S (1987) Spontaneous hepatitis B e antigen to antibody seroconversion and reversion in Chinese patients with chronic hepatitis B virus infection. Gastroenterology 92: 1839–1843
27. Lok ASF, Lai C-L (1990) Acute exacerbations in Chinese patients with chronic hepatitis B virus (HBV) infection. J Hepatol 10: 29–34
28. Perrillo RP, Campbell CR, Sanders GE, Regenstein FG, Bodicky CJ (1984) Spontaneous clearance and reactivation of hepatitis B virus infection among male homosexuals with chronic type B hepatitis. Ann Intern Med 100: 43–46
29. Raimondo G, Rodinò G, Smedile V, Brancatelli S, Villari D, Longo G, Squadrito G (1990) Hepatitis B virus (HBV) markers and HBV-DNA in serum and liver tissue of patients with acute exacerbation of chronic type B hepatitis. J Hepatol 10: 271–273
30. Sheen I-S, Liaw Y-F, Tai D-I, Chu C-M (1985) Hepatic decompensation associated with hepatitis B e antigen clearance in chronic type B hepatitis. Gastroenterology 89: 732–735
31. Smith HM, Lau JYN, Davies SE, Daniels HM, Alexander GJM, Williams R (1992) Significance of serum IgM anti-HBc in chronic hepatitis B virus infection. J Med Virol 36: 16–20
32. Sjogren M, Hoofnagle JH (1985) Immunoglobulin M antibody to hepatitis B core antigen in patients with chronic type B hepatitis. Gastroenterology 85: 252–258
33. Tong MJ, Sampliner RE, Govindarajan S, Co RL (1987) Spontaneous reactivation of hepatitis B in chinese patients with HBsAg-positive chronic active hepatitis. Hepatology 7: 713–718
34. Weinstein MC, Fineberg HV (1980) Clinical decision analysis. W.B. Saunders, Philadelphia, pp 601–609
35. Youden WJ (1950) Index for rating diagnostic tests. Cancer 3: 32–35
36. Zoulim F, Mimms L, Floreani M, Pichoud C, Chemin I, Kay A, Vitvitski L, Trepo C (1992) New assays for quantitative determination of viral markers in management of chronic hepatitis B virus infection. J Clin Microbiol 30: 1111–1119

Authors' address: Dr. F. Bonino, Department of Gastroenterology, Molinette Hospital, Corso Bramante 88, I-10126 Torino, Italy.

Arch Virol (1993) [Suppl] 8: 213–218

Serum IgM antibodies to hepatitis C virus in acute and chronic hepatitis C

S. Brillanti, Caterina Masci, M. Miglioli, and **L. Barbara**

Department of Internal Medicine and Gastroenterology, University of Bologna, Italy

Summary. A standardized commercially available immunoassay is not available for detection of IgM antibodies against hepatitis C virus antigens (IgM anti-HCV). Therefore, different "in-house" enzyme immunoassays have been assessed. These assays vary greatly in sensitivity, but specificity seems satisfactory in all of them. A typical IgM antibody response to HCV antigens is usually found in nearly all patients with acute hepatitis C. This antibody response rarely precedes the appearance of IgG anti-HCV, and it persists for a few months at high titer. Low titers of IgM anti-HCV are detectable in 50–80% of cases with chronic hepatitis C. IgM anti-HCV reactivity is typically found during acute exacerbation of chronic hepatitis C. Furthermore, many patients with chronic active hepatitis C without acute exacerbation also have IgM anti-HCV. In these patients a correlation exists between the titer of IgM anti-HCV and the biochemical parameters of liver disease. When alpha interferon therapy induces a sustained remission of liver disease activity, positivity for IgM anti-HCV disappears in more than 70% of cases. In contrast, patients who do not respond to therapy rarely loose IgM anti-HCV. In conclusion, serum IgM antibodies to HCV antigens are reliable markers of active HCV-induced liver disease both in acute and in chronic HCV infection.

Introduction

After the encounter of specific B lymphocytes with hepatitis virus antigens antibodies of the IgM class are the first formed in serum. IgM antibodies generally have only short half-life, characterizing the acute-phase of viral hepatitis. Using sensitive immunoassays, IgM antibodies to viral antigens have been shown to persist for a long time after the acute phase of the disease, in patients with chronically evolving hepatitis B and D [1, 2]. Otherwise, asymptomatic "healthy" carriers of hepatitis viruses are generally negative for IgM antibodies. These data show that the presence of IgM antibodies correlates with the activity of chronic virus-induced liver disease.

HCV infection is diagnosed serologically by detection of IgG antibodies to HCV antigens, but antibody testing cannot reveal whether the patient has recovered from the infection or is still carrying the virus. HCV antigen cannot be directly detected in serum by current assays, and only very low titers of serum hepatitis C viral RNA sequences (HCV RNA) are present. Therefore, to confirm the presence of active HCV infection, reverse transcription and amplification of serum HCV RNA must be performed by polymerase chain reaction (PCR) technique [3].

For the last three years we have assessed and standardized a specific enzyme immunoassay for the detection of IgM antibodies to HCV - derived antigens (IgM anti-HCV). In this brief review, we summarize and discuss the results on the significance of IgM anti-HCV in acute and chronic hepatitis C.

Detection of IgM anti-HCV

To detect serum IgM antibodies to HCV antigens, solid-phase enzyme-immunoassays have been developed. Results on IgM anti-HCV testing were first reported in abstracts published in 1990 [4, 5], but the first full-length paper on IgM anti-HCV was published in July 1991 by Spanish researchers [6]. The authors used a modification of the commercially available first generation anti-HCV test to detect circulating IgM antibodies against the non-structural HCV antigen C100-3. In late 1991, researchers from Abbott Laboratories, published a similar method to detect IgM antibodies to structural and non-structural HCV antigens [7]. Finally, in June 1992, we published the method we have been using in our laboratory since 1989 [8]. In all three of the studies an antigen-coated solid phase and labeled anti-μ antibody was used to detect IgM anti-HCV. Before incubating serum specimens with the solid phase, the sample dilution used varied between the studies: samples were diluted 1:125 in PBS buffer by the Spanish researchers, 1:441 in a diluent containing goat anti-human IgG by the American authors, and in our laboratory we tested specimens diluted 1:21 in PBS buffer, after specific removal of serum IgG, using an IgG removal method with immobilized, recombinantly engineered, protein G.

Since specificity and sensitivity of IgM antibody testing are influenced by both IgG depletion and serum sample dilution, it is difficult to compare the results obtained with different solid-phase antibody capture assays. In addition, researchers from Abbott Laboratories and from Taiwan used immunoblot assays to test serum for IgM anti-HCV [9, 10].

Results

Serum samples from patients with both acute and chronic HCV infections were tested for IgM anti-HCV. Results are summarized in Table 1. There was a high prevalence of IgM antibodies to the core HCV antigen in patients with acute hepatitis C. The appearance of IgM anti-HCV usually occurred coincidentally with appearance of IgG anti-HCV. Six months after the acute phase, almost all patients with resolved hepatitis became IgM anti-HCV negative. Otherwise, the prevalence of IgM antibodies to non-structural HCV antigens, observed in patients with acute hepatitis C, varied greatly between studies.

In patients with chronic HCV infection, different figures of IgM anti-HCV positivity were reported. There was a correlation between the titer of IgM anti-HCV and the biochemical activity of the liver disease. In addition, positivity for IgM anti-HCV frequently appeared during acute exacerbation of chronic hepatitis C. IgM anti-HCV was, however, not

Table 1. Comparison of IgM anti-HCV results by five groups in IgG anti-HCV positive patients

First author	Quiroga JA	Chau KH	Brillanti S	Clemens JM	Chen P-J
Method	ELISA	ELISA	ELISA	immunoblot	immunoblot
Removal of serum IgG	no	yes	yes	yes	yes
Dilution of serum	1:125	1:441	1:21	1:100	1:500
Number tested acute hepatitis C	13	6	0	15	9
% positive NS3/NS4	100	100	–	7	ND
% positive core	ND	100	–	67	89
Number tested chronic hepatitis C	88	29	40	15	47
% positive NS3/NS4	51	ND	82	13	ND
% positive core	ND	8	ND	33	45
Number tested asymptomatic anti-HCV+	0	0	10	0	0
% positive NS3/NS4	–	–	0	–	–
% positive core	–	–	ND	–	–

NS-3, *NS*-4, *core* Putative HCV genomic sequences encoding for non-structural and structural polypeptides
ND Not done

found in those patients chronically infected with HCV, but without increased aminotransferase levels.

The significance of IgM anti-HCV in monitoring interferon therapy for chronic hepatitis C was established. After interferon therapy, patients who did not respond to therapy also did not lose IgM anti-HCV. Similarly, more than 80% of patients with a temporary response to interferon remained positive for IgM anti-HCV after therapy. Otherwise, IgM anti-HCV definitely disappeared in more than 70% of cases with a long-term response to interferon treatment. Furthermore, most of these cases completely lost IgM anti-HCV by the end of therapy.

Discussion

In patients with acute post-transfusion hepatitis C, the presence of IgM anti-HCV indicates active HCV infection and anti-HCV seroconversion, since IgM anti-HCV is not passively transferred. High titers of IgM antibodies against the core HCV antigen are found, and such titers generally wane after a few months [9]. Discordant results on the prevalence of IgM antibodies against non-structural HCV antigens are likely related to the sensitivity of the different assays [6, 7, 9]. In summary, these data indicate that IgM anti-HCV, at high titer, is a useful marker of acute HCV-induced hepatitis. IgM anti-HCV response rarely precedes the IgG response, so testing for IgM anti-HCV is of limited value in IgG anti-HCV negative patients with acute non-A, non-B hepatitis.

Positivity for IgM anti-HCV is not limited to the acute phase of hepatitis C. During reactivation of chronic hepatitis C, IgM anti-HCV reactivity is generally detected [10]. Using more sensitive assays, IgM anti-HCV may be also found in many patients with chronic hepatitis C without reactivation. In these patients, a correlation exists between IgM anti-HCV titers and biochemical evidence of active liver disease [6, 8, 10]. Therefore, IgM anti-HCV may be present in both acute and chronic hepatitis C, but with different titers, and sensitive immunoassays are required to detect IgM anti-HCV in chronically infected patients. In contrast, in HCV-infected subjects with normal ALT levels IgM anti-HCV is rarely found [8]. These data indicate that serum IgM anti-HCV correlates with persistent HCV infection and biochemical evidence of active liver disease.

During alpha interferon therapy for chronic hepatitis C, IgM anti-HCV levels do not change significantly in patients who do not respond to treatment. Similarly, in patients with a temporary response to treatment followed by relapse, IgM anti-HCV levels do not fall to undetectable levels. In contrast, IgM anti-HCV falls to undetectable levels in most patients with a sustained response to interferon therapy [8, 11]. These data indicate that testing of sequential serum specimens for IgM anti-HCV in patients treated with alpha interferon may reveal which patients will have a long-term response. In contrast, detection of serum HCV RNA by PCR is not able to give similar clinical information Furthermore, before treatment, patients who will respond to interferon therapy are more likely to be positive for IgM anti-HCV than the non-responder patients.

In conclusion, serum IgM antibodies to HCV antigens are reliable markers of active HCV-related liver disease both in acute and in chronic HCV infection. Testing for IgM anti-HCV identifies among IgG anti-HCV positive subjects those with an ongoing HCV infection and an active immune response to the virus. During antiviral therapy, resolution of HCV-related liver damage correlates with disappearance of IgM anti-HCV.

References

1. Sjogren M, Hoofnagle JH (1985) Immunoglobulin M antibody to hepatitis B core antigen in patients with chronic type B hepatitis. Gastroenterology 89: 252–258
2. Govindarajan S, Gupta S, Valinluck B, Redeker G (1989) Correlation of IgM anti-hepatitis D virus (HDV) to HDV RNA in sera of chronic HDV. Hepatology 10: 34–35
3. Farci P, Alter HJ, Wong D, Miller RH, Shih JW, Jett B, Purcell RH (1991) A long-term study of hepatitis C virus replication in non-A, non-B hepatitis. N Engl J Med 325: 98–104
4. Brillanti S, Miglioli M, Di Febo G, Barbara L (1990) Serum IgM antibody to hepatitis C virus in subjects with acute and chronic non-A, non-B hepatitis. Ital J Gastroenterol 22: 165
5. Brillanti S, Masci C, Biasco G, Miglioli M, Barbara L (1990) Detection of serum IgM antibody to hepatitis C virus in chronic non-A, non-B hepatitis. J Hepatol 11 [Suppl 2]: S12
6. Quiroga JA, Campillo ML, Catillo I, Bartolomé J, Porres JC, Carreño (1991) IgM antibody to hepatitis C virus in acute and chronic hepatitis C. Hepatology 14: 38–43
7. Chau KH, Dawson GJ, Mushahwar IK, Gutierrez RA, Johnson RG, Lesniewski RR, Mattson L, Weiland O (1991) IgM antibody response to hepatitis C virus antigens in acute and chronic post-transfusion non-A, non-B hepatitis. J Virol Methods 35: 343–352
8. Brillanti S, Masci C, Ricci P, Miglioli M, Barbara L (1992) Significance of IgM antibody to hepatitis C virus in patients with chronic hepatitis. C Hepatology 15: 998–1001
9. Clemens JM, Taskar S, Chau K, Vallari D, Shih JW-K, Alter HJ, Schleicher JB, Mimms LT (1992) IgM antibody response in acute viral infection. Blood 79: 169–172
10. Chen P-J, Wang J-T, Hwang L-H, Yang Y-H, Hsieh C-L, Kao J-H, Sheu J-C, Lai M-Y, Wang T-H, Chen D-S (1992) Transient immunoglobulin M antibody

response to hepatitis C virus capsid antigen in posttransfusion hepatitis C: putative serological marker for acute viral infection. Proc Natl Acad Sci USA 89: 5971–5975

11. Quiroga JA, Bosch O, Gonzalez R, Marriott E, Castillo I, Bartolome J, Carreño V (1992) Immunoglobulin M antibody to hepatitis C virus during interferon therapy for chronic hepatitis C. Gastroenterology 103: 1285–1289

Authors' address: Dr. S. Brillanti, Istituto di Clinica Medica e Gastroenterologia, Policlinico S. Orsola, Via Massarenti 9, I-40138 Bologna, Italy.

Arch Virol (1993) [Suppl] 8: 219–228

Isolate antibody to hepatitis C virus core antigen (C22) by RIBA-2: correlation with HCV-RNA and anti-NS5

G. Taliani, M.C. Badolato, R. Lecce, R. Bruni, C. Clementi, F. Grimaldi, C. Furlan, M. Manganaro, F. Duca, A. Bozza, G. Poliandri, and **C. De Bac**

Institute of Tropical Disease, "La Sapienza" University, Policlinico Umberto I, Rome, Italy

Summary. The presence of circulating hepatitis C virus genome (HCV-RNA), elevated ALT levels and antibodies to an NS5-derived synthetic peptide have been examined in 13 subjects with isolate positivity for antibodies to the HCV core antigen (C22) on RIBA-2 testing. All subjects were followed up for 8–18 months (mean 12.4 months). In seven subjects (54%), intermittent or persistent viremia was associated with abnormal ALT levels (6 subjects) and with positivity for antibodies to NS5-peptide (6 subjects). On the other hand, in 6 out of 13 subjects (46%) no viral replication, no liver cytonecrosis and no antibodies to NS5 were found. It is concluded that isolate reactivity to C22 by RIBA-2 is a heterogeneous condition that corresponds to two distinct categories of subjects: those with active HCV infection and those without evidence of virus replication. Although HCV-RNA determination is the most reliable means of identifying HCV carriers, antibodies to NS5 can be a useful marker of virus activity. In fact, antibodies to NS5 were detected in 6 out of 7 viremic patients, compared to 0 out of 6 non-viremic patients ($P = 0.004$). It remains to be elucidated whether the isolate reactivity to core antigen found in non-viremic subjects represents a specific, HCV-induced antibody response, or is an unrelated crossreactivity.

Introduction

Complete molecular cloning of the hepatitis C virus (HCV) genome and extended expression of viral antigens have led to the development of second generation ELISAs for blood screening and serodiagnosis of HCV infection. These tests employ expression products from the non-structural (NS) regions NS3/C33, NS4/C100 and the putative nucleocapsid (C22) region of the hepatitis C genome [8–10, 20]. The examination of ELISA reactive sera with second-generation recombinant immunoblot assay (RIBA-2, Chiron Corp., Emeryville, CA, U.S.A.) allows recognition of the reactivity to each viral antigen that is present on a nitrocellulose strip as a separate band [14, 22]. By this test, some of the anti-HCV positive sera exhibit an isolate positivity for antibodies to the core antigen (C22) [1, 2, 16].

Isolate anti-C22 positivity is at present difficult to interprete. In particular, it has not been unequivocally defined as to whether it corresponds to the early seroconversion during acute HCV infection [3], whether it is a late remnant of a recovered HCV infection, or is due to the infection by a highly divergent HCV strain [23] which does not cross-react with the NS antigens or the PCR primers. Alternatively, it could be a false-positive reaction [18].

To better understand the meaning of this serologic profile and to evaluate its evolution, we prospectively followed up 13 subjects with isolate positivity for anti-C22 by RIBA-2 in whom aminotransferase levels were serially measured and serum HCV-RNA was tested by the polymerase chain reaction (PCR). Moreover, to assess whether anti-C22 was a true isolate antibody response, antibodies to a synthetic peptide derived from the NS5 non-structural region of HCV were examined by an immunoblot assay.

Materials and methods

Patients

Sera from 256 subjects who tested repeatedly positive for antibodies to hepatitis C virus (HCV) by second-generation ELISA (Ortho Diagnostics Inc., Raritan, NJ, U.S.A.) were examined by second-generation recombinant immunoblotting assay (RIBA-2, Chiron Corporation, Emeryville, CA, U.S.A.) to confirm the ELISA reactivity. Sera from 13 subjects (5%) were found to be reactive to the structural core antigen (C-22) but were negative for antibodies to the non-structural NS3 (C-33) and NS4 (C-100) antigens, and therefore were considered indeterminate as suggested by the producer. These 13 subjects (8 males, mean age 46.7 years, range 20–59 years) were included in the study and were followed up for 8–18 months (mean 12.4 months).

In these patients, alanine aminotransferase (ALT) levels were measured every 4–6 weeks. Further RIBA-2 examination was performed in all subjects at the end of the follow up. Moreover, paired sera obtained from each patient at the beginning and at the end of the follow up, were kept frozen in aliquots at −80°C and examined for circulating HCV-RNA by nested PCR, and for antibodies to HCV by a different immunoblotting assay that employs synthetic peptides derived from the NS4, NS5 and core region of HCV (Liatek HCV, Organon Teknika B.V., Boxtel, The Netherlands). The NS5 derived peptide is not present in the RIBA-2 assay, consequently antibodies to this region could be detected only by Liatek-HCV. Moreover, in this assay, four different core peptides are employed, therefore the reactivity to core peptides was scored from 1 to 4.

Reverse transcription and PCR

Total RNA was extracted from 100 μl of serum by guanidinium isothiocyanate-phenol-chloroform according to Chomczynsky and Sacchi [4]. In brief, 100 μl of serum were added to 500 μl of 4 mol/l guanidinium isothiocyanate, 25 mmol/l sodium citrate (pH 7.0), 0.5% sodium lauryl sarcosinate, 0.1 mol/l β-mercaptoethanol. After phenol-chlorophorm-isoamyl alcohol extraction, 20 μg of glycogen (Boehringer Mannheim Biochemicals, Mannheim, Federal Republic of Germany), one-tenth vol of 3 mol/l sodium acetate (pH 6.0) and 10 vol of absolute ethanol were added and the mixture was left at −80°C overnight. After centrifugation (20 min, 15 000 g at +4°C) the pellet RNA was washed twice with 80% ethanol, dried and resuspended in 20 μl of RNase-free water containing 40 U of RNase inhibitor (Promega Corp., Madison, WI, U.S.A). PCR was performed by combining the reverse transcription step with the first PCR reaction in the same tube. Ten microliters of RNA (corresponding to 50 μl of serum) were added to 90 μl of the PCR buffer (10 mmol/l Tris-HCl, pH 8.4, 50 mmol/l KCl, 3 mmol/l MgCl$_2$, 0.01% gelatin) containing 50 pmol of each outer primer deduced from the 5′ untranslated region (sense, nucleotides 19 to 38, 5′GCGACACTCCACCATGAATC3′, antisense, nucleotides 343 to 321, 5′ATGGTGCACGGTCTACGAGACC 3′), 0.2 mmol/l of each of the four deoxynucleotides, and 6.25 units of avian myeloblastosis virus reverse trancriptase (Promega Corp., Madison, WI, U.S.A.). After 60 min at 43°C for the reverse transcription step, and 5 min denaturation at 94°C, 2.5 units of Taq DNA polymerase (Promega Corp., Madison, WI, U.S.A.) were added and 30 cycles of PCR consisting of 94°C for 1 min, 55°C for 2 min and 72°C for 3 min were carried out in a DNA thermal cycler (MJ Research, Watertown, MA, U.S.A.). At the end of the first PCR reaction, a 10 min extension at 72°C was carried out. For the second PCR reaction, 10 μl of the first reaction were added to 90 μl of PCR buffer containing deoxynucleotides and Taq polymerase as in the first reaction and 50 pmol each of the internal sets of primers (sense, nucleotides 60 to 83, 5′GTCTTCACGCAGAAAGCGTCTAG3′; antisense, nucletides 268 to 251, 5′CCCAACACTACTCGGCTAGC3′). Thirty cycles were carried out as in the first PCR set. Individual 10 μl aliquots of the second amplification mixture were subjected to electrophoresis in 2% agarose gels. Bands of the expected size (209 nucleotides) were visualized by ethidium bromide staining. All samples were assayed at least twice with reproducible results.

Hybridization of PCR product

Hybridization of the final PCR product was performed by the DEIA assay [13, 14] kindly provided by Sorin Biomedica (Saluggia, Italy). The assay employs an antisense oligonucleotide probe derived from a region internal to the PCR product and not overlapping the primers used for the amplification (oligonucleotide position from 185 to 143). The probe is bound to wells of a microtiter plate in which the denatured PCR mixture is incubated. The hybridization is performed in 1 × SSC, 2 × Denhardt's solution, 10 mM Tris-HCl, pH 7.5, 1 mM EDTA for 1 h at 50°C and it is then revealed by means of a monoclonal antibody that selectively recognizes double but not single-stranded DNA. The monoclonal antibody is finally detected by horseradish peroxidase (HRP)-labelled rabbit anti-mouse IgG antibody followed by incubation with the chromogen-substrate solution and reading of the absorbance values at 450 nm. The cut-off value was calculated by adding 2 standard deviations to the mean value of the negative controls.

Results

The ALT values were persistently within the normal range (i.e. <40 U/L) in 7/13 subjects (53.8%), they fluctuated (ranging from 24 U/L to 148 U/L) in 4 subjects (30.7%), while in the remaining 2 subjects (15.4%) they were persistently raised (Table 1). The determination of HCV-RNA at enrollment and at the end of the follow-up showed 5 (38.5%) and 7 (53.8%) positive serum samples, respectively (Table 1). The examination of the PCR positive samples by the DEIA test confirmed in all samples the specificity of the PCR product through the hybridization of the amplified sequences to the immobilized capture HCV-DNA oligonucleotide. The optical density of the positive samples ranged from 0.314 to 2.692 (cut-off value = 0.120), while the optical density of the negative samples never exceeded 0.08 (range 0.014–0.078). On the whole, circulating HCV-RNA was demonstrated once during the observation period in 2 patients (15.4%), on both examinations in 5 patients (38.5%) and was never found in 6 patients (46.6%, Table 1). Of the 7 patients with viremia in one or both examined samples, 6 had fluctuating or persistently raised ALT levels, while the remaining 6 non-viremic patients showed normal ALT levels during the whole observation period (Fisher's exact test, $P = 0.004$; Table 2). Immunoblotting examination by RIBA-2 of sera obtained at the end of the follow-up confirmed the presence of isolate positivity for antibodies to the core antigen (C22) in all patients.

By Liatek HCV assay, in which four different core peptides are present as separate antigen bands, the core reactivity was scored from 1 to 4 reactive bands: 7 patients reacted to 1 or 2 core bands and the remaining 6 patients reacted to 3 or 4 core bands. However, no correlation was found between the number of reactive core bands and viremia (Table 2). Interestingly, by Liatek examination, antibodies to the NS5-derived peptide were found in 6 patients (46.1%), and the reactivity to NS5 band was significantly associated to the presence of circulating HCV-RNA. In fact all patients with antibodies to NS5 were viremic in one or both serum samples, compared to one out of seven anti-NS5 negative subjects (Fisher's exact test, $P = 0.004$; Table 2). None of the

Table 1. Biochemical, serological and virological characteristics of the 13 enrolled patients with isolate antibody to C22 antigen by RIBA-2

Patient no.	Sex	Age yrs	Follow-up months	Enrollment					End follow-up				
					antibodies to synthetic peptides					antibodies to synthetic peptides			
				ALT U/L	NS4	NS5	CORE	HCV RNA	ALT U/L	NS4	NS5	CORE	HCV RNA
1	f	46	10	30	–	–	1[a]	–	27	–	–	1[a]	–
2	m	48	10	28	–	–	1	–	16	–	–	1	–
3	m	39	12	22	–	–	2	–	20	–	–	2	–
4	m	49	13	24	–	–	2	–	19	–	–	2	–
5	m	55	8	20	–	–	4	–	16	–	–	4	–
6	m	45	18	14	–	–	4	–	13	–	–	4	–
7	m	27	14	31	–	+	4	–	55	–	+	4	+
8	m	58	8	23	–	+	4	–	47	–	+	4	+
9	f	58	16	53	–	–	1	+	38	–	–	1	+
10	f	59	18	11	–	+	1	+	12	–	+	1	+
11	f	48	12	48	–	+	4	+	18	–	+	4	+
12	f	20	13	74	–	+	2	+	50	–	+	2	+
13	m	56	10	86	–	+	4	+	85	–	+	4	+

[a] 1–4 indicates the number of core reactive bands

Table 2. Serum HCV-RNA: correlation with ALT increase and positivity for antibodies to NS5- and core-derived synthetic peptides, in 13 subjects

	HCV-RNA +[a]	HCV-RNA −[b]	Fisher's exact test
ALT increase (fluctuating or persistent)	6	0	P = 0.004
Anti-NS5	6	0	P = 0.004
Anti-core reactivity score 1/2	3	4	
			P = n.s.
Anti-core reactivity score 3/4	4	2	

[a] In at least one sample
[b] In both samples

patients had antibodies to the NS4-derived peptide. Finally, the serologic pattern of each patient was unchanged during the follow-up.

Based on the results of HCV-RNA detection by RT-PCR, on the behavior of ALT levels and on the serologic profile obtained by Liatek-HCV, two groups of patients could be recognized: the first group included 6 subjects (Table 1, patients 1 to 6) without circulating HCV-RNA, with persistently normal ALT levels and isolate reactivity to the core peptides. The second group included 7 patients (Table 1, patients 7 to 13) with persistent or intermittent viremia, of whom 6 had fluctuating or persistently raised ALT levels and 6 had antibodies to NS5.

Discussion

The aim of this study was to evaluate the meaning and clinical evolution of the isolate positivity for antibodies to the hepatitis C virus core antigen (C-22) detected by RIBA-2. This antibody profile is, at present, not completely understood [1, 2, 6, 16]. In particular, the study was designated to evaluate whether isolate anti-C22 represents a transient seroconversion phase, to assess the behavior of ALT levels on repeating examination, and to check the presence of viremia and of antibodies to an NS5-derived peptide in subjects followed-up for 8–18 months.

By both RIBA-2 and Liatek examination, the serologic pattern of each subject was shown to remain unchanged during the follow-up. In particular, no seroconversion to NS3- or NS4-derived antigens was found over an observation period of 8–18 months, and although antibodies to NS5 peptide were detected in 6 patients, they were already present at the time of the first examination. This result seems to indicate that the examined subjects were not in the early phase of HCV infection [8, 9].

Based on our findings it was not possible to determine whether anti-NS3, anti-NS4 and anti-NS5 antibodies (in the 7 negative subjects) were never produced, or whether they were lost [5]. However, it was apparent that isolate reactivity to C-22 antigen by RIBA-2 is a heterogeneous condition that implies two distinct categories of patients: those with inactive or absent and those with replicating HCV infection. In fact, in 6 out of 13 subjects (46.1%; Table 1, cases 1–6) no viremia and no cytonecrosis were observed. In these subjects the anti-C22 sero-conversion, without ongoing viral replication, could be due to a favorable host-virus balance [19], or to infection by a less virulent HCV strain [12], or to infection by an HCV strain that was undetectable by our primers because of a mutation in the 5' non-coding region [23]. Moreover, since in these 6 subjects only antibodies reacting to the HCV core antigen and peptides were detected, the possibility of a non-specific crossreactivity cannot be ruled out. An autoantibody, anti-GOR, was recently described which is directed to a host nuclear antigen that shares several amino acids with the HCV core antigen [18]. However, anti-GOR seems to be induced by HCV infection and it is frequently found in patients with HCV infection and autoimmune hepatitis [17]. In the subjects that we describe, however, no anti-nuclear (ANA) nor liver-kidney-microsomal (LKM-1) antibodies have been found (data not shown). Therefore, if the core reactivity was a cross reactivity, it could possibly be due to a different antibody.

In the remaining 7/13 subjects (53.8%; Table 1, case 7–13) the positivity of HCV-RNA in at least one examination indicates the presence of an active HCV infection that in 6 patients (86%) was associated with intermittent or continuous liver cytonecrosis. Although raised ALT levels and viremia tended to be concomitant ($P = 0.004$, Table 2), in 4 out of 12 HCV-RNA positive serum samples the ALT values were within the normal range (Table 1), and one viremic patient had persistently normal ALT levels over an 18 month observation period (Table 1, case 10). Therefore, normal ALT levels do not necessarily imply absence of viremia. Finally, the intensity of the core reactivity, indicated by the number of reactive core bands on Liatek HCV, was not significantly associated with a positive PCR result (P = n.s., Table 2), and it could not be regarded as an indicator of viremic status [21]. These data indicate that PCR examination remains the only reliable means of identifying HCV-RNA carriers.

Interestingly, the presence of antibodies to the NS5-derived synthetic peptide was significantly associated with viral replication, being positive in 6 out of 7 PCR positive subjects compared to 0 out of 6 PCR negative patients (P = 0.004, Table 2). The NS5 gene of HCV encodes for a protein containing the RNA-dependent RNA-polymerase [7], therefore it is not surprising that patients with replicating HCV infection produce antibodies to this protein, which must be well expressed during replication.

It has been recently reported that antibodies to C-33 antigen, encoded by the NS3 gene, are significantly associated with the presence of circulating HCV-RNA sequences, probably because this protein is associated with a helicase enzyme activity necessary for viral replication [15]. Our results indicate that antibodies to the NS5-derived peptide can be detected even in patients with an incomplete antibody response to HCV, who lack antibodies to C33 antigen. Among these patients, anti-NS5 antibodies allow discrimination of those with viral replication from those with inactive infection. Therefore, the detection of anti-NS5 can be of great diagnostic utility.

In conclusion, the isolate positivity for anti-C22 by RIBA-2 can be a rather stable serologic pattern, not associated to a rapidly evolving seroconversion. It seems to correspond to two distinct conditions: one is characterized by persistent or fluctuating viremia and liver cytonecrosis on repeated examinations and presence of antibodies to NS5-derived peptide. The other one, on the contrary, is non-viremic and lacks cytonecrosis and anti-NS5. It remains to be elucidated whether the second condition is the result of an HCV-related seroconversion or is due to a non-specific cross reactivity.

Acknowledgement

This work was supported, in part, by a grant from MURST, project "Cirrosi Virali".

References

1. Boudard D, Lucas JC, Adjouu C, Muller JY (1992) HCV confirmatory testing of blood donors. Lancet 339: 372
2. Chan SW, Simmonds P, McOmish F, Lee Yap P, Mitchell R, Dow B, Follet E (1991) Serological responses to infection with three different types of hepatitis C virus. Lancet 338: 1391
3. Chiba J, Ohba H, Matsuura Y, Watanabe Y, Katayama T, Kikuchi S, Saito I, Miyamura T (1991) Serodiagnosis of hepatitis C virus (HCV) infection with an HCV core protein molecularly expressed by a recombinant baculovirus. Proc Natl Acad Sci USA 88: 4641–4645
4. Chomczynsky P, Sacchi N (1987) Single-step method of RNA isolation by acid guanidinium thiocyanate-phenol-chloroform extraction. Anal Biochem 162: 156–159

5. Farci P, Alter HJ, Wong D, Miller RH, Shin JW, Jett B, Purcell RH (1991) A long-term study of hepatitis C virus replication in Non-A, Non-B hepatitis. N Engl J Med 325: 98–104

6. Follet EA, Dow BC, McOmish F, Lee Yap P, Crawford RJ, Mitchell R, Simmonds P (1992) HCV confirmatory testing of blood donors. Lancet 339: 928

7. Houghton M, Weiner A, Han J, Kuo G, Choo QL (1991) Molecular biology of the hepatitis C viruses: implications for diagnosis, development and control of viral disease. Hepatology 14: 381–388

8. Hutchinson Mc JG, Person JL, Govindarajan S, Valinluck B, Gore T, Lee SR, Nelles M, Polito A, Chien D, DiNello R, Quan S, Kuo G, Redeker AG (1992) Improved detection of hepatitis C virus antibodies in high-risk population. Hepatology 15: 19–25

9. Katayama T, Mazda T, Kikuchi S, Harada S, Matsuura Y, Chiba J, Ohba H, Saito I, Miyiamura T (1992) Improved serodiagnosis of non-A, non-B hepatitis by an assay detecting antibody to hepatitis C virus core antigen. Hepatology 15: 391–394

10. Kuo G, Choo Q-L, Alter HJ, Gitnick GL, Redeker AG, Purcell RH, Miyamura T, Dienstag JL, Alter MJ, Stevens CE, Tegtmeier GE, Bonino F, Colombo M, Lee WS, Kuo C, Berger K, Shuster JR, Overby LR, Bradley DW, Houghton M (1989) An assay for circulating antibody to a major etiologic virus to human non-A, non-B hepatitis. Science 244: 362–364

11. Imberti L, Cariani E, Bettinardi A, Zonaro A, Albertini A, Primi D (1991) An immunoassay for specific amplified HCV sequences. J Virol Methods 34: 233–243

12. Irving WL, Day S, Eglin RP, Bennet DP, Jones DA, Nuttall P, James V (1992) HCV and PCR negativity. Lancet 339: 1425

13. Mantero G, Zonaro A, Bertolo P, Primi D (1991) DNA enzyme immunoassay (DEIA): a general method for detecting polymerase chain reaction products based on anti-DNA antibody. Clin Chem 37: 422–429

14. Marcellin P, Martinot-Peignoux M, Boyer N, Pouteau M, Aumont P, Erlinger S, Benhamou JP (1991) Second generation (RIBA) test in diagnosis of chronic hepatitis C. Lancet 337: 551–552

15. Martinot-Peignoux M, Marcellin P, Xu LZ, Bernuau J, Erlinger S, Benhamou JP, Larzul D (1992) Reactivity to c 33c antigen as a marker of hepatitis C virus multiplication. J Infect Dis 165: 595–596

16. Mateos M, Ballestero S, Polanco AM, Camarero C (1992) Diagnostic difficulties of HCV serological tests. Lancet 340: 116

17. Michel G, Ritter A, Gerken G, Meyer Zum Büschenfelde KH, Decker R, Manns MP (1992) Anti-GOR and hepatitis C virus in autoimmune liver diseases. Lancet 339: 267–269

18. Mishiro S, Takeda K, Hoshi Y, Yoshikawa A, Gotanda T, Itoh Y (1991) An autoantibody cross-reacting to hepatitis C virus core and a host nuclear antigen. Autoimmunity 10: 269–273

19. Pozzato G, Moretti M, Franzin F, Crocè LS, Tiribelli C, Masayu T, Kaneko S, Unoura M, Kobayashi K (1991) Severity of liver disease with different hepatitis C viral clones. Lancet 338: 509

20. Sugitani M, Inchauspè G, Shindo M, Prince AM (1992) Sensitivity of serological assays to identify blood donors with hepatitis C viremia. Lancet 339: 1018–1019

21. Tong CYM, Codd AA (1992) RIBA-2-band intensity and PCR in HCV infection. Lancet 340: 117

22. VanDerPoel CL, Cuypers HTM, Reesink HW, Weiner AJ, Quan S, DiNello R, van Boven JJP, Winkel I, Mulder-Folkerts D, Exel-Oehlers PJ, Schaasberg W,

Leentwaar-Kuypers A, Polito A, Houghton M, Lelie PN (1991) Confirmation of hepatitis C virus infection by new four-antigen recombinant blot assay. Lancet 337: 317–319
23. Su LX, Martinot-Peignoux M, Marcellin P, Benhamou JP, Larzul D (1992) Lack of polymerase chain reaction amplification in the 5′ region of a hepatitis C virus isolate. J Infect Dis 165: 1164–1165

Authors' address: Dr. G. Taliani, Institute of Tropical Diseases, "La Sapienza" University, Viale Regina Elena 324, I-00161 Rome, Italy.

Arch Virol (1993) [Suppl] 8: 229–234

Is HCV transmitted by the vertical/perinatal route?

E. Tanzi[1], C. Stringhi[2], S. Paccagnini[3], M.L. Profeta[1], and A.R. Zanetti[4]

[1] Institute of Virology, University of Milan, [2] Department of Pediatric Infectious Diseases, Hospital Niguarda, Milan, [3] IV Department of Pediatric, Hospital L. Sacco, Milan; and [4] Department of Hygiene, University of Camerino, Camerino, Italy

Summary. Hepatitis type C is the major aetiological cause of both parenterally transmitted and cryptogenic, sporadic or community acquired nonAnonB hepatitis. The lack of known parenteral risk factors in a consistent number of cases with nonAnonB hepatitis has stimulated the search of other possible modes of viral transmission. The aim of this report is to review the evidence both for and against vertical/perinatal transmission of HCV from anti-HCV positive mothers to infants.

Introduction

Hepatitis type C virus (HCV) can be efficiently transmitted by parenteral route as demonstrated by the high prevalence of anti-HCV antibody found among blood and blood product recipients, haemodialysis patients and intravenous drug users [1, 4, 9]. Moreover it is known that HCV is a major aetiological cause of the so-called sporadic, cryptogenic or community acquired nonAnonB (NANB) hepatitis [4, 9].

The failure to detect known parenteral risk factors in a vast proportion of acute and chronic NANB hepatitis cases has raised the hypothesis that HCV infection can be transmitted by other routes than overt percutaneous exposure. Thus, it is reasonable to speculate that HCV, like other blood borne viruses such as hepatitis B virus (HBV) and human immunodeficiency virus (HIV), can be transmitted by sexual or by vertical/perinatal routes. However data reported in published studies are so controversial that they leave the issue unresolved. The aim of this brief article is to review the evidence both for and against transmission of HCV from anti-HCV positive mothers to infants.

Evidence of vertical/perinatal transmission

Several reports show that babies are at higher risk of vertical/perinatal infection when the mother is HCV and HIV co-infected. For example, Giovannini et al. [6] found serological evidence of HCV infection in 11 out of 25 babies (44%) born to HCV/HIV positive mothers. All HCV- positive babies had simultaneously acquired HIV infection. Perez Alvarez et al. [13] showed that one of the 22 children (4.5%) born to HIV/HCV co-infected mothers acquired hepatitis C and HIV infection. In agreement with Giovannini et al. [6], these authors concluded that co-infection by HIV and HCV in children may be associated with a worse evolution of HIV infection.

A further demonstration that the risk of vertical/perinatal transmission of HCV may be enhanced by the contemporary presence of HIV in maternal blood was provided by the study of Weintrub et al. [18] in which 4 out of 43 infants (9%) born to HIV/HCV co-infected mothers developed active anti-HCV antibodies, 3 with elevated ALT levels. Again, all babies who became HCV infected were also infected with HIV. In agreement with these findings, Novati et al. [12] detected by nested Polymerase Chain Reaction (PCR) the presence of HCV-RNA in 4 out of 8 babies born to mothers with HCV and HIV co-infection. One case of vertical hepatitis C infection was reported in Sweden by Wejstal et al. [19] on a cohort of 8 babies born to anti-HCV positive intravenous drug addicted mothers, followed up for 1 year. No data were available concerning HIV infection among these mother/infant pairs.

By reverse transcription and PCR using primers in the highly conserved 5'-untranslated region of the viral genome, Thaler et al. [17] found HCV-RNA in 8 babies born to anti-HCV positive mothers. The persistence of viral RNA from birth up to 19 months of follow-up in the absence of active production of anti-HCV antibodies must be considered extremely important. The fact that transmission of HCV occurred in 5 babies born to mothers with anti-HCV alone suggests that HIV/HCV co-infection is not an absolute requirement for vertical transmission of HCV. Similarly, in Taiwan Joung et al. [8] found HCV-RNA in 11 of 13 infants born to PCR positive mothers with elevated serum ALT. In some children HCV-RNA persisted for up to 5 years in absence of anti-HCV and with normal serum aminotransferases. A high prevalence of HCV-RNA (up to 85%) in infants born to HCV infected mothers has also been reported in Japan by Kuroki et al. [10]. During follow-up all babies remained anti-HCV negative and showed normal ALT values.

Vertical/perinatal transfer of HCV has also been suggested by an anecdotical but well-planned molecular biology study which demonstrated HCV transmission from mother to child through two generations [7].

Failure to detect HCV vertical transmission

In contrast with the above mentioned studies, Fortuny et al. [5] found no serological and clinical evidence of HCV infection in 37 infants born to HIV/HCV positive mothers over an average follow-up period of 16 months. Passively acquired anti-HCV was lost by the age of 6 months and no active antibody production was detected in any of these babies followed from 2 to 32 months (median 15.4 m). However, since babies were not examined for HCV-RNA presence, the authors cannot exclude the presence of a silent HCV infection as seen by others [8, 10, 17].

In the same direction, a retrospective study conducted on stored sera by Stevens et al. [16] showed that none of the 17 infants born to anti-HCV positive women had hepatitis or was anti-HCV positive at 12 months of age or beyond.

Similarly, Lin et al. [11] reported no serological evidence for HCV perinatal transmission or spouse infection in Taiwan. By PCR, using primers of the 5′-non coding region, the same group [3] found no viral RNA in 8 babies born to anti-HCV positive mothers, although 6 of them were positive for HCV-RNA. In agreement with the above reported findings, Roudot-Thoraval [15] in France did not detect HCV-RNA or active antibodies to HCV in any of the 7 babies born to anti-HCV positive mothers followed-up for 12 months. Reinus et al. in the U.S.A. [14] failed to detect evidence of HCV infection in a cohort of 24 babies born to anti-HCV positive mothers, 4 of them HIV co-infected. In this study, HCV-RNA measured by two independent laboratories was found in the cord blood from one baby only. The child showed a transient ALT elevation at 5 months of age in the absence of markers of HAV and HBV infection. Negative PCR on samples collected at 1 and 8 months of age together with the absence of anti-HCV antibody after a 2 year follow-up led the authors to conclude that the occurrence of HCV-RNA in cord blood might have been the result of maternal blood contamination during delivery.

Similarly, in Milan (manuscript in prep.) we have found no evidence of HCV infection among 32 babies born to asymptomatic HCV positive women, although 18 of them (56%) were HCV-RNA positive. However, we have found transmission of HCV infection in 7 out of 15 (47%) infants born to mothers simultaneously infected with HCV (PCR positive) and HIV.

Discussion

The above reported studies provide conflicting evidence of vertical/ perinatal transmission of HCV. According to some investigators, transfer of HCV from mother to infant seems to be a common event while according to others the occurrence seems to be rare. The clinical setting where there is an almost universal agreement on the risk for vertical HCV transmission is when the mother/baby pairs are simultaneously infected with HCV and HIV. Available data suggest a possible synergic interaction between the two viruses, probably because the immune suppression due to HIV may favour HCV replication in the mother and, at the same time, may increase susceptibility of the baby to the infection. This fact may have important medical and public health implications expecially in the urban areas with a high prevalence of intravenous drug abusers where both HCV and HIV infections are endemic.

Another point of agreement is that almost all babies born to HCV infected mothers are anti-HCV positive at birth and that antibodies generally tend to disappear over time, suggesting a passive transfer rather than an "ex novo" antibody production in response to HCV infection.

Although clearance of maternal antibody seems to provide substantial evidence against vertical/perinatal transmission, no definitive conclusions can be drawn since in most studies babies were followed-up too shortly. In other words, due to the short observation periods, we cannot exclude that, at least in some instances, delayed anti-HCV seroconversions might have been missed. The most striking divergencies in the outcomes of different studies have been found when infection in the infant was determined by the detection of HCV-RNA using PCR. Discrepancies among different laboratories, which have been found to range from no viral RNA detection to very high frequency of HCV-RNA detection in anti-HCV seronegative infants, are so wide that they are difficult to understand. Differences in maternal populations studied, in PCR technology employed as well as the existence of multiple sub-types of HCV [2] have all been considered as possible variables in the search for an explanation.

However, the dilemma is not merely whether HCV can be transmitted by vertical/perinatal route, but more important, it concerns the epidemiological relevance of such a mode of transmission in maintaining the world-wide reservoir of HCV carriers.

If transfer of HCV from mother to infant is an uncommon event, we can assume that the large number of adult patients with chronic cryptogenic hepatitis C has acquired infection through unknown mechanisms, whereas if it occurs frequently, it is believable that a consistent number of such cases may have been generated through vertical/perinatal infection and subsequent development of a long-lasting silent chronic carrier state.

Further epidemiological studies are needed in order to better clarify this important issue.

Acknowledgements

The Study Group on vertical transmission of HCV in Lombardy includes: S. Bresciani, M.G. Marin, A. Rodella, D. Padula, G. Pizzocolo (Brescia); G. Cambiè, FF. D'Agostino (Lodi); I. Bulgarelli, M.L. Caccamo, F. Chiodo, E. D'Amico, G. Gubertini, E. Magliano, G. Miotto, M.L. Muggiasca, R. Pozzoli, L. Romanò (Milano); L. Vecchi, B. Zapparoli (Monza).

References

1. Alter HJ, Purcell RH, Shih JW, Melpolder JC, Houghton M, Choo QL, Kuo G (1989) Detection of antibody to hepatitis C virus in prospectively followed transfusion recipients with acute and chronic non-A, non-B hepatitis. N Engl J Med 321: 1494–1500

2. Cha TA, Beall E, Irvine B, Kolberg J, Chien D, Kuo G, Urdea MS (1992) At least five related, but distinct, hepatitis C viral genotypes exist. Proc Natl Acad Sci USA 89: 7144–7148

3. Chen DS, Lin HH, Chang MH, Chen PJ, Sung JL (1991) Mother-to-child transmission of hepatitis C virus. J Infect Dis 164: 428–429

4. Esteban JI, Esteban R, Viladomiu L, Lopez-Talavera JC, Gonzalez A, Hernandez JM, Roget M, Vargas V, Genesca J, Buti M, Guardia J, Houghton M, Choo QL, Kuo G (1989) Hepatitis C virus antibodies among risk groups in Spain. Lancet ii: 294–297

5. Fortuny C, Ercilla MG, Barrera JM, Gil C, Reverter M, Bruguera M, Jimenez R, Castillo R, Rodes J (1991) Vertical transmission of hepatitis C virus (HCV): a prospective study in infants born to HCV-seropositive mothers. In: Hollinger FB, Lemon SM, Margolis HS (eds) Viral hepatitis and liver disease. Williams and Wilkins, Baltimore, pp 418–419

6. Giovannini M, Tagger A, Ribero ML, Zuccotti G, Pogliani L, Grossi A, Ferroni P, Fiocchi A (1990) Maternal-infant transmission of hepatitis C virus and HIV infections: a possible interaction. Lancet 335: 1166

7. Inoue Y, Miyamura T, Unayama T, Takahashi K, Saito I (1991) Maternal transfer of HCV. Nature 353: 609

8. Joung MK, Park CK, Buskell-Bales ZJ, Seef LB, Houghton M, Kuo G, Chien D, Han JH (1991) Transmission of HCV from infected mother to infant. Proceedings of the Third International Symposium on HCV, Strasbourg, p 96

9. Kuo G, Choo QL, Alter HJ, Gitnick GL, Redeker AG, Purcell RH, Miyamura T, Dienstag JL, Alter MJ, Stevens CE, Tegtmeier GE, Bonino F, Colombo M, Lee WS, Kuo C, Berger K, Shuster JR, Overby LR, Bradley DW, Houghton M (1991) An assay for circulating antibodies to a major etiologic virus of human nonA-nonB hepatitis. Science 244: 362–364

10. Kuroki T, Nishiguchi S, Fukuda K, Shiomi S, Monna T, Murata R, Isshiki G, Hayashi N, Shikata T, Kobayashi K (1991) Mother-to-child transmission of hepatitis C virus. J Infect Dis 164: 427–428

11. Lin HH, Hsu HY, Chang MH, Hong KF, Young YC, Lee TY, Chen PJ, Chen DS (1991) Low prevalence of hepatitis C virus and infrequent perinatal or spouse infections in pregnant women in Taiwan. J Med Virol 35: 237–240

12. Novati R, Thiers V, D'Arminio Monforte A, Maisonneuve P, Principi N, Conti M, Lazzarin A, Brechot C (1992) Mother-to-child transmission of hepatitis C virus detected by nested polymerase chain reaction. J Infect Dis 165: 720–723
13. Perez Alvarez L, Gurbindo MD, Hernandez-Sampelayo T, Casado C, Leon P, Garcia Salz A, De Andres R, Najera R (1992) Mother to infant transmission of HIV and hepatitis C infections in children born to HIV-seropositive mothers. AIDS 6: 427–428
14. Reinus JF, Leikin EL, Alter HJ, Cheung L, Shindo M, Jett B, Piazza S, Shih JW (1992) Failure to detect vertical transmission of hepatitis C virus. Ann Intern Med 117: 881–886
15. Roudot-Thoraval F, Guillot F, Huraux C, Girollet PP, Deforges L, Thiers V, Dhumeaux D (1991) Serological and biochemical follow-up of newborns from mothers with hepatitis C virus infection. Proceedings of the Third International Symposium on HCV, Strasbourg, p 85
16. Stevens CE, Taylor PE (1991) Perinatal and sexual transmission of hepatitis C virus: a preliminary report. In: Hollinger FB, Lemon SM, Margolis HS (eds) Viral hepatitis and liver disease. Williams and Wilkins, Baltimore, pp 407–410
17. Thaler MM, Park CK, Landers DV, Wara DW, Houghton M, Veereman-Wauters G, Sweet RL, Han JH (1991) Vertical transmission of hepatitis C virus. Lancet 338: 17–18
18. Weintrub PS, Veereman-Wauters G, Cowan MJ, Thaler MM (1991) Hepatitis C virus infection in infants whose mothers took street drugs intravenously. J Pediatr 119: 869–874
19. Wejstal R, Hermodsson S, Iwarson S, Norkrans G (1990) Mother to infant transmission of hepatitis C virus infection. J Med Virol 30: 178–180

Authors' address: Dr. A.R. Zanetti, Department of Hygiene, University of Camerino, V.le E. Betti 3, I-62032 Camerino (MC), Italy.

VI. Clinical course and therapy of chronic viral hepatitis

Arch Virol (1993) [Suppl] 8: 237–248

Multiple viral infections in HIV-infected children with chronically-evolving hepatitis

G. Nigro[1], **G. Taliani**[2], **A. Krzysztofiak**[1], **S. Mattia**[1], **U. Bartmann**[1], **A. Petruccelli**[1], **P. Falconieri**[1], **E. Fridell**[3], **P. Cinque**[3], **A. Linde**[3], **H. Dahl**[3], and **B. Wahren**[3]

[1] Pediatric Institute and [2] Institute of Tropical and Infectious Diseases, "La Sapienza" University, Rome, Italy
[3] National Bacteriological Laboratory, Karolinska Institute, Stockholm, Sweden

Summary. Hepatic involvement was investigated in 31 children with perinatal HIV-1 infection, who were followed for 2–82 months (mean 30.5). Liver disease, as revealed by increased aminotransferase levels, liver biopsy or necroscopy, was diagnosed in 18 children (58%), of which 7 (22.5%) had acute hepatitis and 11 (35.5%) showed chronic liver disease. Overall, 40 persistently active or recurrent viral infections, as demonstrated by positive culture and/or detection of serum DNA, specific IgM, IgA and high levels of IgG, were revealed in the children with liver disease, while 12 similar infections were detected in 13 children without liver disease (p < 0.001). In particular, the children with liver disease showed a significantly (p < 0.002) higher incidence of cytomegalovirus (CMV) infections than children without liver disease (13 versus 3). Moreover, hepatitis C and B virus infections were revealed only in children with liver disease (5 and 1 patients, respectively). Clinical outcome showed a significantly (p < 0.001) higher mean survival in the children without liver disease than those with liver disease (47.5 versus 18.2 months). In fact, nine of the children with liver disease (50%) died, as opposed to only one of the children without liver disease (7.7%; p = 0.01). Based on these findings, liver disease is indicative of a poor prognosis in children with HIV infection, being related to the presence of multiple active viral infections.

Introduction

A wide range of clinical manifestations has been reported in children with human immunodeficiency virus type 1 (HIV-1) infection [6, 16, 17,

22, 26]. Hepatomegaly or increase of serum aspartate aminotransferase (AST), alanine aminotransferase (ALT) or other enzymes, as well as macro- or microscopic changes of the liver at autopsy, are frequently observed in adult patients with AIDS [18, 20]. In children with perinatal HIV-1 infection, hepatitis has been recently associated with a shorter survival [26].

The aim of our study was to investigate the prevalence and outcome of liver disease in children with perinatal HIV-1 infection, together with the occurrence of infections by viral agents other than HIV, which could also be responsible for the progression of HIV disease.

Materials and methods

Patients

From February 1985 to October 1992, 31 Italian children born to HIV-1-infected mothers were followed up for two to 82 months (mean 30.5). Patients underwent clinical and laboratory examinations on admission and then every 1–2 months, in accordance with standard procedures. After a mean of 14 months, 28 patients were given zidovudine, following the development of clinical and immunologic features typical of stage P2A according to the rating scale of the Centers for Disease Control [5].

Laboratory tests

AST, ALT, alkaline phosphatase (ALP), lactate dehydrogenase (LDH) and gamma-glutamyl transferase (gGT) levels were examined to evaluate disease activity and liver function. Acute hepatitis was defined by a high and sharp increase of ALT levels returning to normal within six months and/or by histologic findings of marked hepatic involvement. Chronic hepatitis was defined by a persistent or fluctuating increase of ALT levels, reaching a twofold increase above upper normal limit at least three times.

Virological investigations

Markers for hepatitis B virus (HBV) and antibodies against hepatitis A virus (HAV) were detected by ELISA and/or RIA from Abbott (North Chicago, IL, U.S.A.) and Sorin (Saluggia, Italy). Antibodies to hepatitis C virus (HCV) were detected by a second generation EIA (Ortho, NY, U.S.A.) and positive results were confirmed by Recombinant Immunoblotting Assay (RIBA). Cytomegalovirus (CMV) infection was diagnosed by viral isolation from urine and blood on primary human embryonic or foreskin fibroblasts and by detection of CMV-DNA by polymerase chain reaction (PCR), complement-fixing (CF) and class-specific CMV antibodies (IgG, IgA, IgM) by commercial (Radim, Pomezia, Italy) and non-commercial enzyme immunoassays [12]. Epstein-Barr virus (EBV) infection was diagnosed by detection of anti-viral capsid (VCA) IgM and IgG antibodies and anti-membrane antigens (MA) IgA antibodies by

EIAs from Dupont (Wilmington, DE, U.S.A.), and by anti-nuclear antigens (NA) IgM and IgG antibodies by EIAs from Incstar (Stillwater, MN, U.S.A.). IgM, IgA and IgG antibodies to herpes simplex viruses (HSV) types 1 and 2 were detected with EIAs from Radim. IgM and IgG antibodies to human herpesvirus 6 (HHV-6) were detected using immunofluorescence assays [11]. B19-DNA in the serum by PCR, and antibodies (IgM, IgG) to parvovirus B19 were detected by two EIAs using synthetic peptide or recombinant protein as coating antigens [8, 24].

Based on the clinical features shown by each of the patients, additional serological and culture investigations for other pathogens were performed. In particular, cultures for viruses other than herpesviruses were carried out on HEp-2 cells, routine virus-specific antibodies were examined by CF test, and IgM and IgG with commercial EIAs. To eliminate interference by the rheumatoid factor (RF) in detecting virus-specific IgM, IgM-positive sera were re-tested after RF was eliminated by direct latex agglutination (Gamma, Houston TX, U.S.A.). Furthermore, to confirm IgM-positive results, μ-capture systems for CMV-IgM (Medac, Hamburg, Federal Republic of Germany), EBV-VCA-IgM (Ismunit, Pomezia, Italy) and HSV-IgM (Poli, Milan, Italy) were also used.

Polymerase chain reaction (PCR)

Different PCR procedures were used for the detection of CMV-DNA, B19-DNA, HBV-DNA and HCV-RNA in serum samples. CMV-DNA was detected by nested PCR using primers from the fourth exon of CMV-IEA1 gene located in EcoRI J fragment of strain Ad169 [4]. The first flanking primer set consisted of 5' TGAGGATAAGCGGGAGATGT 3' (C; nucleotides 1729-1748 complementary to the antisense strand) and 5' ACTGAGGCAAGTTCTGCAGT 3' (D; nucleotides 1951-1970 complementary to the sense strand). The nested primer set consisted of 5' AGCTGCATGATGTGAGCAAG 3' (A; nucleotides 1767-1786 complementary to the antisense strand) and 5' GAAGGCTGAGTTCTTCTTGGTAA 3' (B; nucleotides 1893-1912 complementary to the sense strand). For the detection of B-19-DNA [9], the outside primer pair consisted of 5' GGACTGTAGCAGATGAAGAG (390; nucleotides 2955-2974) and 5' TATGGGACTGATGGTG (490; nucleotides 3364-3349), the inside pair consisted of 5' GGGTTTCAAGCACAAGTAG (190; 3002-3220) and 5' CCTTATAATGGTGCTCTGGG (290; 3291-3272). HBV-DNA was detected using primers (10) for the beginning and the end of "S" region (positions 254-273: 5' ACTCGTGGTGGACTTCTCTC 3'; positions 705-686: 5' GAACCACTGAAC-AAATGGCA 3') and "C" region (positions 1778-1800: 5' GGAGGCTGTAGGCA-TAACTTTTT 3'; positions 2290-2270: 5' TAGGGGCATTTGGTGGTCTGT 3'). Finally, HCV-RNA was detected (25) using outside primers (sense, nucleotides 19 to 38, 5' GCGACACTCCACCATGAATC 3'; antisense, nucleotides 342 to 321, 5' ATGGTGCACGGTCTACGAGACC 3') and inside primers (sense, nucleotides 60 to 38, 5' GTCTTCACGCAGAAAGCGTCTAG 3'; antisense, nucleotides 268 to 251, 5' CCCAACACTACTCGGCTAGC).

Statistical analysis

P-values were calculated by two-tailed Student's T-test, Fisher's exact test and chi-square test.

Results

Prevalence and outcome of liver disease

Out of 31 patients that were followed, 7 (22.5%) had acute hepatitis and 11 (35.5%) showed chronic liver disease. Among a total of 18 children (58%) with liver disease, 9 (50%) died, as opposed to only one (7.7%; p < 0.001) of the children without liver disease.

Clinical outcome and occurrence of viral infections in children with acute hepatitis

Of 7 children with acute hepatitis, 5 died at a mean age of 6.4 months. They all showed signs of active CMV infection, as demonstrated by detection of CMV-DNA, CMV-specific IgM and IgA antibodies (Table 1). Of these, two patients with active CMV infection shown by autoptic multi-organ CMV inclusions or positive CMV culture had detectable

Table 1. Active viral infections detected in 7 HIV-infected children with acute hepatitis

Patient no.	Follow-up (months)	Viral infections	Positive parameters	Outcome HIV disease
1	7	absent		lost to follow-up
2	14	HCV	antibodies	dead
		CMV	DNA, IgA, autoptic CMV inclusions	
3	7	HCV	antibodies	lost to follow-up
4	3	HCV	antibodies	dead
		CMV	culture, DNA, IgM, autoptic CMV inclusions	
		EBV	VCA-IgM, MA-IgA, EBNA-IgM	
		B19	DNA, IgM	
5	3	CMV	IgA, autoptic CMV inclusions	dead
6	2	CMV	IgA, autoptic CMV inclusions	dead
7	9	adenovirus	culture, CF-seroconversion histologic examination	dead
		CMV	culture, IgA	

HCV Hepatitis C virus, *CMV* cytomegalovirus, *EBV* Epstein-Barr virus, *VCA* virocapsid antigens, *EBNA* Epstein-Barr nuclear antigens, *MA* membrane antigens, *B19* parvovirus B19, *CF* complement fixation

CMV-specific IgA antibodies but negative IgG and IgM. Antibodies to HCV were detected in three children aged 2, 6 and 13 months. Three children also showed antibodies to B19, while adenovirus and EBV infections were revealed in one patient each. No active viral infections were detected in a 5-month-old boy, who showed increased AST (174 U/L) and ALT (209 U/L) levels for about one month. Both surviving children were lost to follow-up at the age of seven and eight months, respectively, so a relapse of the liver disease might have occurred.

Of 4 autopsied children, 3 showed disseminated CMV infection, in one of which CMV inclusions were also found in the liver, and the remaining one had CMV pneumonitis. In addition, one patient also had *Pneumocistis carinii* pneumonitis and *Candida albicans* esophagitis, while another patient also showed *Toxoplasma gondii* encephalitis and pneumonitis. The only patient for whom autopsy was not carried out underwent liver biopsy, when acute hepatitis occurred. In spite of the presence of bilateral retinitis and CMV-positive throat culture, only adenovirus hepatitis was demonstrated by positive culture, CF-seroconversion (from $<1:8$ to $1:64$) and typical histopathologic changes.

Clinical outcome and occurrence of viral infections in children with chronic liver disease

Out of 11 patients with chronically-evolving liver disease, 4 died at a mean age of 16 months, while the remaining patients were followed for a mean of 30.3 months (Table 2). All children who died showed active CMV-B19 coinfection. In addition, 2 children also had signs of reactivated EBV infection. Autoptic examination, permitted in 2 patients, showed CMV inclusions in the liver of one. Major histologic changes included lymphocytic infiltration, cholestasis, necrosis, fibrosis and steatosis.

Of the surviving 7 patients (mean age: 31.3 months), 5 showed persistently or recurrently active CMV infection associated with EBV (4 cases), B19 (4 cases), HHV-6 (2 cases), HCV (1 case) or HSV-1 (1 case). The other two children showed HBV infection or HCV associated with persistently active EBV infection. However, since non-specific reactivity in the detection of HHV-6 antibodies could not be firmly established, the prevalence of HHV-6 infections was probably higher than revealed. Liver biopsy, performed on 3 children, showed histologic signs of viral hepatitis due to CMV infection in one and inflammation and steatosis in the other two patients. No correlation was found between the number of viral infections, mean aminotransferase levels and the clinical outcome ($p > 0.05$).

Table 2. Active viral infections occurring in 11 HIV-infected children
with chronic liver disease

Patient no.	Follow-up (months)	Viral infections	Positive parameters	Outcome HIV disease
1	18	CMV	culture, IgM	dead
		EBV	VCA-IgM, MA-IgA	
		B19	DNA, IgM	
2	26	CMV	culture, DNA, IgA	P2AD1D3
		EBV	VCA-IgM, MA-IgA, EBNA-IgM	
		HSV-1	IgM, IgA	
		B19	DNA, IgM	
3	23	CMV	DNA, IgA	dead
		EBV	VCA-IgM, EBNA-IgM	
		B19	DNA, IgM	
4	14	CMV	DNA, autoptic CMV inclusions	dead
		B19	IgM	
5	27	CMV	culture, IgA	P2ABD1F
		HHV-6	IgM	
		B19	DNA, IgM	
6	41	CMV	culture, IgA	P2AD1
		EBV	VCA-IgM, EBNA-IgM	
7	24	HBV	HBsAg, HBeAg, DNA	P2A
8	11	HCV	RNA, antibodies	P2AC
		EBV	MA-IgA, EBNA-IgM	
9	72	HCV	RNA, antibodies	P2AD1
		CMV	culture, IgA	
		EBV	MA-IgA, EBNA-IgM	
		B19	IgM	
10	9	CMV	culture, IgM, autoptic CMV inclusions	dead
		B19	DNA, IgM	
11	19	CMV	culture, DNA, IgA	P2ABD1-3
		EBV	VCA-IgM, MA-IgA	
		HHV-6	IgM	
		B19	DNA, IgM	

HBV Hepatitis B virus, *HCV* hepatitis C virus, *CMV* cytomegalovirus, *EBV* Epstein-Barr virus, *VCA* virocapsid antigens, *MA* membrane antigens, *EBNA* Epstein-Barr nuclear antigens, *HSV-1* herpes simplex virus type 1, *HHV-6* human herpesvirus 6, *B19* parvovirus B19

Three patients, showing persistently increased AST (ranging between 382 and 1,970 U/L) and ALT (between 482 and 692 U/L levels for about 2 years, concomitant with positive CMV culture and/or serum CMV-DNA, were given ganciclovir at a dosage of 10 mg/kg/day for several courses of one to three weeks. Virological response and clinical improvement occurred in two patients, who temporarily showed aminotransferases dropping to values near to normal and decreased size of the liver. The third patient, who was treated with ganciclovir, continued to show positive CMV-DNA detection and high ALT levels until death occurred at 13 months. The child with chronic HBV hepatitis was given interferon, which was followed by immediate although temporary decrease of the aminotransferase levels.

Infections in children without liver disease

Of 13 children, who did not show liver disease, three still have asymptomatic HIV disease (Table 3). They all showed a significantly ($p < 0.001$) higher mean survival (47.4 months) than patients with liver disease ($p < 0.001$), both acute ($p < 0.0001$) and chronic ($p = 0.01$). Only one patient died at 51 months, showing concomitant infections with CMV, B19, EBV and HSV-1. Two other children showed persistently active CMV-B19 coinfection. In the remaining 5 symptomatic HIV-infected children, 4 active infections caused by B19, 3 by EBV and one by CMV were revealed. The 3 children who are asymptomatically HIV-infected are seronegative for the tested viruses. Therefore, a significantly ($p < 0.001$) lower number of active viral infections (15 versus 43) was detected in the children without liver disease than in those with acute or chronic liver disease.

Discussion

Our study showed that liver disease is a frequent and serious complication in children with HIV infection, being often chronically-evolving. In fact, of 18 children with liver disease, 5 died during the acute stage and 11 developed chronic disease, which was associated with fatal evolution in a further 4 patients. Multiple active virus infections appeared to be frequently associated with acute or chronic liver disease, based on the detection of specific serum DNA, IgM and IgA antibodies as well as culture. While the specificity of viral DNA or IgA antibodies is generally reliable, positive IgM antibodies may be due to non-specific factors or polyclonal B-cell stimulation by different agents. However, in our

244 G. Nigro et al.

Table 3. Active viral infections occurring in 13 HIV-infected children without liver disease

Patient no.	Follow-up (months)	Viral infections	Positive parameters	Outcome HIV disease
1	45	CMV	culture	P2ABD2
2	49	absent		P1A
3	44	EBV B19	MA-IgA, EBNA-IgM DNA	P2AD1
4	82	B19	DNA, IgM	P2A
5	18	B19	DNA, IgM	P2A
6	12	absent		P1A
7	50	B19	IgM	P2A
8	54	EBV	MA-IgA, EBNA-IgM	P2A
9	67	EBV	MA-IgA, EBNA-IgM	P2A
10	18	CMV B19	culture, DNA IgM	P2AC
11	61	absent		P1B
12	51	CMV EBV HSV-1 B19	culture, DNA VCA-IgM, EBNA-IgM IgM, IgA IgM	dead
13	66	CMV B19	culture, IgA DNA, IgM	P2A

CMV Cytomegalovirus, *EBV* Epstein-Barr virus, *VCA* virocapsid antigens, *MA* membrane antigens, *EBNA* Epstein-Barr nuclear antigens, *HSV-1* herpes simplex virus type 1, *B19* parvovirus B19

patients, IgM antibodies were recurrent and generally not associated with IgM antibodies to other viruses. Moreover, to avoid cross-reactions and interference by RF we also tested for the presence of IgM antibodies to CMV, EBV and HSV-1 with μ-capture EIA systems. In two children with liver disease, selective stimulation of B cells related to primary or reactivated HHV-6 infection is supported not only by positive IgM antibodies but also by the concomitant seroconversion. In these patients, possible cross-reactions between HHV-6 and CMV antibodies can be ruled out, since both children had only CMV-specific IgA and IgG antibodies at levels lower than those generally shown by HHV-6-negative children. However, since many children showed non-interpretable HHV-6 results, due to non-specific reactivity, as can occur in patients with AIDS [11], some cases of HHV-6 infection could have been over-looked. Finally, to avoid false positive results in the detection of B19 antibodies, different EIA systems based on two non-overlapping epitopes of the viral proteins, were used [8, 24].

In children with liver disease, CMV was the most frequently implicated virus and was occasionally detected just before the patient's death or only at autopsy [7]. In fact, viral cultures on samples from patients with AIDS can be rather laborious, CMV-DNA is not easily detectable, and antibody response may not be prompt or complete. The clinical spectrum of CMV hepatitis in childhood is broad: from self-limited outcome in immunocompetent children, although severe cases have occasionally been reported [13, 15], to highly severe or fatal outcome in immunodeficient patients [3, 7]. In our HIV-infected children, CMV-associated liver disease was frequent and severe, being present in 9 out of 10 patients who died. Ganciclovir appeared to induce a decrease of the aminotransferase levels in two of three children with chronic liver disease, but withdrawal of the drug was again followed by an increase of the liver enzymes.

A possible mechanism for the hepatic involvement of herpesvirus infections in our patients is an immune-mediated action, apart from a direct action [21, 23]. In fact, all herpesviruses, in particular CMV, are associated with immunosuppression [2]. The progressive impairment of T, B and NK immune cells, with subsequent inhibition of the induction of gamma-interferon, could have allowed these viruses to produce a more prolonged and intense pathogenic effect on the hepatocytes than usual. Subsequently, HIV could have been enhanced in its pathogenic role, with accelerated progression of HIV disease. Other infections with herpesviruses, such as EBV, HHV-6 and HSV-1, were detected in 8, 2 and one child(ren) with liver disease, respectively. There was, however, no significant ($p > 0.05$) difference from the frequency of these infections in the children without liver disease.

Of the typical hepatitis viruses, HBV infection was revealed in a 1-year-old boy who developed chronic hepatitis, while HCV infection was confirmed in 4 children who showed specific antibodies after the age of 6 months. In an infant, who died at three months of CMV pneumonitis associated with EBV and B19 infections, HCV infection can be only presumed. In fact, the test for HCV-RNA by PCR was negative, as in 3 others, but the HCV RNA could have been degraded since the children's sera had been thawed and frozen several times. The development of chronic HBV hepatitis in one patient was rather surprising, since this infection is generally associated with transient and slight increase of the aminotransferase levels in patients with AIDS, due to immunological impairment [20]. Chronic HBV evolution in an HIV-infected child, who did not show concomitant infections, may be an indirect demonstration of the immunosuppressive role of herpesviruses and possibly B19.

As reported previously [14], we found a high prevalence of persistently positive or recurrent B19-IgM antibodies in our HIV-infected children. However, there was no significant difference ($p > 0.05$) in the rate of B19 infections between the patients with liver disease and in those without. Although B19 is a virus having specific tropism for bone marrow [1], a possible role of B19 in the development of liver disease in HIV-infected children cannot be ruled out.

In conclusion, our study showed that liver disease is an important clinical manifestation of HIV disease in children, probably as a consequence of the frequent involvement of viral agents, capable of inducing immunosuppression and activation of HIV [19]. Therefore, for the outcome of liver disease in HIV-infected children, a combined antiviral therapy is necessary.

Acknowledgements

The authors thank Dr. J. Booth (St. George's Hospital, London) for the detection of CMV-DNA and Prof. M.L. Zerbini for the detection of B19-DNA in some serum samples, Dr. P. Pisano, Dr. E. Bevivino, Dr. A. Modica, T. Mango, A. Porcaro, L. Simonelli and S. Valia for their excellent collaboration. English was revised by P. Byrne. This work was supported by the National Council of Research (CNR), project FATMA (contract no. 9103613), and the Ministry of University and Scientific Research (60% funds).

References

1. Anderson MJ (1991) Parvoviruses. In: Belshe RB (ed) Textbook of human virology. Mosby Year Book, St Louis, pp 1012–1020
2. Banks TA, Rouse BT (1992) Herpesviruses – immune escape artists? Clin Infect Dis 14: 933–941
3. Bronsther O, Makowka L, Jaffe R, Demetris AJ, Breinig MK, Ho M, Esquivel CO, Gordon RD, Iwatsuki S, Tzakis A, Marsh JW Jr, Mazzaferro V, Van Thiel D, Starzl TE (1988) Occurrence of cytomegalovirus hepatitis in liver transplant patients. J Med Virol 24: 423–434
4. Brytting M, Sundqvist V-A, Stalhandske P, Linde A, Wahren B (1991) Cytomegalovirus DNA detection of an immediate early protein gene with nested primer oligonucleotides. J Virol Methods 32: 127–138
5. Centers for Disease Control (1987) Classification system for human immunodeficiency virus (HIV) infection in children under 13 years of age. MMWR 36: 225–236
6. Falloon J, Eddy J, Wiener L, Pizzo PA (1989) Human immunodeficiency virus infection in children. J Pediatr 114: 1–30

7. Frenkel LD, Gaur S, Tsolia M, Scudder R, Howell R, Kesarwala H (1990) Cytomegalovirus infection in children with AIDS. Rev Infect Dis 12: S820–S826
8. Fridell E, Cohen BJ, Wahren B (1991) Evaluation of a synthetic peptide enzyme-linked immunosorbent assay for immunoglobulin M to human parvovirus B19. J Clin Microbiol 29: 1376–1381
9. Fridell E, Bekassy AN, Larsson B, Eriksson BM (1992) Polymerase chain reaction with double primer pairs for detection of human parvovirus B19 induced aplastic crises in family outbreaks. Scand J Infect Dis 24: 275–282
10. Garbuglia AR, Manzin A, Budkowska A, Taliani G, Clementi M, Delfini C, Carloni C (1991) Levels of pre-S antigens and HBV DNA in sera from high and low viremic HBV carriers. J Med Virol 35: 273–282
11. Linde A, Dahl H, Wahren B, Fridell E, Salahuddin Z, Biberfeld P (1988) IgG antibodies to human herpesvirus-6 in children and adults both in primary Epstein-Barr virus and cytomegalovirus infections. J Virol Methods 21: 117–123
12. Nigro G, Mattia S, Midulla M (1989) Simultaneous detection of specific serum IgM and IgA antibodies for rapid serodiagnosis of congenital or acquired cytomegalovirus infection. Serodiagn Immunother Infect Dis 3: 808–813
13. Nigro G, Bartmann U, Mattia S, Nastasi V, Torre V, Manganaro M, Perrone T, Tucciarone L (1992) Acute hepatitis in childhood: virological, immunological and clinical aspects. Biomed Pharmacother 46: 155–160
14. Nigro G, Luzi G, Fridell E, Ferrara M, Pisano P, Castelli Gattinara G, Mezzaroma I, Söderlund M, Rasnoveanu D, Aiuti F (1992) Parvovirus infection in children with AIDS: high prevalence of B19-specific immunoglobulin M and G antibodies. AIDS 6: 679–684
15. Nigro G, Mattia S, Vitolo R, Bartmann U, Midulla M (1992) Hepatitis in pre-school children: prevalent role of cytomegalovirus. Arch Virol [Suppl] 4: 268–272
16. Oleske J, Minnefor A, Cooper R, Thomas K, dela Cruz A, Ahdieh H, Guerrero I, Joshi VV, Desposito F (1983) Immune deficiency syndrome in children. JAMA 249: 2345–2349
17. Oxtoby MJ (1990) Perinatally acquired human immunodeficiency virus infection. Pediatr Infect Dis J 9: 609–619
18. Prufer-Kramer L, Kramer A, Weigel R, Rogler G, Fleige B, Krause PH, Hahn EG, Riecken EO, Pohle HD (1991) Hepatic involvement in patients with human immunodeficiency virus infection: discrepancies between AIDS patients and those with earlier stages of infection. J Infect Dis 163: 866–869
19. Rando RF, Pellett PE, Luciw PA, Bohan CA, Srinivasan A (1987) Transactivation of human immunodeficiency virus by herpesviruses. Oncogene 1: 13–18.
20. Rustgi VK, Hoofnagle JH, Gerin JL, Gelmann EP, Reichert CM, Cooper JN, Macher AM (1984) Hepatitis B virus infection in the acquired immunodeficiency syndrome. Ann Intern Med 101: 795–797
21. Sacks SL, Freeman HJ (1984) Cytomegalovirus hepatitis: evidence for direct hepatic viral infection using monoclonal antibodies. Gastroenterology 86: 346–354
22. Scott GB, Buck BE, Leterman JG, Bloom FL, Parks WP (1984) Acquired immunodeficiency syndrome in infants. N Engl J Med 310: 76–81
23. Snover DC, Horwitz CA (1984) Liver disease in cytomegalovirus mononucleosis: a light microscopical and immunoperoxidase study of six cases. Hepatology 4: 408–415
24. Söderlund M, Brown KE, Meurman O, Hedman K (1992) Prokaryotic expression of a VP1 polypeptide antigen for diagnosis by a human parvovirus B19 antibody enzyme immunoassay. J Clin Microbiol 30: 305–311

25. Taliani G, Badolato MC, Lecce R, Bruni R, Clementi C, Grimaldi F, Furlan C, Manganaro M, Duca F, Bozza A, Poliandri G, De Bac C Isolate antibody to hepatitis C core antigen (C22): correlation with HCV-RNA and anti-NS5. Arch Virol [Suppl] 8: 219–228

26. Tovo PA, De Martino M, Gabiano C, Cappello N, D'Elia R, Loy A, Plebani A, Zuccotti GV, Dallacasa P, Ferraris G, Caselli D, Fundaro C, D'Argenio P, Galli L, Principi N, Stegagno M, Ruga E, Palomba E and the Italian Register for HIV infection in children (1992) Prognostic factors and survival in children with perinatal HIV-1 infection. Lancet 339: 1249–1253

Authors' address: Dr. G. Nigro, Pediatric Institute, "La Sapienza" University, Viale Regina Elena 324, I-00161 Rome, Italy.

Arch Virol (1993) [Suppl] 8: 249–255

Lymphoblastoid interferon in chronic hepatitis C patients who were "non responders" to recombinant interferon alpha (rIFN alpha)

G. Budillon, L. Cimino, B. D'Ascoli, and **P.V. Napolitano**

Cattedra di Gastroenterologia, Second School of Medicine, Università di Napoli
"Federico II", Napoli, Italy

Summary. The aim of this study was to investigate whether there is a variable responsiveness of the hepatitis C virus and/or of the affected host to different types of interferon. We treated 21 patients affected by chronic hepatitis C who failed to respond to recombinant alpha interferon, after a four-month interval, with lymphoblastoid interferon.

Alanine transaminase (ALT) serum level was normal in 9 patients (43%) at the end of treatment. At the end of a 12 months treatment-free follow-up serum ALT remained normal in 3 patients, 4 patients relapsed and 2 dropped-out.

As yet there are no reports of differences in the therapeutic efficacy of the two types of interferon used in our study. The good response to the lymphoblastoid interferon in our non responders to recombinant alpha interferon may be due to the immune modulation induced by the four-month wash-out interval between the two therapies or to a different virus sensitivity to this interferon.

Introduction

Interferon (IFN) is well known to be efficacious in the treatment of chronic viral hepatitis C. However, the dosage of IFN and length of treatment have yet to be clearly established [1–8]. A response to the drug is usually obtained in 50–70% of cases at a dose of 3–6 MU. About 20–30% of all patients do not respond to IFN from the start of treatment, while about 50% of initial "responders" relapse after cessation of treatment. Therefore, a stable response is obtained in about 20–30% of cases [9, 10].

Various factors have been considered to affect the response to IFN: age, gender, length of disease, initial level of cytolysis, presence of siderosis, the initial latency of the response. However, factors that are clearly predictive of a positive response to the drug have not yet been identified [11–15].

The biological mechanism underlying the virus-host interaction are not completely known. In particular, still obscure are the direct cytopathic effect of the virus, the various sites of viral replication, and the role played by the host immune system.

Probably, different mechanisms occur in infections which are resistent to IFN from the onset of treatment compared with those which relapse after the first course. Immediate non-response and relapse after an initial course of IFN can, in theory, be attributed to different factors: incomplete inhibition of viral replication or delayed access to the drug by the cells or tissues in which replication takes place, mutants unaffected by the drug, formation of anti-IFN antibodies, triggering of an autoimmune mechanism of maintenance of the liver lesion, interference of other, previously latent viruses, and lastly, a different sensitivity of the virus or of the host to the action of different types of IFN. This study was designed to investigate the last point.

Materials and methods

We studied 21 patients affected by chronic hepatitis C, with or without cirrhosis, who did not respond to a previous 6-month cycle of rIFN-alpha (16 patients had received 3 MU three times a week for four months, followed by 6 MU t.i.w. for two months; and five patients were given 6 MU three times a week for six months).

Table 1 shows the data of each of the patients studied. All patients were treated for eight weeks with lymphoblastoid IFN at a dose of 6 MU three times a week, followed by 3 MU three times a week for 16 weeks.

In all patients, tests for autoantibodies (ANA, AMA, SMA, and LKM) conducted before and after the beginning of treatment were negative. None of the patients had a history of alcohol abuse, and tests for alpha 1 anti-trypsin or ceruloplasmin abnormalities and HIV were negative.

The criterion for a positive response to treatment was the normalization of ALT serum level during treatment.

Results

Nine subjects (43%) of the initial 21 non-responders had a positive response to treatment with normalization of ALT serum level. The time of the response was variable, but was always within the first three months (see Fig. 1). The characteristics of responders and those of patients who remained resistent to treatment are shown in Table 2. Data obtained during the follow-up of the nine patients who responded to the new treatment is reported in Table 3.

Table 1. Characteristics of the 21 patients who were non responders to recombinant alpha interferon

N	Gender	Age	Diagnosis	Initial ALT
1	M	47	CAH	154
2	M	46	CAH	275
3	M	42	CAH	153
4	F	60	CIR	120
5	M	48	CIR	135
6	F	62	CIR	270
7	F	53	CIR	139
8	M	32	CAH	179
9	F	67	CAH	121
10	F	55	CIR	114
11	M	47	CAH	112
12	M	54	CIR	212
13	M	62	CIR	94
14	M	53	CIR	144
15	F	38	CIR	300
16	F	57	CIR	160
17	M	46	CAH	156
18	M	50	CAH	144
19	F	47	CAH	92
20	F	53	CAH	280
21	M	48	CIR	300

CAH Chronic active hepatitis, *CIR* cirrhosis

Four patients relapsed, two were lost to follow-up and only three were definitively cured by the new cycle of treatment.

Discussion

Our study shows that in a group of patients with C virus hepatitis resistent to alpha-rIFN, a 43% subset may become positive responders during a second course of a different type of interferon, as indicated by normalizing of the serum transaminase levels.

Obviously, because of the small sample studied, we are unable to reach any definite conclusion. However, some preliminary considerations are called for. As far as we are aware, this study is the first to describe a different efficacy of the various interferons available, with particular reference to rIFN-alpha and/or lymphoblastoid-IFN as used in our study.

It is feasible that anti-IFN antibodies may be induced by the treatment, and this may bring about a decreasing effect of the drug in

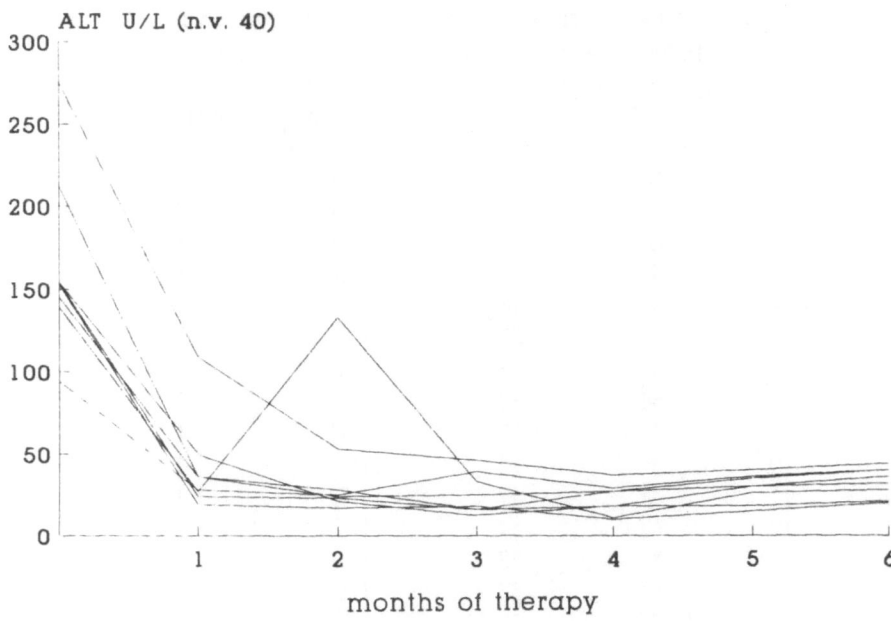

Fig. 1. Serum alanine aminotransferase (*ALT*) time-course in the 9 patients who were "responders" to the second treatment with lymphoblastoid IFN

Table 2. Characteristics of responders to lymphoblastoid IFN versus non responders

	Responders to the new treatment	Non responders confirmed	p
N. Patients	9	12	n.s.
Cirrhosis	5	6	n.s.
Severe CAH	0	1	n.s.
Mod/mild CAH	4	5	n.s.
ANA positive	0	0	n.s.
AMA positive	0	0	n.s.
SMA positive	0	0	n.s.
LKM positive	0	0	n.s.
Iron > 150 mcg/dl	2	1	n.s.

ANA Antinuclear antibody, *AMA* antimitochondrial antibody, *SMA* smooth muscle antibody, *LKM* liver kidney microsomal antibody, *n.s.* not significant

patients who initially responded well to therapy. It is well known that r-IFN alpha tends more than other types of IFN to induce specific antibodies (6–16% against 1% of lymphoblastoid) ([16], see also Antonelli and Dianzani, this volume). However, this is unlikely in our study, because it was performed on patients that were non responders "ab

Table 3. Follow-up of 9 patients who were responders
to lymphoblastoid IFN

Patient no.	Follow-up
1	ALT still normal 13 months after cessation of treatment
2	Lost to follow-up
3	ALT still normal 9 months after cessation of treatment
6	Lost to follow-up
7	Relapsing 4 months after cessation of treatment
12	ALT still normal 4 months after cessation of treatment
13	Relapsing 4 months after cessation of therapy
14	Relapsing 2 months after cessation of therapy
17	Relapsing at cessation of treatment

initio" to r-IFN alpha and not on subjects who relapsed after an initial positive response. Another possible variable that can affect the clinical response is the wash-out period between the first and the second course of therapy, as has been described in a single instance [17]. Should this be the case, the therapy-free interval could exert an immunomodulating effect.

Finally, it cannot be excluded that there is a different responsiveness of the virus to the drug, at the beginning of the treatment or after selection of a genomic variant. However, it should be stressed that we are in the realm of hypothesis, and this speculation finds no support for any of the hepatitis viruses known so far.

The data that emerge from this study do not weigh in favor of one or another of the various above-described postulates; probably an immunobiologic investigation could establish the mechanism of action of the phenomenon observed in our study.

In conclusion, further studies of this type should be performed on a larger population, and investigating in greater detail the potential biologic variables that could interfere with the responsiveness of patients with chronic hepatitis C to IFN.

References

1. Peters M, Davis GL, Dooley JS, Hoofnagle JH (1986) The interferon system in acute and chronic viral hepatitis. In: Popper H, Schaffer F (eds) Progress in liver disease, vol 8. Grune and Stratton, New York, pp 453–467
2. Davis GL, Hoofnagle JH (1986) Interferon in viral hepatitis: role in pathogenesis and treatment. Hepatology 1038–1041

3. Davis GL, Balart LA, Schiff ER, Lindsay KL, Bodenheimer HC, Perrillo RP, Carey W, Jacobson IM, Payne J, Dienstag JL, Van Thiel DH, Tamburro C, Lefkowitch J, Albrecht J, Meschievitz C, Ortego TJ, Gibas A, The Hepatitis Interventional Therapy (HIT) Group (1989) Treatment of chronic hepatitis C with recombinant alpha interferon. A multicenter randomized, controlled trial. N Engl J Med 321: 1501–1506

4. Di Bisceglie AM, Martin P, Kassianides C, Lisker-Melman M, Murray L, Waggoner J, Goodman Z, Banks SM, Hoofnagle JH (1989) Recombinant interferon alpha therapy for chronic hepatitis C. N Engl J Med 321: 1506–1510

5. Hoofnagle JH, Mullen KD, Jones DP, Rustgi V, Di Bisceglie A, Peters M, Waggoner JG, Park Y, Jones A (1986) Treatment of chronic non-A non-B hepatitis with recombinant human alpha interferon. N Engl Med 315: 1575–1578

6. Thomson AJ, Doran M, Lever AML, Webster ADB (1987) Alpha interferon therapy for non-A non-B Hepatitis transmitted by gammaglobulin replacement therapy. Lancet 1: 539–541

7. Jacyna MR, Brooks MG, Loke RHT, Main J, Murray-Lyon IM, Thomas HC (1989) Randomized controlled trial of interferon alfa (lymphoblastoid interferon) in chronic non-A non-B hepatitis. Br Med J 298: 80–82

8. Gòmez-Rubio M, Porres JC, Castillo I, Quiroga JA, Moreno A, Carreno V (1990) Prolonged treatment (18 months) of chronic hepatitis C with recombinant alfa-interferon in comparison with a control group. J Hepatol 11 [Suppl 1]: S63–S67

9. Tine F, Magrin S, Craxì A, Pagliaro L (1991) Interferon for non-A non-B chronic hepatitis. A meta-analysis of randomized clinical trials. J Hepatol 13: 192–199

10. Marcellin P, Boyer N, Giosta E, Degotte C, Courouce AM, Degos F, Coppere H, Cales P, Couzigou P, Benhamou JP (1991) Recombinant human alfa-interferon in patients with chronic non A, non B hepatitis: a multicenter randomized controlled trial from France. Hepatology 13: 393–397

11. Cimino L, Nardone G, Citarella C, Perna E, Capuano G, Budillon G (1991) Treatment of chronic hepatitis C with recombinant interferon alfa. Ital J Gastroenterol 23: 399–402

12. Causse X, Godinot H, Chevallier M, Chossergross P, Zoulim F, Ouzan D, Heyraud JP, Fontanges, Albrecht J, Meschievitz C, Trepo C (1991) Comparison of 1 or 3 MU of interferon alfa-2b and placebo in patients with chronic non-A, non-B hepatitis. Gastroenterology 101: 497–502

13. Di Bisceglie AM, Hoofnagle JH (1991) Therapy of chronic hepatitis C with alpha-interferon: the answer? Or more questions? Hepatology 13: 601–603

14. Davis GL, Lindsay K, Albrecht J, Bodenheimer HC, Balart LA, Perrillo RP, Dienstag JL, Tamburro J, Schiff ER, Carey W, Payne J, Jacobson IM, Van Thiel DH, Lefkowitch J, The Hepatitis Interventional Therapy (HIT) Group (1990) Predictors of response to recombinant alpha interferon (rIFN) treatment in patients with chronic hepatitis C (Abstract). Hepatology 12: 905

15. Lindsay KL, Davis GL, Bodenheimer HC, Albrecht J, Balart LA, Perrillo RP; Dienstag JL, Tamburro C, Schiff ER, Carey W, Payne J, Jacobson IM, Van Thiel DH, Lefkowitch J, The Hepatitis Interventional Therapy Group (1990) Predictors of relapse and response to re-treatment in patients with an initial response to recombinant alpha interferon (rIFN) therapy for chronic hepatitis C (Abstract). Hepatology 12: 847

16. Antonelli G, Currenti M, Turriziani O, Dianzani F (1991) Neutralizing antibodies to interferon alpha: relative frequency in patients treated with different interferon preparation. J Infect Dis 163: 882–885

17. Marcellin P, Boyer N, Castelnau C, Degos F, Martinot M, Loriot MA, Erlinger S, Benhamou JP (1991) Retreatment with recombinant alpha interferon in patients with chronic hepatitis C (Abstract). Gastroenterology 100: 771

Authors' address: Dr. G. Budillon, Cattedra di Gastroenterologia II, Nuovo Policlinico, via Pansini 5, I-80131 Napoli, Italy.

Arch Virol (1993) [Suppl] 8: 257–263

Non responders to interferon therapy among chronic hepatitis patients infected with hepatitis C virus

F. Piccinino, F.M. Felaco, L. Aprea, V. Messina, G. Pasquale, and **E. Sagnelli**

Institutes of Tropical and Infectious Diseases, School of Medicine,
2nd University of Naples, Naples, Italy

Summary. We studied a series of 268 chronic hepatitis C patients (31 chronic persistent hepatitis CPH, 69 mild chronic active hepatitis CAH, 125 severe CAH, and 43 active cirrhosis) enrolled from 1988 to 1991 in different therapeutic protocols using lymphoblastoid or recombinant interferon (IFN) at a dosage of 3 mega units (M.U.), three times a week for 12 months. Of these patients 54.8% showed a complete response (normalization of aminotransferases), 14.2% a partial response (decrease in aminotransferases of over 50%), 27.6% no response, and 3.4% a substantial progressive increase in the liver enzymes during IFN (becoming worse). The prevalence of non responders was lower in CPH (9.7%) than in CAH patients (31.9% in the mild form and 20.8% in the severe), and significantly higher in patients with cirrhosis (53.5%). No correlation was observed between non response and the baseline aminotransferase level or the patient's sex. Patients under 35 had a better response to IFN when compared with patients 36–50 years. This is probably due to the higher prevalence of CPH patients with a good response to IFN in the youngest group. No effect was gained in non responders by increasing the dose or shifting from recombinant to lymphoblastoid IFN; three patients were then treated with steroids, but only one benefitted. For 5 of the 9 patients who became worse, steroids were started after discontinuation of IFN therapy, and they induced a favorable response only for the 3 who had developed autoantibodies during IFN treatment.

Introduction

In hepatitis C virus (HCV) positive chronic hepatitis (HCV-CH) patients, α interferon (IFN) treatment is reported to be effective in about 50% of cases while therapy lasts. Another 20% of patients show a partial response to therapy with a decrease in the aminotransferase values which, however, never reach the normal range during treatment. For the remaining 30% of patients IFN treatment has no beneficial effect [3–8]. Predicting factors of response to IFN treatment have not yet been definitely identified, with the exception of the duration of the disease, age of patient, and presence of cirrhosis, as reported by some authors [1, 2]. It is reasonable to hypothesize that these three factors are not completely independent of each other, since often older patients have a longer duration and a more severe disease.

In the last four years we have treated with αIFN a large series of patients with HCV-CH enrolled in different therapeutic protocols using recombinant or lymphoblastoid αIFN. This study reports on those patients who failed to respond to IFN treatment, and focuses on the causes of the non response and on the measures taken to improve response.

Materials and methods

A total of 275 consecutive biopsy-proven HCV-CH patients observed from 1988 to 1991 in the Institutes of Infectious and Tropical Diseases of the 2nd University of Naples were treated with αIFN. Drug addicts or anti-HIV positive patients were not included in the present series. Also excluded from the treatment were the anti-HCV positive patients with serum markers of autoimmunity. The median age of our patients was 49 years, (range 16–69), the male to female ratio 2.3.

Almost all patients had symptom-free community acquired HCV-CH. For these patients the disease was identified after a slight increase in the aminotransferase values was noticed or because of anti-HCV positivity. We consider it almost impossible to date the onset of the disease in such patients, as the liver enzymes may only be slightly abnormal and an asymptomatic disease is almost always the rule.

Fifty-three patients received recombinant α2A-IFN, 112 patients received recombinant α2B-IFN, and 110 patients received lymphoblastoid αIFN. All patients were administered 3 MU of IFN intramuscularly 3 times a week for 12 months.

The response to treatment was predefined as "complete" when aminotransferases (ALT) normalized during treatment, "partial" when a decrease in the aminotransferase levels of more than 50% was achieved, "no response" when no variation in the enzyme levels was observed during treatment.

Results

Because of severe side effects in the first 6 months of therapy, IFN was discontinued for 7 patients (4 showed substantial weight loss, 2 developed urticaria, and one showed a sharp decrease in platelet count); these patients were subsequently excluded from the evaluation.

Table 1 summarizes the characteristics of the remaining 268 patients before IFN treatment. Out of these patients, 31 had CPH, 69 mild CAH,

Table 1. Clinical data on 268 HCV-CH patients on entering the study

Histology	No. of patients	Age median (range)	Sex M/F	ALT M + S.D.	Platelets M + S.D.
CPH	31	41 (22–66)	22/9	128 ± 75	213 ± 45
Mild CAH	69	44 (26–63)	58/11	127 ± 71	217 ± 63
Severe CAH	125	50 (16–69)	85/40	220 ± 154	194 ± 50
Active cirrhosis	43	55 (35–69)	22/21	197 ± 182	148 ± 50

Table 2. Response to IFN in relation to liver histology

Histology	No. of patients	% of patients with			
		complete response	partial response	no response	worsening
CPH	31	80.7	6.4	9.7	3.2
Mild CAH	69	52.2	13.0	31.9	2.9
Severe CAH	125	56.8	17.6	20.8	4.8
Active cirrhosis	43	34.9	11.6	53.5	–
Total	268	54.8	14.2	27.6	3.4

125 severe CAH and 43 had active cirrhosis. As expected, the median age was progressively higher from CPH to active cirrhosis patients. The male to female ratio was 2.3. The higher mean values of ALT were observed in the more severe forms of chronic disease. Out of the 268 HCV-CH patients, 147 (54.8%) showed a complete response at the end of the 12th month of treatment, 38 (14.2%) a partial response and 74 (27.6%) no response. Nine patients (3.4%) showed a substantial progressive increase in the ALT levels after beginning IFN treatment. For these patients, defined as "worsening", IFN was discontinued after 3–6 months of treatment because of the clear evidence of deterioration of condition. On comparing the response to IFN with the histology before treatment, we observed that the prevalence of non responders was lower in CPH patients (9.7%) than in CAH (31.9% for the mild form and 20.8% for the severe) and significantly higher in patients with active cirrhosis (53.5%). Also when we considered the partial response, we observed a lower prevalence of patients in the CPH (6.4%) than in the other groups (13.0% in mild CAH, 17.6% in severe CAH, and 11.6% in cirrhosis) (Table 2).

Table 3. Predicting factors of response to IFN therapy for 225 HCV-CH
without cirrhosis

		No. of patients	Complete response	Partial response no response worsening	Significance level (P)
ALT initial levels					
	mean ± S.D.	225	194 ± 163.1	173.5 ± 114.2	N.S.
Sex	M	165	62.4%	37.6%	N.S.
	F	60	48.4%	51.6%	
Age	≤35	36	75%*	25%*	*P < 0.01
	36–50	111	49.5%*	50.5%*	
	>50	78	64.1%	35.9%	
Age in CAH	≤35	25	64%	36%	
	36–50	96	46.9%	53.1%	N.S.
	>50	73	63%	37%	

If we consider the partial and non responders together as patients with a rather poor clinical response to IFN, we observe that the prevalence of patients who show little or no benefit from such treatment increases from 16.1% for CPH to 44.9% and 38.4% respectively for mild and severe CAH, and to 65.1% for liver cirrhosis patients. On the other hand, it is interesting to note that the highest prevalence of complete response was observed among CPH patients.

The patients who showed worsening were evenly distributed among the CPH group and CAH groups, none was in the cirrhosis group.

Since cirrhosis is a very reliable predictor of non response to IFN therapy it would seem reasonable to exclude patients with cirrhosis from the evaluation of the predicting factors besides histology.

Since the term "non responders" no longer includes the cirrhosis patients, the evaluation now concerns 225 patients.

Table 3 reports the predicting factors of response to IFN treatment in our series of HCV-CH patients without cirrhosis.

No correlation was observed between non response and the baseline ALT level or the patient's sex. Patients under 35 had a better response to IFN compared to those in the 36–50 age group. (P < 0.01). Patients over 50 showed a complete response in 64% of cases. When the evaluation was restricted to CAH patients, no significant difference was observed as regards age. This is probably due to the higher prevalence of CPH patients with a good response to IFN in the youngest age group.

Almost all patients who had a complete response to IFN, showed normalization of ALT within the first 3 months of treatment; in a few

Table 4. Measures to improve the response rate in non responders

	No. of patients	No. of patients with		
		complete response	partial response	no response
Increasing dose	22	1	2	19
Shifting from recombinant to lymphoblastoid IFN	10	1	1	8
Steroids	3	1	–	2

cases the ALT normalized between the 3rd and 6th month of therapy. Prolonging the treatment in non responders from 6 to 12 months did not achieve further improvement.

For patients with no response to IFN therapy some attempts were made to improve the response rate. First we tried increasing the IFN dose, then shifting from recombinant to lymphoblastoid IFN, finally using steroids if the disease was severe (Table 4).

Increasing the dose of IFN for non responders in the 12th month of treatment is virtually ineffective, as only 1 patient out of 22 showed a complete response after increasing dose.

Similarly, when we tried shifting from recombinant to lymphoblastoid IFN, a complete response was attained by only one patient out of 10.

In 3 patients with severe CAH who were non responders to a high dose of IFN, even though no marker of autoimmunity was present, we tried using steroids. One out of the three patients showed normalization in the aminotransferases.

In our series of chronic hepatitis patients, 9 showed a substantial progressive increase in the aminotransferase levels during IFN treatment (Table 5) and were defined as worsened. After discontinuation of therapy 4 of these patients showed a decrease in the ALT which reached pretreatment levels, 4 patients showed persistently high transaminase levels, and no follow-up is available for the last one.

One of the 9 patients became positive for anti-nuclear antibodies (ANA) and 2 liver kidney membrane (LKM) antibody-positive during IFN treatment. In one, the increase in the ALT levels was symptomatic with the appearance of jaundice and fever. The 3 patients who developed autoantibodies during IFN therapy subsequently showed a response to steroids and 2 of them also showed persistent normalization of ALT levels. In contrast, no response was observed in the other 2 patients

F. Piccinino et al.

Table 5. HCV chronic hepatitis patients with substantial progressive increase in aminotransferase levels during IFN

Patients	Sex	Histology	Autoantibodies during IFN	AMT[a] after discontinuation of IFN	Steroids
1	F	CPH	absent	High	No
2	M	CAH	absent	decreased[b]	No
3	F	CAH	absent	unknown	No
4	F	CAH	absent	decreased	No
5	F	CAH	absent	high	Yes (no response)
6	F	CAH	absent	decreased	Yes (no response)
7	F	CAH	ANA +	decreased	Yes (response +)
8	M	CAH	LKM +	high	Yes (response +++)
9	M	CAH	LKM +	high	Yes (response +++)

[a] *AMT* Aminotransferases
[b] Decreased to pre-treatment values

treated with steroids who showed no development of autoantibodies during IFN (Table 5).

Conclusion

The presence of liver cirrhosis is the most important factor in predicting non response to IFN treatment; in contrast, having CPH or being under 35 seem to be predictors of a good response. Six months is sufficient to evaluate the efficacy of IFN therapy, after which an alternative approach must be considered for non responders. Increasing the dose or shifting to the other types of αIFN for non responders appears to be ineffective.

Autoantibodies, including LKM, must be determined before and during IFN treatment. In fact, some patients may develop autoantibodies during IFN therapy and clinical symptoms will actually develop in some cases. For these patients steroids seem to be effective, as seen with our three cases.

References

1. Alberti A, Chemello G, Diodati G (1992) Treatment of chronic hepatitis C with different regimens of interferon alpha-2a (IFN-2a). Hepatology 16: 75A
2. Cammà C, Craxí A, Tinè F (1992) Predictors of response to alpha-interferon (IFN) in chronic hepatitis C: a multivariate analysis on 361 treated patients. Hepatology 16: 131A

3. Davis GL, Balart LA, Schiff ER, Lindsay KL, Bodenheimer HC, Perrillo RP, Carey W, Jacobson IM, Payne J, Dienstag JL, Van Thiel DH, Tamburro C, Lefkowitch J, Albrecht J, Meschievitz C, Ortego TJ, Gibas A, The Hepatitis Interventional Therapy (HIT) Group (1989) Treatment of chronic hepatitis C with recombinant alpha interferon. A multicenter randomized, controlled trial. N Engl J Med 321: 1501–1506
4. Di Bisceglie AM, Martin P, Kassianides C, Lisker-Melman M, Murray L, Waggoner J, Goodman Z, Banks SM, Hoofnagle JH (1989) Recombinant interferon alpha therapy for chronic hepatitis C. N Engl J Med 321: 1506–1510
5. Hoofnagle JH, Mullen KD, Jones DP, Rustgi V, Di Bisceglie A, Peters M, Waggoner JG, Park Y, Jones A (1986) Treatment of chronic non-A non-B hepatitis with recombinant human alpha interferon. N Engl Med 315: 1575–1578
6. Jacyna MR, Brooks MG, Loke RHT, Main J, Murray-Lyon IM, Thomas HC (1989) Randomized controlled trial of interferon alfa (lymphoblastoid interferon) in chronic non-A, non-B hepatitis. Br Med J 298: 80–82
7. Marcellin P, Boyer N, Giostra E (1991) Recombinant human alpha interferon in patients with chronic non-A, non-B hepatitis: a multicenter randomized controlled trial from France. Hepatology 13: 393–397
8. Tiné F, Magrin S, Craxí A (1991) Interferon for non-A, non-B chronic hepatitis. J Hepatol 13: 192–199

Authors' address: Dr. F. Piccinino, Istituto di Malattie Tropicali, c/o Ospedale Gesù e Maria, Via D.Cotugno 1, I-80135 Napoli, Italy.

Arch Virol (1993) [Suppl] 8: 265–269

The relationship between Le^Y antigen and the therapeutic efficacy of interferon in chronic hepatitis C

M. Sata[1], H. Nakano[1], T. Hino[1], T. Kosedo[2], M. Adachi[2], and K. Tanikawa[1]

[1] The Second Department of Medicine, Kurume University School of Medicine, Asahi-machi, Kurume, Fukuoka, [2] Japan Immunoresearch Laboratories, Sakae-machi, Takasaki, Gunma, Japan

Summary. The expression of Le^Y antigen (Fuc α 1 $\rightarrow$ 2Gal β 1 $\rightarrow$ 4 [Fuc $\alpha \rightarrow$ 3] GlcNAc β 1 $\rightarrow$ R), recognized by the monoclonal antibody BM-1, was studied in peripheral blood T-lymphocyte subpopulations in patients with viral hepatitis, and in liver tissue of patients with acute viral hepatitis. The relationship between the expression of Le^Y antigen on peripheral blood T-lymphocytes and the effects of interferon (IFN) therapy for chronic hepatitis type C were also evaluated. Le^Y antigen is not markedly expressed in B or T-lymphocytes of healthy individuals. However, it was strongly expressed in CD8 and CD4 T-lymphocytes in patients with viral hepatitis. In the acute phase of acute viral hepatitis, the expression of Le^Y antigen was more markedly expressed on peripheral CD8 T-lymphocytes than on CD4 T-lymphocytes. In chronic hepatitis type B and type C, it was significantly expressed more often on CD4 T-lymphocytes. In the liver tissues of patients with acute viral hepatitis, Le^Y antigen was expressed on hepatocytes and infiltrating lymphocytes. IFN therapy for chronic active hepatitis type C proved most effective when Le^Y antigen was more markedly expressed on the patient's CD4 and CD8 peripheral blood T-lymphocytes before treatment. Further studies are needed to clarify the relationship between the mechanisms of hepatic cell injury and Le^Y antigen.

Introduction

Aberrant glycosylation, which is related to oncogenic transformation, has been studied extensively, and the basic chemistry of such changes has been established fairly well, but little is known about the role of carbohydrate antigens and their changes in viral hepatitis [5]. Recent reports of the expression of Le^Y carbohydrate antigen (Le^Y antigen) on CD4 T-lymphocytes infected by HIV [2] and Le^X antigen expression on cytomegalovirus-infected teratocarcinoma cells [3], however, suggest that aberrant glycosylation occurs in cells infected by virus or is triggered by viral infection in some cells. Based on the results that Le^Y antigen is expressed on peripheral blood T-lymphocytes of patients with HIV infection, we studied the relationship between the expression of Le^Y antigen on peripheral CD4 and CD8 T-lymphocytes and the therapeutic effect of IFN against chronic hepatitis C virus infection.

Materials and methods

The expression of Le^Y antigen in peripheral blood T-lymphocyte subpopulations was studied in patients with viral hepatitis by flow cytometry with two-color analysis using the monoclonal antibody BM-1 [1], which recognizes Le^Y antigen. Lymphocyte fractions separated from peripheral blood of patients were reacted with FITC-labeled anti-Le^Y monoclonal antibody (FITC-labeled BM-1 monoclonal antibody) and with the phycoerythrin-labeled antibody CD4, CD8, or CD3 and were analyzed with a flow cytometer (EPICS-C, Coulter Electronics Inc., Hialeah, FL, U.S.A.). The percentage of lymphocytes positive for Le^Y in CD4-positive and CD8-positive cells (BM-1$^+$ and CD4$^+$/CD4$^+$ $\times$ 100, BM-1$^+$ and CD8$^+$/CD8$^+$ $\times$ 100) were calculated.

Le^Y antigen in the liver was studied in tissues obtained by liver biopsy from patients with acute hepatitis using the direct fluorescence antibody method with FITC-labeled anti-Le^Y monoclonal antibody (FITC-labeled BM-1 monoclonal antibody).

Natural IFN-α or recombinant IFN-α 2a was administered i.m. daily at 3–9 MIU/day for 2 weeks and 3 times a week for 14 weeks thereafter in 23 patients with chronic active hepatitis C.

HBsAg, IgM class anti-HBc, total anti-HBc, HBV-DNA, anti-HCV, and HCV-RNA were measured for the diagnosis of viral hepatitis.

Results

In 124 healthy individuals (mean age 42.8 years), the percentage of lymphocytes that expressed Le^Y antigen was 7.1 $\pm$ 3.2 (0.9–13.3%) in CD4 T-lymphocytes and 4.8 $\pm$ 2.7 (0–14.5%) in CD8 T-lymphocytes. In 41 patients in the acute phase of acute viral hepatitis, the percentages of lymphocytes expressing Le^Y antigen in CD4 and CD8 T-lymphocytes were significantly higher than in the healthy individuals, and the increase in the percentage of Le^Y antigen-positive lymphocytes in CD8 T-lymphocytes was especially marked. In the convalescent phase, the percentages of Le^Y antigen-positive lymphocytes in CD4 and CD8 T-lymphocytes were reduced, and the values were similar to those in healthy subjects. Such changes were observed regardless of the type

Table 1. Le^Y antigen expression (defined by monoclonal antibody BM-1) in CD4- and CD8-positive peripheral blood lymphocytes of normal subjects and patients with viral hepatitis

Subjects	Le^Y expression	
	CD4	CD8
Normal control (n = 124)	7.1 ± 3.2	4.8 ± 2.7
Acute hepatitis (n = 41)	14.2 ± 6.6	24.5 ± 9.5
Chronic hepatitis (n = 159)	13.4 ± 5.3	9.7 ± 5.1

* P < 0.05

Table 2. Ratio of Le^Y antigen expression in CD4- and CD8-positive peripheral blood lymphocytes of patients with acute and chronic viral hepatitis

Subjects	% Le^Y in CD4/% Le^Y in CD8	
	≧1	<1
Acute hepatitis (n = 41)	5/41	36/41
Chronic hepatitis (n = 159)	145/159	14/159

of hepatitis virus. In patients with chronic hepatitis B or C, the expression of Le^Y antigen in CD4 and CD8 T-lymphocytes was significantly higher than that in healthy individuals, but its expression on CD8 T-lymphocytes was less than that in patients with acute hepatitis (Table 1). The ratio between the percentage of Le^Y-antigen-positive lymphocytes in CD4 T-lymphocytes and that in CD8 T-lymphocytes (%Le^{Y+} in CD4/%Le^{Y+} in CD8) was 1.0 or greater in many patients with chronic hepatitis but was less than 1.0 in many patients with acute hepatitis (Table 2).

The expression of Le^Y antigen was studied in the liver tissue of 7 patients with acute hepatitis A (n = 2), B (n = 2), and C. The antigen was expressed in the cytoplasm of hepatocytes in all 7 patients, on the cell membrane of hepatocytes in 2, and in infiltrating mononuclear cells in 4 of the 6 patients examined.

Of the 23 patients with chronic active hepatitis C, IFN therapy was effective, i.e. the ALT and AST values were normalized for 6 months or longer after the end of the administration of IFN, in 11 (responders) while 12 were non responders. The percentages of Le^Y antigen-positive CD4 T-lymphocytes and CD8 T-lymphocytes determined within 2 weeks

Table 3. Relationship between Le^Y antigen expression on CD4- and CD8-positive peripheral blood lymphocytes before therapy, and the therapeutic efficacy of interferon in chronic hepatitis C

Le^Y expression	IFN treatment	
	responders (n = 11)	nonresponders (n = 12)
CD4 (+)	17.3 ± 6.73%	11.3 ± 3.24%
	$P < 0.05$	
CD8 (+)	20.2 ± 8.83%	12.3 ± 5.24%
	$P < 0.05$	

before the beginning of the IFN therapy were significantly higher in the responders (17.3 ± 6.73 and 20.2 ± 8.83%, respectively) than in the non responders (11.3 ± 3.24 and 12.3 ± 5.24%, respectively) (Table 3).

Discussion

Hepatic cell injury by hepatitis virus is explained as a result of destruction of virus-infected hepatocytes by cytotoxic T-lymphocytes rather than a result of the cytopathic effect of the virus ([4, 6, 9] see also Ferrari et al., this volume).

Le^Y antigen, studied here, was shown to be expressed in hepatocytes, intrahepatic infiltrating lymphocytes, and peripheral blood CD4 and CD8 T-lymphocytes. Moreover, the percentages of Le^Y antigen-positive lymphocytes in peripheral blood CD4 and CD8 T-lymphocytes were different between acute hepatitis and chronic hepatitis, and according to the stage of the disease. These findings suggest that Le^Y antigen is an important carbohydrate antigen, the expression of which changes with immunological responses of the host. Therefore, this carbohydrate antigen may serve as a pathological marker of viral hepatitis, in addition to many markers of the hepatitis viruses themselves.

IFN has attracted attention as a possible treatment for chronic hepatitis C. The HCV genotype [7] and the amount of HCV-RNA [8] are reported to be useful as indices for prediction of the effects of IFN therapy. In addition to these viral factors, histological findings of the liver and the duration of illness are known to be predictive factors. According to our results, the therapeutic effect was greater when more

Le^Y antigen was expressed on CD4 and CD8 T-lymphocytes. These observations suggest that the percentage of CD4 and CD8 T-lymphocytes expressing the Le^Y antigen is useful for selection of patients for IFN therapy, determination of the time of its beginning, and prediction of its therapeutic effects.

References

1. Abe K, McKibbin JM, Hakomori S (1983) The monoclonal antibody directed to difucosylated type 2 chain (Fuc α 1 → 2Gal β 1 → 4 [Fuc α 1 → 3] GlcNAc β 1-R; Y determinant). J Biol Chem 258: 11793–11797
2. Adachi M, Hayami M, Kashiwagi N, Mizuta T, Ohta Y, Gill MJ, Matheson DS, Tamaoki T, Shiozawa C, Hakomori S (1988) Expression of Le^Y antigen in human immunodeficiency virus-infected human T cell lines and in peripheral lymphocytes of patients with acquired immune deficiency syndrome (AIDS) and AIDS-related complex (ARC). J Exp Med 167: 323–331
3. Andrews PW, Gonczol E, Fenderson BA, Holmes EH, O'Malley G, Hakomori S, Plotkin S (1989) Human cytomegalovirus induced stage-specific embryonic antigen 1 in differentiating human teratocarcinoma cells and fibroblasts. J Exp Med 169: 1347–1359
4. Eddleston ALWF, Williams R (1974) Inadequate antibody response to HBAg or suppressor T-cell defect in development of active chronic hepatitis. Lancet 2: 1543–1545
5. Hakomori S (1985) Aberrant glycosylation in cancer cell membranes and focused on glycolipids: overview and perspectives. Cancer Res 45: 2405–2414
6. Naumov NV, Mondelli M, Alexander GJM (1984) Relationship between expression of hepatitis B virus antigens in isolated hepatocytes and autologous lymphocyte cytotoxicity in patients with chronic hepatitis B virus infection. Hepatology 4: 63–68
7. Okamoto H, Sugiyama Y, Okada S, Kurai K, Akahane Y, Sugai Y, Tanaka T, Sato K, Tuda F, Miyakawa Y, Mayumi M (1992) Typing hepatitis C virus by polymerase chain reaction with type-specific primers: application to clinical surveys and tracing infections sources. J Gen Viol 73: 673–679
8. Takada N, Takase S, Enomoto T, Takada A, Date T (1992) Clinical backgrounds of the patients having different types of hepatitis C virus genomes. J Hepatol 14: 35–40
9. Vallbracht A, Gabriel P, Maier K (1986) Cell-mediated cytotoxicity in hepatitis A virus infection. Hepatology 6: 1308–1314

Authors' address: Dr. M. Sata, Second Department of Medicine, Kurume University School of Medicine, 67 Asahi-machi, Kurume, Fukuoka 830, Japan.

Arch Virol (1993) [Suppl] 8: 271–277

Antibodies to interferon alpha in patients

G. Antonelli[1] and **F. Dianzani**[2]

[1] Department of Biomedicine, Sec. Virology, University of Pisa, Pisa, [2] Institute of Virology, University "La Sapienza", Tiburtina, Rome, Italy

Summary. A massive amount of information on interferons (IFNs) has been gathered since their original description as antiviral agents in 1957. Human IFNs have now been used clinically for over a decade and their therapeutic efficacy has been well established for some human neoplasias and viral diseases. During these studies, it has also been documented that some of the patients treated with IFNs can develop antibodies to IFNs which can affect their therapeutic efficacy. Here, it is summarized what is currently known on the biological and clinical aspects of these antibodies.

Introduction

The interferons (IFNs) are members of a family of proteins produced by animal cells in response to a variety of stimuli [9]. They represent one of the body's natural defenses. Although originally described as potent antiviral agents, they were subsequently found to affect many other cellular functions and now they are considered pleiotropic hormonal effectors [9]. Because of these properties and characteristics, including low toxicity, broad efficacy, and high specific activity, the IFNs have been introduced into clinical trials and various interventional strategies.

A variety of human (hu) IFNs are now commercially available as therapy for infectious and neoplastic diseases. Some are produced by recombinant DNA technology (rIFN) in eukaryotic or prokaryotic cells, while others, the natural IFNs (nIFN), are purified from supernatants of leukocytes or cultured human cells after induction by various stimuli.

Three IFN antigenic types have now been described: IFN alpha, produced by leukocytes; IFN beta, produced mainly by fibroblasts; IFN gamma, produced mainly by T-lymphocytes [9, 22].

Unlike IFN beta and gamma, IFN alpha consists of different subtypes coded for by several non-allelic genes. There are least 18 different huIFN alpha I genes and 6 huIFN alpha II genes. The homology between regions coding for the different IFN alpha genes is about 90% among the alpha I genes and 70% between IFN alpha I and alpha II [12, 22]. To date, most biological and clinical studies have been based on products of the huIFN alpha I genes, since these have been most extensively characterized.

Despite their high homology, the biological activities of the various IFN alpha subtypes may present some differences such as the lower specific activity of the alpha 1 compared to alpha 2, the inability of IFN alpha 2 to induce MHC antigens on monocytes or of IFN alpha J to display NK induction activity at low concentration [12]. Moreover, it has been shown that different IFN alpha preparations may demonstrate different activities against HIV-replication, whereby IFN alpha D is less active than IFN J/C and IFN alpha A [26].

IFN-induced antibodies

IFNs, although homologous proteins, have been reported to be immunogenic when administered to patients. Further, low titers of natural antibodies to IFNs may be present in normal individuals [7, 25]. An explanation for this phenomenon has not yet been provided; however, it is clear that antibodies to other cytokines or hormones (such as GM-CSF or insulin) can occasionally be found in healthy individuals or can be induced by therapeutic treatment in patients.

The first observation of anti-IFN antibodies dates to 1981, when Vallbracht identified the presence of neutralizing antibodies to IFN in a patient treated with IFN beta [29]. Since 1981, many other reports have been published concerning development of anti-IFN antibodies in patients treated with IFN alpha and beta, while IFN gamma appears to be an exception [8, 10, 11, 13, 14, 16, 17, 21, 23, 27].

At least three assays are currently available for detecting antibodies to IFN. These include an enzyme immunoassay (EIA), a radioimmunoassay (RIA), and an antibody-neutralizing biossay (ANB).

Of the three, only ANB is capable of detecting antibodies to functional epitopes, those involved in generating the observed biological activity of IFNs. The ANB assay permits the detection of neutralizing antibodies directed against any type of IFN. Such neutralizing antibodies can inhibit the antiviral activity of the IFNs as well as their antiproliferative and immunomodulatory activities.

Although EIA and RIA tests may detect neutralizing antibodies to IFN alpha, they also detect non-neutralizing or binding antibodies, i.e. an antibody directed against epitopes that are not involved in the biological activity of IFN alpha. Although the clinical significance of binding antibodies has yet to be fully established, some preclinical evidence suggests that this type of antibody can prevent normal clearance and degradation of administered IFN alpha, therefore modifying its pharmacokinetics [4, 24].

As would be expected, the frequency of seroconversion measured by EIA or RIA is always higher than the seroconversion frequency revealed with ANB [11]. Thus, one critical factor to be considered in calculating the incidence of seroconversion is the assay type that is being employed. Other factors such as dosage, route of administration, treatment duration, or underlying disease can affect seroconversion incidences.

In fact, the percentage of patients developing antibodies during or after IFN treatment varies considerably, from total absence in a group of hairy cell leukemia patients [14] to as high as 56% seroconversion in one group of melanoma patients [10]. Additionally, it was tempting to speculate that also the type of IFN administered could influence the frequency of both binding and non-neutralizing antibody development. Indeed, differences have been observed between the frequency of seroconversion to IFN in patients treated with rIFN alpha 2a, rIFN alpha 2b and IFN alpha N1, a natural mixture of different IFN alpha subtypes [33] (Table 1) [1–3, 18]. Interestingly, antibody produced in rIFN alpha 2 treated patients recognizes IFN alpha 2 (IFN alpha A) and possibly other individual subtypes of IFN alpha but fails to neutralize nIFN alpha (Table 2) [2]. These results obtained in an in vitro assay may imply that the treatment of patients with nIFN alpha could overcome the resistance to rIFN alpha due to the development of neutralizing antibodies. Indeed several papers have been published concerning the successful use of

Table 1. Incidence of antibodies in patients treated with recombinant interferons (rIFN-alpha 2a or rIFN-alpha 2b) or lymphoblastoid interferon (IFN-alpha N1)

Treatment	Neutralizing antibody no. pos./no. tested (%)	Non-neutralizing antibody no. pos./no. tested (%)
rIFN-alpha 2a	15/74 (20.2)	33/74 (44.6)
rIFN-alpha 2b	10/144 (6.9)[a]	21/144 (14.6)[a]
rIFN-alpha N1	1/78 (1.2)[b]	7/74 (9.4)[a]

[a] $p < 0.01$
[b] $p < 0.001$
Data from [2, 3]

Table 2. Neutralizing activity of sera derived from rIFN alpha 2 treated patients against different subtypes of IFN alpha

Sera no.	Titer (NU)[a] against IFN alpha					
	2a	A	D	A/D[b]	B/D[c]	N1[d]
1	355	266	66	<5	ND	<5
2	266	166	22	<5	ND	<5
3	88	33	<5	<5	ND	<5
4	178	67	<5	<5	ND	<5
5	667	ND	ND	ND	<5	<5
6	167	ND	ND	ND	<5	<5
7	333	ND	ND	ND	<5	<5
8	167	ND	ND	ND	<5	<5

[a] *NU* Neutralization units (see [11])
[b] *A/D* Hybrid IFN alpha
[c] *B/D* Hybrid IFN alpha kindly provided by Dr. H. Hockeppel (Ciba Geigy, Basel)
[d] N_1 IFN alpha N1 is a mixture of different IFN alpha subtypes produced by Namalwa cells
ND Not done

nIFN alpha to restore clinical response in patients who developed antibodies to rIFN alpha [5, 6, 30, 32].

Taken together, these results raise new and important questions, the answers to which have still to be elucidated. First, it is hard to explain why only a minority of patients develops antibodies, or why some types of IFN induce a higher percentage of seroconversion. Furthermore, it remains to be ascertained why in some patients these antibodies are produced only transiently during therapy and why they have a well-defined specificity.

Certainly, one of the key issue of the anti-IFN antibody development field is to establish whether these antibodies can affect the therapeutic efficacy of the administered IFN.

Resistance to the antitumor or antiviral effects of recombinant IFNs associated with development of neutralizing antibodies has been found in chronic myelocytic leukaemia [31], hairy cell leukaemia (HCL) [28], cryoglobulinaemia [5] or hepatitis patients [19], whereas other reports indicate that antibodies to IFN do not appear to be associated with any adverse clinical sequelae [11, 15]. The failure of IFN therapeutic activity may not only be due to the "qualitative" presence of antibodies to IFN but also to the amount and neutralizing efficacy of such antibodies. In this respect it is interesting to mention, as already stated, that patients who become resistant to rIFN alpha treatment can be effectively treated with natural IFN alpha preparations. Data currently available do not, however, allow definite conclusions to be made, but in the light of some recent results it is tempting to speculate that at least in some patients the development of neutralizing antibodies may correlate with loss of the therapeutic effect. For instance, neutralizing antibodies to rIFN alpha could be detected in 3 out 4 cryoglobulinemia patients who relapsed during maintenance therapy while none of the patients with persistent response showed detectable amounts of antibody to IFN [5]. Furthermore, in a trial conducted in Germany on HCL patients, neutralizing antibodies have been found in all relapsing patients (9/59) while none of the antibody-negative patients relapsed [32].

Moreover, in a recent study on hepatitis C patients, the seroconversion rate was significantly lower in responder patients than in non-responders. Specifically, the therapeutic failure due to antibody formation might be attributed to hepatitis patients who after an initial response showed disease reactivation concomitantly with anti-IFN antibody appearance, while no significance could be assigned to the antibody development in patients who never showed ALT-reduced levels during treatment [20].

However, although these findings strongly suggest that, at least in some patients, anti-IFN antibody production may affect the response to IFN, the clinical significance of seroconversion during IFN treatment remain to be ascertained in detail.

Acknowledgements

This work was supported by a grant from CNR, Progetto Finalizzato "Biotecnologie e biostrumentazione". We thank S. Tamburrini for help in manuscript preparation.

References

1. Antonelli G, Currenti M, Turriziani O, Dianzani F (1990) Incidence of anti-IFN alpha neutralizing antibodies in hepatitis patients treated with different IFN alpha preparations. Antiviral Res 1: S104
2. Antonelli G, Currenti M, Turriziani O, Dianzani F (1991) Neutralizing antibodies to interferon-alpha: relative frequency in patients treated with different interferon preparations. J Infect Dis 163: 882–885
3. Antonelli G, Currenti M, Turriziani O, Riva E, Dianzani F (1992) Relative frequency of non-neutralizing antibodies to interferon in hepatitis patients treated with different IFN alpha preparations. J Infect Dis 165: 593
4. Bendtzen K, Svenson M, Jonsson V, Hippe E (1990) Autoantibodies to cytokines: friends or foes? Immunol Today 11: 167–169

5. Casato M, Laganà B, Antonelli G, Dianzani F, Bonomo L (1991) Long-term results of interferon therapy in essential mixed cryoglobulinemia. Blood 78: 1–6

6. Catani L, Gugliotta L, Zauli G, Bagnara GP, Antonelli G, Mattioli Belmonte M, Vianelli N, Bonsi L, Brunelli MA, Tura S (1992) In vitro inhibition of interferon alpha-2a antiproliferative activity by antibodies developed during treatment for essential thrombocythaemia. Haematology 77: 318–321

7. De Maeyer-Guignard J, De Maeyer E (1986) Natural antibodies to interferon-alpha and interferon-beta are a common feature of inbred mouse strains. J Immunol 136: 1708–1711

8. Dianzani F, Antonelli G, Amicucci P, Cefaro A, Pintus C (1990) Low incidence of neutralizing antibody formation to interferon in human recipients. J Interferon Res 9: S33–S36

9. Dianzani F, Antonelli G, Capobianchi MR (1990) The biological basis for clinical use of interferon. J Hepatol 11: S5–S10

10. Dummer R, Muller W, Nestle F, Wiede J, Dues J, Lechner W, Haubitz I, Wolf W, Bill E, Burg G (1991) Formation of neutralizing antibodies against natural interferon-beta, but not against recombinant interferon-gamma during adjuvant therapy for highrisk malignant melanoma patients. Cancer 67: 2300

11. Figlin RA, Itri LM (1989) Anti-interferon antibodies: a perspective. Semin Hematol 25: 9–15

12. Finter NB (1991) Why are there so many subtypes of alpha-interferons? J Interferon Res [Special issue, January] 185

13. Galton JE, Bedford P, Scott JE, Brand CM, Nethersell ABW (1990) Antibodies to lymphoblastoid interferon. Lancet 2: 572–573

14. Golomb HM, Fefer A, Golde DW, Ozer H, Portlock C, Silber R, Rappeport J, Ratain MJ, Thompson J, Bonnem E, Spiegel R, Tensen L, Burke JS, Vardiman JW (1988) Report of a multi-institutional study of 193 patients with hairy cell leukemia treated with interferon-alfa 2b. Semin Oncol 15: 7

15. Itri IM, Campion M, Dennin RA, Polleroni AV, Gutterman JU, Groopman JE, Trown PW (1987) Incidence and clinical significance of neutralizing antibodies in patients receiving recombinant interferon alpha 2a by intramuscular injection. Cancer 59: 668–674

16. Jacobs S, Friedman RM, Nagabhushan TL (1989) Detection of serum neutralizing antibodies to interferon alfa-2b (Intron A) by means of an enzyme immunoassay. J Interferon Res 9: S292

17. Leavitt R, Ratanantharathorn V, Ozer H, (1987) Alpha-2b interferon in the treatment of Hodgkin's disease and non-Hodgkin's lymphoma. Semin Oncol 14: 18–23

18. Liao MJ, Axelrod HR, Kuchler M, Yip YK, Kirkbright E, Testa D (1992) Absence of neutralizing antibodies to interferon in condyloma acuminata and cancer patients treated with natural human leukocyte interferon. J Infect Dis 165: 757–760

19. Lok SF, Lai CL, Leung KY (1990) Interferon antibodies may negate the antiviral effects of recombinant alpha-interferon treatment in patients with chronic hepatitis B virus infection. Hepatology 12: 1266

20. Milella M, Antonelli G, Santantonio T, Currenti M, Monno L, Mariano N, Angarano G, Dianzani F, Pastore G (1993) Neutralizing antibodies to recombinant alpha-interferon and response to therapy in chronic hepatitis C virus infection. Liver 13: 146–150

21. Mogensen KE, Daubas PH, Gresser I, Sereni D, Varet B (1981) Patient with circulating antibodies to alpha interferon. Lancet ii: 1227–1228

22. Pestka S (1986) Interferon from 1981 to 1986. Methods Enzymol 119: 3–23
23. Quesada JR, Hawkins M, Horning S, Alexantian R, Borden E, Merigan T, Adams F, Gutterman JU (1984) Collaborative phase I–II study of recombinant DNA-produced leukocyte interferon (clone A) in metastatic breast cancer, malignant lymphoma and multiple myeloma. Am J Med 77: 427–435
24. Rosenblum MG, Unger BW, Gutterman JU, Hersh M, David GS, Frincke JM (1985) Modification of human lekocyte interferon pharmacology with a monoclonal antibody. Cancer Res 45: 2421
25. Ross C, Hansen MB, Schyberg T, Berg K (1990) Autoantibodies to crude human leucocyte interferon (IFN), native human IFN, recombinant human IFN-alpha 2b and human IFN-gamma in healthy blood donors. Clin Exp Immunol 82: 57–62
26. Sperber SJ, Gocke DJ, Haberzettl C, Kuk R, Schwartz B, Pestka S (1992) Anti-HIV-1 activity of recombinant and hybrid species of interferon-alpha. J Interferon Res 12: 363–368
27. Spiegel J, Spicehandler JR, Jacobs SL, Oden EM (1986) Low incidence of serum neutralizing factors in patients receiving recombinant alpha-2b interferon (Intron A). Am J Med 80: 223–228
28. Steis RG, Smith JW, Urba WJ, Clark JW, Itri LM, Evans LM, Schoenberber C, Longo DL (1988) Resistance to recombinant interferon alpha-2a in hairy cell leukemia associated with neutralizing anti-interferon antibodies. N Eng J Med 318: 1409–1413
29. Vallbracht A, Treuner T, Flehming B, Joester KE, Miethammer D (1981) Interferon neutralizing antibodies in a patient treated with human fibroblast interferon. Nature 287: 496–498
30. von Wussow P, Hartman F, Freund M, Poliwoda H, Deicher H (1988) Leukocyte derived interferon-alpha in patients with antibodies to recombinant IFN-alpha 2b. Lancet i: 882–883
31. von Wussow P, Freund M, Block B, Dietrich H, Poliwoda H, Deicher H (1987) Clinical significance of anti-IFN alpha antibody during interferon therapy. Lancet ii: 635–636
32. von Wussow P, Pralle H, Hochkeppel HK, Jakschies D, Sonnen S, Schmidt H, Muller-Rosenau D, Franke M, Haferlach T, Zwingers T, Rapp U, Deicher H (1991) Effective natural interferon-alpha therapy in recombinant interferon-alpha-resistant patients with hairy cell leukemia. Blood 78: 38–43
33. Zoon KC, Miller D, Bekisz J, zur Nedden D, Enterline JC, Nguyen NY, Ren-qui H (1992) Purification and characterization of multiple components of human lymphoblastoid interferon-alpha. J Biol Chem 267: 15210–15216

Authors' address: Dr. F. Dianzani, Institute of Virology, University "La Sapienza", V.le di Porta Tiburtina 28, I-00185 Roma.

VII. Liver transplantation and chronic viral hepatitis

Arch Virol (1993) [Suppl] 8: 281–289

_Archives_____
Virology
© Springer-Verlag 1993

Patterns and mechanisms of hepatitis B/hepatitis D reinfection after liver transplantation

Anna Linda Zignego[1], **D. Samuel**[2], **P. Gentilini**[1], and **H. Bismuth**[2]

[1] Istituto di Clinica Medica II, University of Florence, Florence, Italy
[2] Hepatobiliary Surgery and Liver Transplantation Research Unit, Hopital Paul Brousse, Villejuif, France

Summary. Viral recurrence is the limiting factor in orthotopic liver transplantation (OLT) for hepatitis B virus (HBV) related liver disease. In fact, high rates of HBV infection of the transplanted liver are reported, followed by the recurrence of liver disease in a high percentage of cases. The importance of reinfection stimulates the study of its modalities and mechanisms in order to better identify preventive measures and better select patients for OLT. In HBV and HDV positive patients, the outcome of liver transplantation appears significantly better than in patients that are solely HBV positive, in spite of a high rate of HDV reinfection. Long-term analysis (5 years) of HBV and HDV infection, using the PCR technique, in 15 patients transplanted for an HBV/HDV positive liver disease and treated with anti-HBs immunoglobulin (HBIG), revealed that all patients experienced an HDV reinfection, but only about 7 were still harboring the virus after four years of follow-up. HDV reinfection was either associated to HBV reinfection or isolated whereas no cases of HBV isolated reinfection was observed. Isolated HDV reinfection was frequent and transient in all but one case that was superinfected by HBV. Infected peripheral blood mononuclear cells seem to be implicated in HBV superinfection of HDV infected liver. Liver damage was observed only in cases of HBV/HDV co-infection, suggesting that, in vivo, HBV is necessary to produce liver damage although it is not essential for HDV absorption to target cells, HDV penetration of these cells or HDV genomic replication. In addition, in isolated HDV infection, transient HDV viraemia and its low levels suggest that, perhaps in these patients HDV uses a very limited presence of HBV or alternative ways which are not efficient enough for envelope production. These data suggest that, particularly in HDV positive patients, antiHBs Ig administration, which has previously been proven to significantly reduce HBV reinfection in HBsAg-positive patients, may be useful in changing the natural history of repetition of the original viral infection and liver disease after OLT.

Introduction

Orthotopic liver transplantation (OLT) is an effective therapy in the treatment of end-stage liver disease. HBsAg-positive chronic hepatitis is among the most common indication worldwide, either because of end-stage liver disease or the development of hepatocellular carcinoma [1]. However, viral hepatitis still represents serious problems for the therapeutic benefits of liver grafting: HBV reinfection is a major cause for long-term liver dysfunction and in some cases, may also lead to cirrhosis in a few months [2, 3]. For this reason, many centers consider the risk of reinfection and mortality too high. The critical importance of reinfection has stimulated the study of its modalities and mechanisms in order to choose more effective preventive measures and to better select patients for OLT.

It has been shown that HBV reinfection of liver graft may be caused both by free virus in plasma and by virus hidden in non-hepatic cells; in particular, the presence of extrahepatic reservoirs of HBV, involving peripheral blood mononuclear cells (PBMC), may explain reinfection observed in patients that showed no evidence of HBV viremia at transplantation [4, 5].

Recent studies show that the outcome of liver transplantation is significantly better in HBV/HDV positive patients, than in those that are HBV-positive alone, in spite of a high rate of HDV reinfection after intervention [6–8].

Studies on HBV/HDV reinfection using classical virological methods

In a study performed by Rizzetto et al. [9] the clinical outcome of 7 patients transplanted for an HBV/HDV positive, HBV-DNA negative cirrhosis was evaluated. During an observation period ranging from 7 weeks to 15 months, the infection recurred in 5 patients: expression of HDAg in the graft was detected very early after intervention and in two cases was not accompanied by expression of HBV in the serum or liver.

From 16 patients transplanted for a HBV/HDV positive and HBV DNA-negative cirrhosis and treated, from the outset, with high doses of HBIG, we observed that only 3 became HBV-positive after OLT, while 13 remained HBsAg-negative during the follow-up without any signs of viral liver damage. It was possible, however, to detect isolated HDV infection in 8 of these patients [10]. HDV reinfection more frequently occurred within the first month and was also detected within 3 days after OLT [11]. In a series of 7 patients that we followed until the 34th month after OLT for both helper and defective virus reinfection, HDV RNA was detected in all seven patients with either persistent (4/7) or transient (3/7) positivity; combined HBV reinfection and recurrence of liver disease were observed only in the first group of patients. In the remaining 3 patients, HDV RNA was detectable despite the absence of markers of HBV infection. HDV RNA positivity before OLT was indicative of an almost immediate reinfection and, after intervention, was the only marker which was constantly detectable [12]. These studies were performed by means of molecular biological methods such as Northern blot, Southern Blot, Spot and Slot hybridization test, of proven efficacy in detecting HDV RNA or HBV DNA sequences in biological samples, but suffering from a limited sensitivity.

Virological analysis of HBV/HDV reinfection using PCR methods

More recently, the availability of very sensitive virological methods for the identification of both HDV and HBV infections (HBV PCR and HDV RT-PCR) allowed a deeper analysis of virological events occurring in transplanted patients for HDV positive liver disease. We followed, for a long period of time (5 years), 15 patients transplanted for an HBV and HDV positive cirrhosis and treated with HBIG. During the first year after transplantation, it was possible to distinguish four main virological groups of patients: subjects without any signs of viral reinfection (group 1: patients 1 and 2, Table 1); patients with HDV reinfection and without markers of HBV reinfection (isolated HDV reinfection) (group 2 = patients 3–10, Tables 1, 2, Fig. 2) patients with markers for both HDV and HBV infection and with a recurrence of liver disease (group 3: patients 11–14, Table 1); patients with markers of both HBV and HDV infection without evidence of liver disease (group 4: patient 15, Table 1). During the second year of follow-up, 6 patients with isolated HDV reinfection (group 2) lost the virus, whereas one was also reinfected by HBV and experienced a recurrence of liver disease (patient 6, Table 2, Fig. 3). Finally, 2 patients (nos. 1 and 2) without any viral reinfection (group 1) were reinfected by both delta and B viruses with the recurrence of liver disease as well. At the end of the second year after LT, group 1 included 6 patients, group 2 one and group 3 five. During the third year after LT, the remaining subject with isolated HDV reinfection (patient 5, group 2) lost the virus. At the

Anna Linda Zignego et al.

Table 1. Serum HDV RNA and HBV DNA before and after liver transplantation (first year of follow-up)

Patient no.	Before transplantation		After transplantation	
	HDV RNA	HBV DNA	HDV RNA	HBV DNA
1	−	−	−	−
2	−	−	−	−
3	+	−	+	−
4	+	−	+	−
5	−	−	+	−
6	+	−	+	−
7	+	−	+	−
8	+	−	+	−
9	+	−	+	−
10	+	−	+	−
11	−	−	+++	+++
12	−	−	+++	+++
13	+	−	++	++
14	−	−	+++	+++
15	+	−	+++	+++

Titers of HDV RNA corresponding to 1:1–1:100 (+); >1:100–1:10000 (++); 1:10000–1:1 000 000 (+++)

end of the third year of follow up only 3 groups (groups 1, 3 and 4) remained. Finally, during the fourth year of followup, one patient with HBV/HDV reinfection (patient 13, group 3) cleared both viruses, whereas the patient with both HBV and HDV infection and no signs of liver damage (patient 15, group 4, Fig. 4) experienced acute hepatitis. No modifications in virological patterns were observed during the fifth year of follow-up. At the end of the period of observation, only two groups of transplanted patients remained: patients without any signs of HBV/HDV infection and no evidence of liver disease (group 1 = 8 patients) and patients with both HBV and HDV infection and evidence of liver disease (group 3 = 7 patients). Thus, we observed a progressive reduction in the type of infection patterns leading, after four years, to only two virological groups of subjects that seem to be definitive, as suggested by the absence of modifications observed during the last year of follow up (Figs. 1, 2, 3 and 4). All patients included in this study experienced HDV reinfection even if very limited in some cases, but at the end of the observation period, only a minority (46.6%) were still harboring HDV. Only patients who became both HDV and HBV positive, 8 out of the 15 subjects followed, experienced a recurrence of liver disease according to data from different series [8, 11]. This was

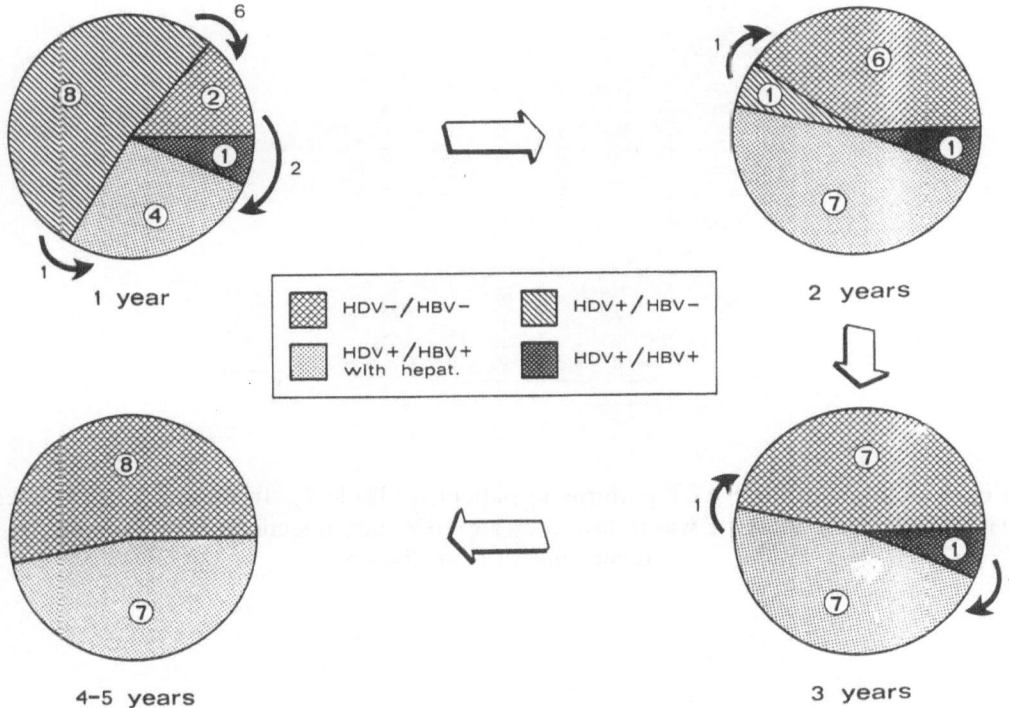

Fig. 1. Patterns of HBV/HDV reinfection after liver transplantation and their evolution during the follow-up period. Black arrows indicate the direction of migration of some subjects from one group to another observed during the second, third and fourth year of follow-up, respectively

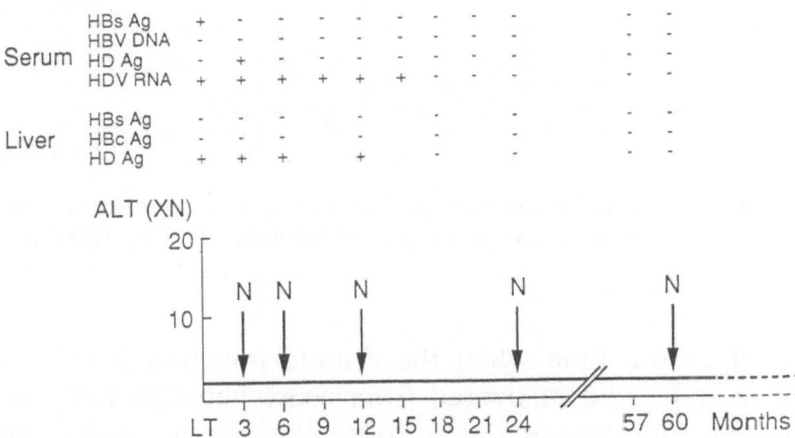

Fig. 2. Virological and ALT patterns in a patient (patient 3; Table 1), who after LT experienced an isolated HDV reinfection. Analogous patterns were observed in other subjects from the same virological group, with the exception of patient 6 (patients 3–5 and 7–10; Table 1)

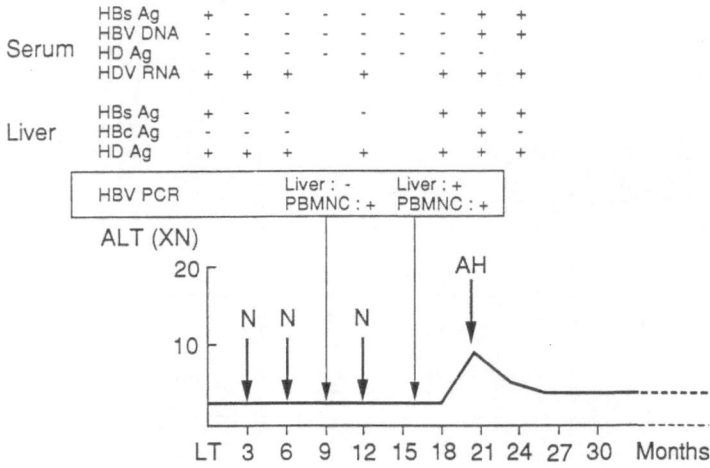

Fig. 3. Virological and ALT patterns in patient 6 (Table 1). In this case, an isolated HDV reinfection after LT was followed by an HBV superinfection of the liver and by a recurrence of liver disease

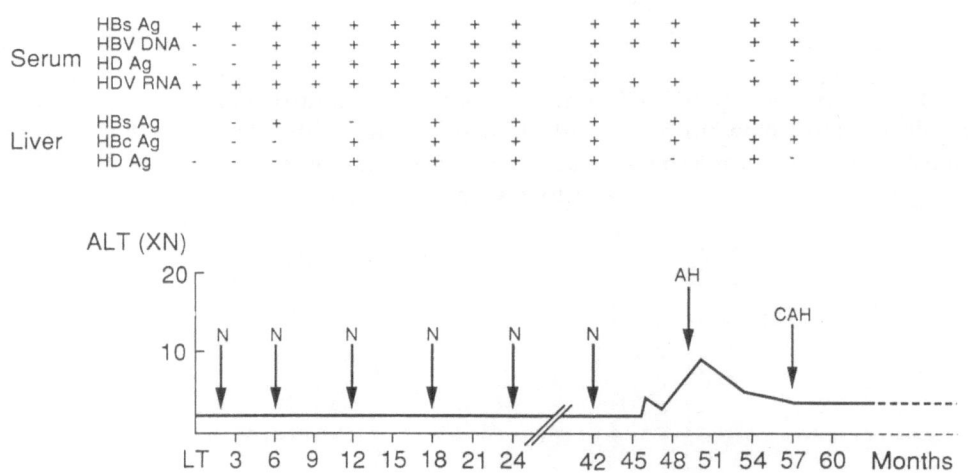

Fig. 4. Virological and ALT patterns in patient 15 (Table 1). This subject maintained a normal liver for more than 3 years in spite of a combined HBV/HDV reinfection

independent of the time when the double infection itself took place. All but one case, who recovered from acute hepatitis B/D, evolved to chronicity. Two mechanisms seem to be effective in double reinfection: an HBV/HDV co-reinfection of the transplanted liver and an HBV superinfection of an HDV isolated infection, whereas no cases of HBV isolated infection, followed by an HDV superinfection, (the most common mechanism in natural infection), were observed. It should be

Table 2. PCR analysis in patients with isolated HDV reinfection: HBV DNA sequences in liver and PBMC (first year of follow-up)

Patients	HBV RNA in liver	HBV DNA in PBMC	Recurrence of liver disease (N/Y)
4	+	−	N
5	−	−	N
6	+	+	Y

noted that from three cases of isolated HDV infection, where we also took into account PBMC infection using PCR, HBV DNA was not detectable in the serum, liver or PBMC samples of two patients who lost the virus. HBV DNA was evident, however, in cells from the third patient, who during the second year of follow up, experienced both liver HBV reinfection and a recurrence of liver disease (Table 2 and Fig. 3). This suggests that the HBV superinfection of an HDV positive liver may be explained by the presence of circulating reservoirs of the helper virus. In contrast, HDV isolated liver reinfection was probably simply mediated by circulating particles. In fact, in all but one patient, HDV viraemia was detected by PCR before OLT (Table 1). The rapidity of liver HDV reinfection in these cases is in agreement with this hypothesis. Delayed HDV reinfection, in patients where serum HDV RNA was negative before OLT (as observed in two patients who were reinfected by both HBV and HDV during the second year of follow-up), is more difficult to explain; in fact, HDV has not yet been detected in extrahepatic sites. More accurate studies will ascertain this possibility.

On the whole, these data indicate that HBV is not essential for HDV absorption to target cells, HDV penetration of these cells or HDV genomic replication, as has been suggested by a series of in vitro studies [13, 14], but it is involved in the generation of liver damage. In addition, we found that in cases of isolated HDV infection, the levels of HDV viremia were low and transient, in contrast with the persistently high levels generally observed when HDV infection was associated with HBV infection (Table 1). This suggests, perhaps, that low-level HDV infection may depend on a very limited presence of HBV or on suboptimal alternative ways for envelope production.

The necessity of HBV/HDV presence for hepatic damage does not necessarily mean that the damage itself immediately follows the association of HDV and HBV infections. In fact, we observed a patient who maintained a contemporary elevated replication of both viruses for a period of more than three years, without any biochemical or histological evidence of liver damage. Factors influencing a different clinical evolution for a double HDV/HBV liver infection are currently unknown. Partial sequencing of the HDV strain infecting the patient described above has not revealed significant genomic differences when compared with previously described sequences. It is possible that other factors, in addition to viral infection, are important in determining liver damage. It is also possible that an HDV plus HBV infection of the liver may be clarified over a long term follow-up, as demonstrated by a patient who was reinfected by both viruses, then experienced hepatitis and finally, during the third year of follow-up, cleared both viruses and resolved the hepatitis.

Prevention

From a practical point of view, these data suggest that particularly in HDV positive patients, HBIG, which has previously been proven to significantly reduce HBV reinfection in HBsAg positive patients [6, 7], may be useful in changing the natural history of the repetition of original viral infection and liver disease after transplantation. This may be deduced from the key role played by HBV reinfection and confirmed by the high percentage of HDV reinfected patients who cleared the virus during follow up (8/15 = 53.3%), as well as by the low percentage of patients who, in turn, also acquired HBV persistent infection (7 patients = 46.6%). In contrast, the percentage of HBV reinfection in similar patients not treated with HBIg appears to be significantly higher (>70%), as indicated by the study of Ottobrelli et al. [8].

An indirect confirmation, in contrast to what was initially believed, was that the presence of HDV viremia before OLT was not correlated with the recurrence of liver disease, but only to the rapidity of HDV liver reinfection after OLT. Thus, HDV RNA determination in serum before intervention cannot be considered a negative prognostic factor.

Acknowledgements

This study was supported by a grant from the Italian CNR (Prevention and Control Disease Factors, subproject SP3, contract no. 91.00168) and the Italian Liver Foundation.

References

1. Bismuth H (1989) Hepatic transplantation in Europe: improved survival. Eur J Gastroenterol Hepatol 1: 79–82
2. Lauchart W, Muller R, Pichlayr R (1987) Immunoprophylaxis of hepatitis B virus reinfection in recipients of human liver allografts. Transplant Proc 19: 2387–2389
3. Starzl TE, Demetris AJ, Van Thiel D (1989) Liver transplantation. N Engl J Med 321: 1092–1099
4. Zignego AL, Samuel D, Gugenheim J, Chardan B, Bismuth H (1988) Hepatitis B virus replication and mononuclear blood cell infection after liver transplantation. In: Zuckerman (ed) Viral hepatitis and liver disease. Williams and Wilkins, Baltimore, pp 808–809
5. Feray C, Zignego AL, Samuel D, Bismuth A, Reynes M, Tiollais P, Bismuth H, Brechot C (1990) Persistent hepatitis B virus infection of mononuclear cells without concomitant liver infection. Transplantation 49: 1155–1158
6. Samuel D, Zignego AL, Bismuth A, Serres C, Feray C, Benhamou J-P, Brechot C, Bismuth H (1990) Outcome of HBV infection after liver transplantation for HBsAg positive patients: experience with long term prophylaxis: In: Hollinger FB, Lemon SM, Margolis H (eds) Viral hepatitis and liver disease. Williams and Wilkins, Baltimore, pp 648–650
7. Samuel D, Bismuth A, Mathieu D, Arulnaden J-L, Reynes M, Benhamou J-P, Brechot C, Bismuth H (1991) Passive immunoprophylaxis after liver transplantation in HBsAg-positive patients. Lancet 337: 813–815
8. Ottobrelli A, Marzano A, Smedile A, Recchia S, Salizzoni M, Cornu C, Lamy ME, Otte JB, De Hemptinne B, Geubel A, Grendele M, Colledan M, Galmarini D, Marinucci G, Di Giacomo C, Agnes S, Bonino F, Rizzetto M (1991) Patterns of hepatitis delta virus reinfection and disease in liver tranplantation. Gastroenterology 101: 1649–1655
9. Rizzetto M, Macagno S, Chiaberge E, Verme G, Negro F, Marinucci G, Di Giacomo C, Alfani D, Cortesini R, Milazzo F, Doglia M, Fassati LR, Galmarini D (1987) Liver transplantation in hepatitis delta virus disease. Lancet ii: 469–471
10. Reynes M, Zignego AL, Samuel D, Fabiani B, Gugenheim J, Tricottet Y, Brechot C, Bismuth H (1989) Graft hepatitis delta virus reinfection after orthotopic liver transplantation in HDV cirrhosis. Transplant Proc 21: 2424–2426
11. Grendele M, Colledan M, Gridelli B (1990) Does the experience with liver transplantation suggest that hepatitis Delta virus is not cytopathic? Transplant Proc 22: 1551–1553
12. Zignego AL, Dubois F, Samuel D, Georgopoulou U, Reynes M, Gentilini P, Bismuth A, Benhamou J-P, Hadziyannis SJ, Bismuth H, Brechot C (1990) Serum hepatitis delta virus RNA in pattients with delta hepatitis and in liver graft recipients. J Hepatol 11: 102–110
13. Taylor J, Mason W, Summers J, Goldberg J, Aldrich C, Coates L, Gerin J, Gowans E (1987) Replication of human hepatitis delta virus in primary cultures of woodchuck hepatocytes. J Virol 61: 2891–2895
14. Mason WS, Taylor JM (1989) Experimental systems for the study of hepadnaviruses and hepatitis delta virus infections. Hepatology 9: 635–645

Authors' address: Dr. Anna Linda Zignego, Istituto di Clinica Medica II, University of Florence viale Morgagni 85, I-50134 Florence, Italy.

Arch Virol (1993) [Suppl] 8: 291–304

Hepatitis C virus infection in liver allograft recipients

**L. Caccamo[1], M. Colledan[1], B. Gridelli[1], G. Rossi[1], M. Doglia[1], S. Gatti[1],
P. Ghidoni[2], A. Lucianetti[1], G. Lunghi[3], U. Maggi[1], G. Paone[1], P. Reggiani[1],
D. Galmarini[1], and L.R. Fassati[1]**

[1] Istituto di Chirurgia Sperimentale e dei Trapianti Centro Trapianto Fegato
[2] Istituto di Anatomia e Istologia Patologica, and [3] Istituto di Igiene, Università di
Milano, Ospedale Maggiore I.R.C.C.S., Milan, Italy

Summary. The impact of HCV infection after liver transplantation remains a topic of discussion. The aims of this study were to define the prevalence of anti-HCV antibodies in liver donors; the risk of acquired HCV infection and HCV re-infection according to the pre-transplant anti-HCV status; the prevalence of HCV infection in post-transplant chronic hepatitis. Sera from 42 recipients with follow up longer than 6 months and their donors were tested for anti-HCV. By results at pre-transplant time patients were classified as follows: donor (D) negative and recipient (R) negative (D−/R−) 31; D−/R+ 9; D+/R− 1; D+/R+ 1. Twenty-one patients with sustained hepatic dysfunction underwent liver biopsy. In group D−/R−, 5 patients showed anti-HCV positivity and 3 (9.7%) of them had acquired HCV hepatitis. In group D−/R+, 6 patients showed persistent anti-HCV positivity and 4 (44.4%) of them had recurrent HCV hepatitis; of these 2 died due to liver failure. The 2 patients of groups D+/R− and D+/R+ had normal liver function. Anti-HCV negative hepatitis was found in 2 patients. The prevalence of anti-HCV positivity in liver donors appeared low (3.2%). Acquired HCV infection rate was 9.7%. Pre-transplant HCV infection led to a high incidence of recurrence (44.4%). HCV was the major etiological agent in post-transplant chronic hepatitis (77.8%).

Introduction

Hepatitis C virus (HCV), an RNA virus that belongs to the family of *Flaviridiae*, has been identified as the principal etiologic agent of post-transfusion hepatitis [13]. Liver cirrhosis in patients with presence of antibodies to HCV (anti-HCV) is one of the major indications for orthotopic liver transplantation (OLT) world-wide [35, 42]. The role of HCV in organ transplantation is still a matter of discussion [8, 18], but the risk of HCV infection after transplantation is beginning to be recognized as being responsible for a large number of liver impairments [7, 9, 11, 17, 27, 28, 32]. A new infection can appear as the result of organ-to-host transmission or, particularly in hepatic transplantation, due to the use of large amounts of blood products. From the experience of liver transplantation in Hepatitis B Virus (HBV) carriers, it is known that post-transplant infection is almost universal [36], if an immuno-prophylaxis program is not maintained during the long-term follow-up [31]. In anti-HCV positive cirrhotic recipients a high risk of reinfection must be expected after liver transplantation because no prophylaxis is available. Nevertheless, the prevalence and the clinical course of HCV reinfection after liver transplantation has still not been well established [8, 10, 37]. The aim of this retrospective study is to evaluate the prevalence of anti-HCV positivity in liver donors, the incidence and clinical course of HCV infection in liver transplant patients and the prevalence of anti-HCV in chronic hepatitis in allografts.

Materials and methods

From our series, HBsAg-negative, IgM antiHBc-negative patients were selected, for whom serum samples of both the donor and the recipient were available. Sera from 60 recipients and their apparently healthy liver donors were therefore tested using a second generation anti-HCV enzyme-linked immunosorbent assay (ELISA); supplementary testing with a second generation 4 antigen recombinant immunoblotting assay (RIBA) was performed in all ELISA positive cases. All sera were stored at $-20°C$ until the time of anti-HCV testing. The post-transplant status was determined using the last available serum sample. Anti-HCV was considered detectable only if confirmed by RIBA. Both anti-HCV assays were carried out according to the manufacturers' instructions. The second generation ELISA (HCV EIA, Abbott Laboratories, North Chicago, IL, U.S.A.) detects antibodies to recombinant HCV antigens derived from two regions (c200 and c22), the non-structural and core regions of the HCV genome. The ELISA cut-off value was determined by the mean of known negative controls plus 0.4 absorbance units. All specimens with an absorbance level less than the cut-off value were considered negative. All specimens that were found to have an absorbance level greater than the cut-off level were rerun to confirm the result, otherwise a negative result was assumed. The second generation RIBA (RIBA HCV Test System, Chiron Corp., Emeryville, CA, U.S.A.) detects antibody to four recombinant antigens (5-1-1, c100, c33 and c22) from the non-structural and core regions of the HCV genome. Using the RIBA test the antigen band intensities were compared to weakly and moderately positive IgG control bands. A response of 1+ or greater to any of the four antigens was interpreted as reactive. RIBA reactivity to neither antigen constituted a negative test; reactivity to only one antigen was considered an indeterminate response; positive results needed reactivity to at least two antigens.

Immunosuppressive treatment was based, when possible according to the white cells count, on triple drug administration (Cyclosporin A, low dose of Prednisone and

Azathioprine). The steroids were gradually reduced and eventually withdrawn during the long-term follow up. Rejection episodes were treated with high doses (0.5–1 gr) of methylprednisolone (bolus). Steroid-resistant rejection was treated with a 10–14 day course of anti-lymphocyte T monoclonal antibodies (Orthoclone OKT3). If rejection was not reversed), retransplantation was performed.

On the basis of anti-HCV determination at the time of the transplantation the recipients were assigned to one of the following four groups: I) donor negative – recipient negative (D−/R−); II) donor negative – recipient positive (D−/R+); III) donor positive – recipient negative (D+/R−); IV) donor positive – recipient positive (D+/R+). Forty-two patients with follow-up periods longer than 6 months were selected according to changes or lack of changes in the anti-HCV status in the last available serum sample.

Blood component exposure for each pre-transplant anti-HCV negative patient was assessed by the total units of packed red cells (RPC) and fresh frozen plasma (FFP) transfused during the hospital stay.

Common liver function tests were regularly determined during the follow-up, and functional graft impairment was defined as persistent when lasting longer than one month, and as transient when recovery was achieved in a month. In the presence of biochemical dysfunction either biliary and vascular technical complications were ruled out by appropriate imaging diagnostic methods. Liver biopsy samples, when available, were evaluated for hepatitis. Diagnosis of hepatitis required the presence of portal inflammation, lobular mononuclear cell infiltrate and hepatocellular necrosis; the appearance of intralobular bridging fibrosis was required to note an active evolution. In order to establish the etiology of the allograft damage, known causes of hepatitis other than HCV were excluded by the serial analysis of patients' serum samples for HBV markers and antibodies (IgM and IgG titer) against Hepatitis A virus, Cytomegalovirus (CMV), Epstein-Barr virus, Herpes simplex virus and Herpes zoster virus. Each liver sample was fixed in buffered formalin and routinely processed. Staining for HBsAg, HBcAg was performed using commercially available monoclonal antisera (Dako Corp., Carpinteria, CA, U.S.A.) and a standard indirect immunoperoxidase technique in paraffin embedded liver sections. CMV detection in liver tissue was performed with the in situ hybridization technique using a streptavidine-alkaline phosphatase-labeled CMV-DNA probe (Human CMV, Kreatech, Amsterdam, The Netherlands).

The presence of hepatitis along with anti-HCV positivity and no evidence of other known causes of hepatitis was defined as HCV hepatitis. The presence of HCV hepatitis in a pre-transplant anti-HCV positive recipient was defined as a recurrence of the primary disease, while HCV hepatitis in a pre-transplant anti-HCV negative patient was defined as a de novo (or acquired) infection. Usually the diagnosis of viral hepatitis led to a reduction or withdrawal of steroid therapy, but no other specific modification in the immunosuppressive regimen or antiviral therapy was undertaken.

Values are presented as mean, standard deviation (SD) and range. Statistical analyses were performed using the Student's t-test with a level of $p < 0.05$ accepted as statistically significant.

Results

The concordance rate between second generation ELISA and RIBA tests in this transplanted population was 100% in the donors and pre-transplant recipients samples, and 97% in the post-transplant sera (Table

L. Caccamo et al.

Table 1. Second generation ELISA and RIBA results in post-transplant follow-up

	RIBA negative	RIBA indeterminate	RIBA positive
ELISA negative	24	0	0
ELISA positive	1	5	12

Table 2. Donors' and recipients' data

	Donors	Recipients
Gender: M/F	40/23	32/28
Age (yrs):		
mean	22.03	30.45
SD	15.65	19.17
range	2–63	1–60
Anti-HCV pos	2	19

1). Sera from 2/63 (3.17%) liver donors and 19/60 (31.9%) pre-transplant liver recipients were found positive for anti-HCV (Table 2). Primary diseases are presented in Table 3. Both the anti-HCV positive donors had normal hepatic function tests and a normal liver appearance at the harvesting procedure. Retransplantation was performed in 3 recipients (OLT 154/167, 158/169, 162/163) using a graft from a donor having the same anti-HCV status as the former. Five anti-HCV negative patients and 4 anti-HCV positive patients died sooner than 6 months after surgery. These 9 recipients, along with 4 other anti-HCV negative patients and 5 other anti-HCV positive patients, all alive with less than 6 months follow-up, were excluded from the present study. Finally, a total of 42 patients were analyzed in the follow-up study (Table 4). At the time of last out-patient control (mean 21.73 months, SD 11.49, range 6–39) 12/42 (28.57%), recipients were found anti-HCV positive and divided as follows: in Group D−/R− 5/31 (16.12%) patients became positive after transplantation (mean time 6.8 months, range 2–16), with sustained liver dysfunction in 4 cases (OLT 86, 97, 122, 165). Another 4 patients (OLT 98, 106, 139, 160) showed an indeterminate RIBA test, with an episode of transient graft dysfunction during the long-term follow up in one case (OLT 139). In group D−/R+ 6/9 (66.66%) patients were found to be anti-HCV positive with long-standing hepatic dysfunction (OLT 80, 92, 117, 133, 143, 157); among these patients, 2 (OLT 80, 133) died because of sub-acute liver failure from recurrent HCV hepatitis after a mean time

Table 3. Primary diseases

	Pre-OLT anti-HCV neg	Pre-OLT anti-HCV pos
Cirrhosis	5	7
Biliary cirrhosis	7	1
Alcoholic cirrhosis	6	2
Fulminant hepatitis	4	0
Biliary atresia	4	2
Sclerosing cholangitis	3	0
Hepatoma	3	5
Byler's disease	2	1
Wilson's disease	2	0
Metabolic disorders	5	1

Hepatoma and cirrhosis were present in 6 patients and primary hepatic cancer was an indication of OLT in 2 patients (one in each group). Metabolic disorders were: cirrhosis in cystic fibrosis (2), Type I Crigler-Najjar disease (2), Alpha-1-anti-trypsin deficiency (1), Cholesterol-ester storage disease (1)

Table 4. Follow-up results

D/R Groups (n)	Post-OLT Anti-HCV pos	Persistent graft dysfunction	HCV hepatitis	Death due to HCV infection
−/− (31)	5	13	3	0
−/+ (9)	6	7	4	2
+/− (1)	0	1	0	0
+/+ (1)	1	0	0	0

of 4 months since acute hepatitis onset. Another recipient (OLT 88) was anti-HCV indeterminate with normal liver function, while 2 other patients (OLT 91, 144) were found anti-HCV negative. The patient of group D+/R− was still negative with normal hepatic parameters 26 months after surgery. The patient of group D+/R+ underwent a long period (11 months) of anti-HCV negativity, but she suddenly showed a transient liver function impairment and thereafter was found to be anti-HCV positive; after that episode she recovered and showed normal graft function 24 months after transplantation.

No significant difference has been demonstrated by the analysis of the blood-product exposure among pre-transplant anti-HCV negative

Table 5. Blood products exposure in D-/R- group

		Post-OLT anti-HCV neg	Post-OLT anti-HCV pos	
RPC:	mean	7.78	3.0	$p = 0.131$
	SD	8.39	3.46	
	range	0–32	0–11	
FFP:	mean	21.3	13.42	$p = 0.304$
	SD	19.49	11.49	
	range	2–80	2–40	

recipients with regard to the post-operative anti-HCV seroconversion (Table 5).

The liver picture of the biopsy taken during the harvesting procedure in both anti-HCV positive donors showed mild hepatocellular swelling.

Persistent abnormal liver function tests were found in 21/42 patients that underwent 29 liver biopsies (Table 6). Among 13 patients in group D−/R− undergoing a biopsy, 3 (9.67%) (OLT 86, 97, 122) had a de novo HCV hepatitis. Among these recipients the clinical course of OLT 86 is unusual because of a double acquired primary virus hepatitis which occurred during the long-term follow up: a HBsAg serum positivity was evident 6 months after surgery, but the peak of serum aminotransferase levels appeared 3 months later as diagnosed by histology and confirmed by immunohistochemistry; steroids were withdrawn; after one year the patient spontaneously cleared HBsAg from her blood, but anti-HCV positivity appeared for the first time and at a graft biopsy control a progression to chronic hepatitis with a negative staining for HBV was found. Five other patients (OLT 83, 96, 136, 145, 165) had a CMV hepatitis which was treated by gancyclovir administration with complete recovery in 3 (OLT 83, 96, 165) and partial response in the other 2 (OLT 136, 145); among these patients one showed a seroconversion for anti-HCV (OLT 165). OLT 149 had an acute rejection episode successfully treated with methylprednisolone boluses. OLT 154 primary disease was an autoimmune cirrhosis with incidental hepatic cancer and soon after surgery she showed severe hepatic damage that was diagnosed as an acute rejection refractory to the medical treatment; the patient was retransplanted (OLT 167) after 4 months, but died in the early post-retransplantation period because of multiorgan failure. OLT 100 received liver replacement for sclerosing cholangitis associated with ulcerative colitis; after surgery, continual deterioration in hepatic function and at biopsy signs of rejection along with lobular inflammation were discovered; only transient and partial response to methylprednisolone

Table 6. Data of patients that underwent liver biopsy for graft dysfunction of medical causes

OLT	Sex/age	Follow-up (months -m-)	Pre-OLT D/R anti-HCV	Onset of graft dysfunction (m)	Peak ALT (U/L)	Peak Tot. Bil. (mg/dl)	Peak GGT (U/L)	Post-OLT anti-HCV	Diagnosis of graft dysfunction	Outcome	Last ALT	Last GGT
80	F/15	6	-/+	1	814	36.8	1164	4 Ag +	Sub-AHF	death	111	425
83	F/47	39	-/-	9	101	2.3	331	-	CMV-H	recovery	22	82
86	F/47	38	-/-	8	487	1.3	669	c22 c33 +	HCV-H	Hepatitis	97	250
91	F/2	36	-/+	1	88	0.3	63	-	CMV-H	recovery	45	25
92	M/58	35	-/+	9	485	5.3	2187	4 Ag +	HCV-H	Hepatitis	24	328
96	F/42	34	-/-	2	345	2.5	801	-	CMV-H	recovery	41	55
97	F/43	33	-/-	6	203	2.4	230	c22 c33 +	(HBV) HCV-H	Hepatitis	196	99
100	M/8	32	-/-	17	411	1.7	522	-	NANBNC-H	Hepatitis	255	479
117	M/44	9	-/+	3	233	2.2	494	c22 c33 +	HCV-H	death	154	494
122	M/49	24	-/-	2	933	2.9	184	4 Ag +	HCV-H	Hepatitis	184	64
124	F/15	24	+/+	7	611	2	608	c22 c33 c100 +	?	recovery	9	6
127	F/4	23	-/-	21	1848	1.2	157	-	?	recovery	24	40
133	M/45	11	-/+	8	1272	84	1350	c22 c33 +	Sub-AHF	death	33	39
136	F/1	19	-/-	1	524	1.6	365	-	CMV-H	recovery	60	8
139	M/5	18	-/-	6	370	2.1	136	c22 +	CMV-H	recovery	27	11
141	F/21	17	-/-	2	790	26.2	724	-	NANBNC-H	Hepatitis	132	146
145	F/12	15	-/-	1	239	2	399	-	CMV-H	recovery	107	41
149	M/47	12	-/-	8	692	1.5	907	-	Acute rejection	recovery	42	222
154	F/28	6	-/-	1	207	60.8	2080	-	Acute rejection	death after Retx	10	100
157	M/60	9	-/+	4	126	1.3	635	4 Ag +	Acute rejection	recovery	32	356
165	F/1	7	-/-	1	385	3.5	379	c100 5-1-1 +	CMV-H	recovery	15	11

ALT Alanine-amino transferase (Normal Value NV <40), *Tot. Bil* total bilirubin (NV <1.2), *GGT* gamma-glutamil transpeptidase (NV <50), *4 Ag +* reactivity to the four antigens of RIBA test, *sub-AHF* sub-acute hepatic failure, *CMV-H* CMV hepatitis, *HCV-H* anti-HCV positive hepatitis, *NANBNC-H* non A non B anti-HCV negative hepatitis, *Retx* retransplantation boluses administration was obtained. OLT 141 presented a severe hepatitis and an active hepatitis was found histologically with a rapid cirrhogenic progression demonstrated after 10 months at her last biopsy; she was anti-HCV negative throughout follow-up. When OLT 113 showed a hepatic damage, biliary anastomotic stenosis was diagnosed and a choledocojejunostomy was successfully performed.

Among 7 patients in group D−/R+ undergoing a liver biopsy, 4 (44.44%) (OLT 80, 92, 117, 133) had recurrent HCV hepatitis; in 2 (22.22%) of these (OLT 80, 133) severe hepatitis was evident, and a rapid progression to liver failure led to death after a mean time of 4 months after onset of acute hepatitis. Among the other 2 anti-HCV positive patients, OLT 117 died after 9 months of follow-up from gastrointestinal hemorrhage having a cancer recurrence, and OLT 92 is still alive but with a tumor recurrence and an active hepatitis. OLT 157 showed an acute rejection which was reversed with medical therapy. OLT 91 had CMV hepatitis that was successfully treated and showed anti-HCV negativization during follow-up. OLT 143 suffered from biliary anastomotic stenosis that was successfully treated by the percutaneous placement of a trans-stenotic metallic expandible stent. The group D+/R− patient (OLT 116) underwent medical therapy for tuberculosis of the lungs and presented a hepatic histological picture of mild cholestasis with an abnormal liver biochemistry being permanently anti-HCV negative.

Transient severe liver function impairment occurred in 3 pediatric patients classified as follows: in group D−/R− 2 recipients (OLT 127, 139) showed an abrupt rise in serum transaminases at a mean of 13.5 months after transplantation. Liver biopsy was performed in OLT 127 showing only minor changes. Both patients rapidly (mean time of 9 days) recovered either spontaneously (OLT 127) or after a single methylprednisolone bolus of 500 mg (OLT 139). In OLT 139 an anti-HCV indeterminate RIBA resulted, beginning 3 months after transplantation, while OLT 127 was found permanently anti-HCV negative. After these episodes, both patients showed normal hepatic function tests (mean time of 7 months follow-up since liver dysfunction time).

High serum aminotransferase levels were observed in the group D+/R+ patient (OLT 124) 7 months after transplantation; at liver biopsy only minor changes were found and spontaneous recovery in 20 days was observed. This patient had a four-antigen RIBA reactivity at pre-transplant time, she was found anti-HCV negative throughout a period of 11 months after surgery, but anti-HCV positivity (without reactivity for the 5-1-1 antigen alone) was found again after the described episode of liver dysfunction.

Discussion

This study enabled us to evaluate the serologic, clinical, and histological features of HCV infection in liver transplant recipients. It has been shown that second generation anti-HCV techniques enhance the sensitivity and specificity of detection of anti-HCV antibodies [11, 14, 19, 21]. The concordance rate between second generation ELISA and RIBA tests in this transplanted population was similar to the experience in chronic hepatitis patients [2] and other liver transplant studies [6, 10]. Ultimately, however, direct detection of HCV-RNA, either in serum or in liver tissue, will serve as the gold standard for presence of HCV infection [6, 7, 32]. At present HCV-RNA detection is still an experimental method and has a limited application for clinical purposes.

Due to the small number of anti-HCV positive donors, no conclusions can be derived from our experience of using such grafts. No persistent liver allograft damage was observed, during a mean follow-up of 25 months, in the 2 recipients of livers harvested from anti-HCV positive donors. In both donors liver-function tests, gross and microscopic appearance were normal. The management of anti-HCV positive organ donors is still controversial in view of the shortage of available grafts which is particularly dramatic in Italy, either because controversy about test significance remains or because evidence for transmission of HCV infection had been demonstrated [24, 25, 35, 40]. The anti-HCV assay is the only available test before the harvesting procedure, but the technique employed is crucial in judging the result because the first generation test has both low specificity and sensitivity [25]. Anti-HCV assays, regardless of the technique, do not reliably distinguish the infectious from the non-infectious state. Recently, Alberti el al. [1] reported that 9/16 (56.25%) anti-HCV positive asymptomatic subjects had histological features of hepatitis and were HCV-RNA positive in the serum; the authors concluded that serum HCV-RNA is a sensitive and specific marker of liver disease in anti-HCV positive individuals, which can be independent of hepatic function tests. The best way to avoid an infectious organ transplantation would be to determine the presence of HCV-RNA sequences; but this technique takes a long time and cannot be performed before the transplant procedure. In order to limit organ waste, a possibility for reducing potential damage might be to transplant macrospically healthy livers from anti-HCV positive donors with normal function tests, only to anti-HCV positive recipients [23, 26]. An objection to this approach can arise from the demonstration of the existence of more than one strain of HCV [12], however since the possibility of intraindividual viral mutations has also been shown [20], we currently follow this policy. A further investigation could be to evaluate a donor liver biopsy for the presence of hepatitis, before the transplantation.

Other than from an infected organ, HCV infection can be transmitted by blood-product transfusions. In transplant patients after parenteral exposure to HCV, anti-HCV seroconversion may not occur due to induced immunosuppression [42], but even serum HCV-RNA sequences can be detected without evidence of liver disease [25]. Finally, hepatitis occurring in long-term follow up has been described independent of blood-product exposure [42]. In this unique group of transplant recipients, prevalence of de novo HCV infection was low without any correlation with blood-product exposure. The clinical course frequently progressed to chronic hepatitis, but did not lead to death or need for retransplantation after a mean of 15 months follow up after onset of hepatitis. It has been speculated that immunosuppression may prevent clearing of the virus and recovery from the acquired liver disease. This hypothesis might be confuted by the clinical course of OLT 86; this patient showed a spontaneous recovery from a de novo HBV infection over 17 months from the first time of HBsAg positive detection; moreover she did not recover from HCV infection one year after the first anti-HCV positivity detection. This can be explained either by the need for a longer follow-up period or by the particular presentation of a two acquired infection by hepatitis viruses. Wang et al. [39] did not find any significant interference between two viruses in 5 chronic hepatitis patients.

In the same manner as HBV positive liver-allograft recipients, pre-transplant anti-HCV positive patients are at a high risk of recurrence since extra-hepatic sites of replication have been demonstrated [15]. As in HBV positive patients [3, 30, 33], recurrence of HCV infection after liver transplantation appears to be a crucial factor affecting the long-term result of transplantation [17, 28, 35]. Poterucha et al. found that after liver transplantation the anti-HCV status was almost always maintained in pre-transplant positive patients and had a correlation with hepatic dysfunction [27]. Among anti-HCV positive patients pre-transplant serum HCV-RNA positivity did not itself lead to a higher rate of re-infection [42]. Forty-four percent of our anti-HCV positive patients with a follow up, longer than 6 months showed hepatitis, while another patient was anti-HCV positive, maintaining normal hepatic function during a follow up of 16 months. In our study, HCV recurrence was the major cause of death in 2 patients with a clinical course of subacute liver failure. Re-infection of the allografts appeared at a mean time of 5.25 months after surgery, but with a wide variability among our 4 patients (range 1–9).

HCV infection was a major cause of chronically impaired graft function after liver transplantation in HBV negative recipients [8, 9, 22, 29]. Anti-HCV serology together with chronic hepatitis occurred in 16.7% (7/42) of recipients; moreover 2 other patients suffered anti-HCV negative hepatitis after a mean of 15 months since hepatitis onset. According to the second generation RIBA, anti-HCV was present in 77.8% (7/9) of patients with chronic hepatitis.

Minimal changes in histological specimens were obtained in the 2 patients that underwent a graft biopsy because of a transient severe biochemical dysfunction. Aminotransferase levels promptly recovered spontaneously in these patients, as occurred in the case reported by Colina et al. [4]. A definitive diagnosis of the cause of the graft damage was not achieved, and anti-HCV positivity occurred in one and an indeterminate result in the other patient.

The management of the post-transplant HCV infection is still controversial. The experience in post-transplant HBV re-infection did not show any effective medical therapy [16, 41]; retransplantation showed poor results as recently reported in an American multicenter study [5], but the role of immunoprophylaxis needs further evaluation as shown by a single successful experience in our study (unpublished observation). With regard to HCV post-transplant infection, some authors have reported on the administration of interferon with only partial or no responses [17, 34, 38, 41], and also retransplantation showed poor results [17, 20].

In conclusion, this study has described: a low prevalence of anti-HCV positivity in liver donors; a low rate of acquired HCV infection after hepatic transplantation; a high risk of recurrent infection in pre-transplant anti-HCV positive patients. Anti-HCV serology evaluation was useful in association with the histopathological findings. Both de novo and recurrent HCV hepatitis have shown the trend to progress to chronic hepatitis. Among causes of chronic allograft impairments after liver transplantation, HCV infection occurred as an almost universal etiologic agent. At present the lack of either a therapy or a prophylaxis against HCV can enhance either the post-transplant morbidity and mortality rates and the need for retransplantation. Moreover HCV infection does not offer any realistic probability of recovery even after retransplantation. According to these results we think that anti-HCV positive cirrhosis might be considered as a relative contraindication to liver transplantation.

Acknowledgements

The authors would like to thank Ms. L. Vento for her invaluable technical assistance. The present study was supported by a grant from Contributo di Ricerca Quota 60%

MURST 1991 -Università di Milano-, and by a grant from Contributo Ricerca Corrente 1991 – Ministero della Sanità – Ospedale Maggiore di Milano IRCCS.

References

1. Alberti A, Morsica G, Chemello L, Cavalletto D, Noventa F, Pontisso P, Ruol A (1992) Hepatitis C viraemia and liver disease in sympton-free individuals with anti-HCV. Lancet 340: 697–698
2. Alter HJ (1992) New kit on the block: evaluation of second-generation assays for detection of antibody to the hepatitis C virus. Hepatology 15: 350–353
3. Colledan M, Grendele M, Gridelli B, Rossi G, Fassati LR, Ferla G, Doglia M, Gislon M, Galmarini D (1989) Long-term results after liver transplantation in B and Delta hepatitis. Transplant Proc 21: 2421–2423
4. Colina F, Mollejo M, Moreno E, Alberti N, Garcia I, Gomez-Sanz R, Castellano G (1991) Effectiveness of histopathological diagnoses in dysfunction of hepatic transplantation. Review of 146 histopathological studies from 53 transplants. Arch Pathol Lab Med 115: 998–1005
5. Crippin JS, Carlen SL, Borcich A, Bodenheimer HC Jr (1992) Retransplantation in hepatitis B: a multicenter experience. Hepatology 16: 277A
6. Donegan E, Wright TL, Kim M, Roberts J, Ascher N, Wiber J, Chan S, Qaun S, Detmer J, Polito A, Urdea M (1992) Diagnosis of hepatitis C viral (HCV) infection in liver transplant recipients. Hepatology 16: 70A
7. Feray C, Gigou M, Samuel D, Thiers V, Paradis V, David MF, Reynes M, Brechot C, Bismuth H (1992) Natural history of hepatitis C virus (HCV) infection after liver transplantation (LT). Hepatology 16: 47A
8. Ferrell LD, Wright TL, Roberts J, Ascher N, Lake J (1992) Hepatitis C viral infection in liver transplant recipients. Hepatology 16: 865–876
9. Freese D, Ho S, Snover D, Bloomer J, Rank J, Sharp H, Knaak M, Allen J, Hamman K, Payne W (1992) Chronic hepatitis after liver transplantation: etiology and outcome. Hepatology 16: 48A
10. Gimeno C, Lumbreras C, Colina F, Fuertes A, Lizasoanin M, Aguado JM, Loinaz C, Iglesias J, Moreno E, Noriega AR (1992) Value of a new four-antigen recombinant immunoblott assay (4-RIBA) in predicting recurrence of hepatitis C virus infection following liver transplantation (LT). The Jean Hamburger memorial congress, XIVth international congress of the Transplantation Society. Symposium and poster abstracts, Paris, August 16–21 1992, p 613
11. Gretch DR, Dela Rosa C, Perkins J, Carithers RL Jr (1992) HCV infection in liver transplantation recipients: chronic reinfection universal, de novo acquisition rare. Gastroenterology 16: 45A
12. Houghton M, Weiner A, Han J, Kuo G, Choo Q-L (1992) Molecular biology of the hepatitis C virus: implications for diagnosis, development and control of viral disease. Hepatology 14: 381–388
13. Kuo G, Choo QL, Alter JH, Gitnik GL, Redeker AG, Purcell RH, Miyamura T, Dienstag JL, Alter MJ, Stevens CE, Tegtmeier GE, Bonino F, Colombo M, Lee WS, Kuo D, Berger K, Shuster JR, Overby LR, Bradley DW, Oughton M (1989) An assay for circulating antibodies to a major etiologic virus of human non-A, non-B hepatitis. Science 244: 362–364

14. Jeffers LJ, Hasan F, De Medina M, Reddy R, Parker T, Silva M, Mendez L, Schiff ER, Manns M, Houghton M, Choo Q-L, Kou G (1992) Prevalence of antibodies to hepatitis C virus among patients with cryptogenetic chronic hepatitis and cirrhosis. Hepatology 15: 187–190
15. Li X, Jeffers L, Reddy R, De Medina M, Parker T, Schiff ER (1992) Detection of HCV-RNA by RT-PCR in lymphocyte subsets. Gastroenterology 102: A842
16. Marcellin P, Samuel D, Loriot MA, Gigou M, Benhamou JP, Bismuth H (1991) Use of ARA-AMP in liver transplant patients with HBV graft infection. 5th Congress of European Society for Organ Transplantation, book of abstracts, Maastricht, October 7–10 1991, p 273
17. Martin P, Munoz SJ, Di Bisceglie AM, Rubin R, Waggoner JG, Armenti VT, Moritz MJ, Jarrell BE, Maddrey WC (1991) Recurrence of hepatitic C virus infection after orthotopic liver transplantation. Hepatology 13: 719–721
18. Martin P, Munoz SJ, Friedman LS (1992) Liver transplantation for viral hepatitis: current status. Am J Gastroenterol 87: 409–418
19. McHutchinson JG, Person JL, Govindarajan, Valinluck B, Gore T, Lee SR, Nelles M, Polito A, Chien D, DiNello R, Quan S, Kuo G, Redeker AG (1992) Improved detection of hepatitis C virus antibodies in high-risk populations. Hepatology 15: 19–25
20. Müller H, Otto G, Goeser T, Arnold J, Pfaff J, Thielmann L (1992) Recurrence of hepatitis C virus infection after orthotopic liver transplantation. Transplantation 54: 743–745
21. Nakatsuji Y, Matsumoto A, Tanaka E, Ogata H, Kiyosawa K (1992) Detection of chronic hepatitis C virus by four diagnostic systems: first-generation and second-generation enzyme-linked immunosorbent assay, second-generation recombinant immunoblot assay and polimerase chain reaction analysis. Hepatology 16: 300–305
22. Nakhleh RE, Schwarzenberg SJ, Bloomer J, Payne W, Snover DC (1990) The pathology of liver allografts surviving longer than one year. Hepatology 11: 465–470
23. New L, Korb S, Brown M, Light JA, Alarif L (1992) The clinical implications of transplanting hepatitis C positive donor organ. The Jean Hamburger memorial congress, XIVth international congress of the Transplantation Society. Symposium and poster abstracts, Paris, August 16–21 1992, p 805
24. Pereira BJG, Milford EL, Kirkman RL, Levey AS (1991) Transmission of hepatitis C virus by organ transplantation. N Engl J Med 325: 454–460
25. Pereira BJG, Milford EL, Kirkman RL, Quan S, Sayre KR, Johnson PJ, Wilber JC, Levey AS (1992) Prevalence of hepatitis C virus in organ donors positive for hepatitis C antibody and in the recipients of their organs. N Engl J Med 327: 910–915
26. Pirsch JD, Belzer FO (1992) Transmission of HCV by organ transplantation. N Engl J Med 326: 412
27. Poterucha JJ, Rakela J, Ludwig J, Taswell HF, Wiesner RH (1991) Hepatitis C antibodies in patients with chronic hepatitis of unknown etiology after orthotopic liver transplantation. Transplant Proc 23: 1495–1497
28. Pouterucha JJ, Rakela J, Lumeng L, Lee C-H, Taswell HF, Wiesner RH (1992) Diagnosis of chronic hepatitis C after liver transplantation by detection of viral sequences by polymerase chain reaction. Hepatology 15: 42–45
29. Quiroga J, Colina I, Demetris JA, Starzl TE, Van Thiel D (1991) Cause and timing of first allograft failure in orthotopic liver transplantation: a study of 177 consecutive patients. Hepatology 14: 1054–1062

30. Rizzetto M, Macagno S, Chiaberge E, Verme G, Negro F, Marinucci G, Di Giacomo C, Alfani D, Cortesini R, Milazzo F, Doglia M, Fassati LR, Galmarini D (1987) Liver transplantation in hepatitis delta virus disease. Lancet ii: 469–471

31. Rossi G, Grendele M, Colledan M, Gridelli B, Fassati LR, Maggi U, Reggiani P, Gatti S, Piazzini A, Lunghi G, Cardone R, Galmarini D (1989) Prevention of hepatitis B virus reinfection after liver transplantation. Transplant Proc 23: 1969

32. Sallie R, Cohen A, Rayner A, O'Grady J, Tan KC, Portman B, Williams R (1992) Hepatitis "C" recurrence following orthotopic liver transplantation: a PCR and histological study. Hepatology 16: 48A

33. Samuel D, Bismuth A, Matheiu D, Arulnaden J-L, Reynes M, Benhamou J-P, Brechot C, Bismuth H (1991) Passive immunoprophylaxis after liver transplantation in HBsAg-positive patients. Lancet 337: 813–815

34. Samuel D, Gigou M, Feray C, David MF, Reynes M, Brechot C, Bismuth H (1992) Recombinant alpha interferon for hepatitis C on the graft in liver transplant patients. Hepatology 16: 48A

35. Shah G, Demetris AJ, Gavaler JS, Lewis JH, Todo S, Starzl TE, Van Thiel DH (1992) Incidence, prevalence and clinical course of hepatitis C in liver transplantation. Gastroenterology 103: 323–329

36. Todo S, Demetris AJ, Van Thiel D, Teperman L, Fung JJ, Starzl TE (1991) Orthotopic liver transplantation for patients with hepatitis B virus-related liver disease. Hepatology 13: 619–626

37. Trainor T, Farr G, Cooper S, Hayes D, Regenstein F, Balart L (1990) Antibody to hepatitis C (anti-HCV) in orthotopic liver transplant (OLTx) recipients. Hepatology 12: 866

38. Villamil F, Yu C-H, Hu K-Q, Makowka L, Podesta L, Sher L, Howard T, Memsic L, Hoffman A, Petrovic L, Aziz H, Vierling J (1992) Alfa 2b-interferon (alfa2b-IFN) therapy of hepatitis C following liver transplantation (LT). Hepatology 16: 48A

39. Wang JT, Wang T-H, Sheu J-C, Lin J-T, Wang C-Y, Chen D-S (1992) Posttransfusion hepatitis revisited by hepatitis C antibody assays and polymerase chain reaction. Gastroenterology 103: 609–616

40. Wreighitt TG, Gray JJ, Allain JP, Porlain J, Garson R, Calne RY, Large S, Wallwork J, Alexander DJ (1992) Frequent transmission of hepatitis C virus by organ donation. Hepatology 16: 47A

41. Wright HI, Gavaler JS, Van Thiel DH (1992) Preliminary experience with alpha-2b-interferon therapy of viral hepatitis in liver allograft recipients. Transplantation 53: 121–124

42. Wright TL, Donegan E, Hsu HH, Ferrel L, Lake JR, Kim M, Combs C, Fennessy S, Roberts JP, Ascher NL, Greenberg HB (1992) Recurrent and acquired hepatitic C viral infection in liver transplant recipients. Gastroenterology 103: 317–322

Authors' address: Dr. L. Caccamo, Centro Trapianto Fegato, Ospedale Maggiore Policlinico, Via F. Sforza 35, I-20122 Milano, Italy.

O.-R. Kaaden, W. Eichhorn,
C.-P. Czerny (eds.)
Unconventional Agents and Unclassified Viruses
Recent Advances in Biology
and Epidemiology

1993. 79 partly coloured figures. VIII, 308 pages.
ISBN 3-211-82480-4
Soft cover DM 260,–, öS 1820,–*
(Archives of Virology / Supplementum 7)

P. P. Liberski
The Enigma of Slow Viruses
Facts and Artefacts

1993. 56 figures. XVI, 277 pages.
ISBN 3-211-82427-8
Soft cover DM 250,–, öS 1750,–*
(Archives of Virology / Supplementum 6)

O. W. Barnett (ed.)
Potyvirus Taxonomy

1992. 57 figures. IX, 450 pages.
ISBN 3-211-82353-0
Soft cover DM 290,–, öS 2030,–*
(Archives of Virology / Supplementum 5)

C. De Bac, W. H. Gerlich, G. Taliani
(eds.)
Chronically Evolving Viral Hepatitis

1992. 72 figures. XIV, 348 pages.
ISBN 3-211-82350-6
Soft cover DM 260,–, öS 1820,–*
(Archives of Virology / Supplementum 4)

B. Liess, V. Moennig, J. Pohlenz,
G. Trautwein (eds.)
Ruminant Pestivirus Infections
Virology, Pathogenesis, and Perspectives
of Prophylaxis

1991. 78 figures. VIII, 271 pages.
ISBN 3-211-82279-8
Soft cover DM 220,–, öS 1540,–*
(Archives of Virology / Supplementum 3)

R. I. B. Francki, C. M. Fauquet,
D. L. Knudson, F. Brown (eds.)
Classification and Nomenclature of Viruses
Fifth Report of the International Committee
on Taxonomy of Viruses
Virology Division of the International Union
of Microbiological Societies

1991. IV, 450 pages.
ISBN 3-211-82286-0
Soft cover DM 110,–, öS 770,–*
(Archives of Virology / Supplementum 2)

C. H. Calisher (ed.)
Hemorrhagic Fever with Renal Syndrome, Tick- and Mosquito-Borne Viruses

1991. 75 figures. VII, 347 pages.
ISBN 3-211-82217-8
Soft cover DM 258,–, öS 1800,–*
(Archives of Virology / Supplementum 1)

* 10 % price reduction for subscribers to the journal
   "Archives of Virology"

Springer-Verlag Wien New York

Sachsenplatz 4–6, P.O.Box 89, A-1201 Wien · 175 Fifth Avenue, New York, NY 10010, USA
Heidelberger Platz 3, D-14197 Berlin · 37-3, Hongo 3-chome, Bunkyo-ku, Tokyo 113, Japan

..ives of Virology

Official Journal of the Virology Division
International Union of Microbiological Societies

Editorial Board:
I. W. Halliburton, H.-D. Klenk, F. A. Murphy (Editor-in-Chief), Y. Nagai,
J. H. Strauss, A. Vaheri, M. H. V. Van Regenmortel, D. O. White,
and an International Advisory Board

Virology Division: M. C. Horzinek
Special Issues: C. H. Calisher

Archives of Virology publishes original contributions from all branches of research on viruses and virus infections of humans, animals, plants, insects, and bacteria. Coverage includes the broadest spectrum of topics, from initial descriptions of newly discovered viruses, to studies of virus structure, composition, and genetics, to studies of virus interactions with host cells, host organisms, and host populations. Multidisciplinary studies are particularly welcome, as are studies employing molecular biologic, molecular genetic, and modern immunologic and epidemiologic approaches. For example, studies on the molecular pathogenesis, pathophysiology, and genetics of virus infections in individual hosts, and studies on the molecular epidemiology of virus infections in populations, are encouraged. Studies involving applied research, such as diagnostic technology development, monoclonal antibody panel development, vaccine development, and antiviral drug development, are also encouraged. However, such studies are often better presented in the context of a specific application or as they bear upon general principles of interest to many virologists. In all cases, it is the quality of the research work, its significance, and its originality which will decide acceptability.

Subscription Information:
1994. Vols. 134-139 (4 issues each):
DM 2100,–, öS 14700,–, plus carriage charges
ISSN 0304-8608. Title No. 705

Springer-Verlag Wien New York

Sachsenplatz 4–6, P.O.Box 89, A-1201 Wien · 175 Fifth Avenue, New York, NY 10010, USA
Heidelberger Platz 3, D-14197 Berlin · 37-3, Hongo 3-chome, Bunkyo-ku, Tokyo 113, Japan

E. Kurstak

Viral Hepatitis

Current Status and Issues

In collaboration with Christine Kurstak, A. Hossain, and A. Al Tuwaijri

1993. 26 figures. X, 217 pages.
Soft cover DM 120,–, öS 840,–
ISBN 3-211-82387-5

Prices are subject to change withouth notice

In the 1990's significant advances in the understanding of viral hepatitis have been observed. In particular, our knowledge of the nature and diversity of viruses causing hepatitis in humans have substantially increased.

"Viral Hepatitis: Current Status and Issues" comprehensively and uniquely presents these valuable information all in a single volume for the utmost benefit of medical practitioners, microbiologists as well as those actively involved in health administration world-wide.

The virological, clinical epidemological, diagnostic, therapeutic, and preventive aspects pertaining to all the types of hepatitis known to date including hepatitis C and E are thoroughly discussed.

From the contents:

Hepatitis A Virus and Disease. - Hepatitis B Virus and Disease. - Hepatitis D Virus and Disease. - Hepatitis C Virus, Hepatitis E Virus and Disease. - Different Forms of Viral Hepatitis.

Springer-Verlag Wien New York

Sachsenplatz 4–6, P.O.Box 89, A-1201 Wien · 175 Fifth Avenue, New York, NY 10010, USA
Heidelberger Platz 3, D-14197 Berlin · 37-3, Hongo 3-chome, Bunkyo-ku, Tokyo 113, Japan